The Socio-Economic Values in Land Resource Management

The Socio-Economic Values in Land Resource Management

Guest Editor

Matjaž Glavan

MDPI

Basel • Beijing • Wuhan • Barcelona • Belgrade • Novi Sad • Cluj • Manchester

Guest Editor
Matjaž Glavan
University of Ljubljana
Ljubljana
Slovenia

Editorial Office
MDPI AG
Grosspeteranlage 5
4052 Basel, Switzerland

This is a reprint of the Special Issue, published open access by the journal *Land* (ISSN 2073-445X), freely accessible at: https://www.mdpi.com/journal/land/special_issues/Socioeconomic_Values_LandResource.

For citation purposes, cite each article independently as indicated on the article page online and as indicated below:

Lastname, A.A.; Lastname, B.B. Article Title. *Journal Name* **Year**, *Volume Number*, Page Range.

ISBN 978-3-7258-2551-6 (Hbk)
ISBN 978-3-7258-2552-3 (PDF)
https://doi.org/10.3390/books978-3-7258-2552-3

Contents

land

ticle

volution Mode, Influencing Factors, and Socioeconomic Value f Urban Industrial Land Management in China

i Xie [1], Shuaibing Zhang [2], Kaixu Zhao [3] and Fengmei Quan [4,5,*]

1 School of Urban Construction and Environment, Dongguan City University, Dongguan 523419, China
2 School of Architecture and Urban Planning, Shenzhen University, Shenzhen 518060, China
3 College of Urban and Environmental Sciences, Northwest University, Xi'an 710127, China
4 School of Architecture and Civil Engineering, Xiamen University, Xiamen 361005, China
5 Hualan Design & Consulting Group Co., Ltd., Nanning 530022, China
* Correspondence: 25220170155168@stu.xmu.edu.cn

Abstract: (1) Background: Accurate measurement of the matching relationship between urban industrial land change and economic growth is of great value for industrialized and re-industrialized countries to perform land resource management in territorial spatial planning. (2) Methods: Based on the combination of the Boston Consulting Group matrix, Geodetector, and decoupling model, we constructed a new method integrating "model evolution + driving mechanism + performance evaluation + policy design" in this paper, and conducted an empirical study on the economic value of urban industrial land management in the Yangtze River Delta. (3) Results: The evolution modes of urban industrial land in the Yangtze River Delta are divided into four types: stars, cows, dogs, and question, distributed in structures ranging from an "olive" shape to a "pyramid" shape, with high spatial heterogeneity and agglomeration and low autocorrelation. The government demand led by driving economic growth and making large cities bigger is the key factor driving the change in urban industrial land and the influence of transportation infrastructure and the business environment has remained stable for a long time. The mechanisms of industrialization, globalization, and innovation are becoming increasingly complicated. Industrial land change and value-added growth in most cities have long been in a state of strong and weak decoupling, with progressive decoupling occurring alongside the unchanged stage and regressive decoupling. The government outperforms the market in terms of urban industrial land management, and the degradation of the synergy between urban industrial land and corporate assets emerges as a new threat to sustainable and high-quality development of the region. (4) Conclusions: This paper establishes a technical framework for zoning management and classification governance of urban industrial land to divide the Yangtze River Delta into reduction-oriented transformation policy zoning, incremental high-quality development zoning, incremental synchronous growth zoning, and reduction and upgrading development zoning. It also proposes an adaptive land supply governance strategy for quantitative and qualitative control, providing a basis for territorial spatial planning and land resource management.

Keywords: urban industrial land; decoupling model; land resource management; China

ation: Xie, F.; Zhang, S.; Zhao, K.; an, F. Evolution Mode, Influencing tors, and Socioeconomic Value of oan Industrial Land Management China. *Land* **2022**, *11*, 1580. https: loi.org/10.3390/land11091580

ademic Editor: Matjaž Glavan

eived: 29 August 2022
cepted: 10 September 2022
olished: 15 September 2022

blisher's Note: MDPI stays neutral h regard to jurisdictional claims in olished maps and institutional affilons.

1. Introduction

1.1. Background

Urban industrial land is the most important spatial carrier for cities to promote the development of the manufacturing industry and real economy, and research on its spatial distribution and evolution patterns has attracted many scholars' attention [1]. Industrial land change is a major feature of the spatial evolution of the real economy of cities, and management and control of the quantity, structure, and layout of industrial land are important elements of territorial spatial planning, industrial economics, and land use planning [2,3].Therefore, it is of great value to analyze the spatio-temporal evolution of urban industrial land, control the scale of urban industrial land in a rational manner, a realize the dynamic balance between land and economy to promote high-quality industr development and sustainable urban development [4].

The Outline of the 14th Five-Year Plan for National Economic and Social Developme and Vision 2035 of the People's Republic of China puts forward the strategy of streng ening manufacturing. Against the background of tight constraints on land resources, i necessary to further strengthen the management of industrial land supply and change a explore the planning technology and management methods of high-quality utilization urban industrial land in order to promote high-quality development of the manufacturi industry. Since the reform and opening up, China has enjoyed rapid industrializati relying on its unique land system and industrial land allocation mode, making it t "world factory" [5,6]. Land has played a vital role in China's industrialization, and t academic circles have widely accepted the role of urban industrial land growth in prom ing urban economic development [7]. However, with the process of industrialization a urbanization, economic, social, and environmental problems brought about by extensi use, low output, standing idle, and low productivity are constantly emerging, includi declined efficiency of land resource allocation [8], increased land rent-seeking and plund ing [9], constant land conflicts [10] and mismatches [11], and deteriorated ecological a environmental quality [12,13].

In territorial spatial planning, it is necessary for us to control the scale of urb industrial land and formulate its spatial allocation plan to realize the mutual matchi between land change and industrial economic growth. It is worth noting that no oth studies have analyzed the dynamics and economic effects of industrial land use and implications for territorial spatial planning. It is worth noting that some papers ha analyzed the relationship between industrial land and manufacturing economic chan (characteristics of industrial land supply and its contribution to economic growth) [14,1 but the analysis results have not been applied to territorial spatial planning and a techni approach to integrate "land dynamics-economic value-management policy" is lacki Therefore, it is of great importance to analyze the characteristics, process, pattern, a performance of industrial land changes in Chinese cities; reveal the land demand a change patterns; and apply the research findings to the process of urban planning, spat planning, industrial planning, and land management policy design to promote industr transformation and upgrade and improve urban functions [16].

1.2. Questions and Framework

This paper conducts an empirical study of the Yangtze River Delta based on GIS to and multiple models, mainly trying to answer the following questions. (1) What are t characteristics of the urban industrial land evolution mode in the Yangtze River Del (2) What factors affect the change in urban industrial land using Geodetector softwa (3) What is the relationship between urban industrial land change and real economic grow in the Yangtze River Delta according to the decoupling model, from the government's a enterprises' views? (4) How can the analysis results be applied to the practical process policy design?

This paper consists of six parts. The first part is the introduction, which explains t background of this study. The second part is the literature review, which analyzes t characteristics and shortcomings of the existing studies, and defines the starting poi for research and the core issues concerned. The third part presents the materials a methods, introducing the study area, research methods and steps, selection of indicato and their data sources. The fourth part is the statement of the results, which is the k component of this paper. This paper analyzes the evolution mode and influencing fact of urban industrial land change in the Yangtze River Delta in detail and measures t coordination of urban industrial land change with urban industrial economic developme and enterprise performance change quantitatively to reveal the spatio-temporal effe of urban industrial land change. The fifth part is the discussion, which is the diffic section of the paper. On the one hand, the core point of this paper is compared with relevant literature to show their similarities and differences and possible reasons for these similarities and differences; on the other hand, based on the analysis results and the regional characteristics of the Yangtze River Delta, this paper puts forward the technical method of zoning management and supply governance of urban industrial land. The last part is the section of conclusions, which systematically summarizes and refines the main findings of this study, with an attempt to put forward the theoretical innovation, international value, and practical contribution of this paper, while explaining the shortcomings of this study, and giving a view of the future research direction.

2. Literature Review

2.1. *Urban Industrial Land Change and Influencing Factors*

Studies on the characteristics of urban industrial land change focus on the analysis of urban industrial land area, proportion, boundaries, use purposes, spatial forms and geo-graphical patterns, and its influencing factors. Zambon [17] found that the expansion of urban industrial land in the peri-urban areas of southern Europe is particularly intense and that industrial development is a principal factor driving the spatial spread of coastal cities. Debela [18] analyzed the impact of industrial investment and industrial land changes on smallholder livelihoods and food production in Ethiopia. Xiong [19] analyzed the process of industrial land expansion and its influencing factors in Shunde, China, finding that industrial land space is characterized by prominent "fragmentation" and that decentralization and marketization have the greatest influence on industrial land change while the influence of globalization and technological progress is not significant. Park [20] found that the loss of industrial land in the central urban area over time is closely related to the suburbanization of FDI manufacturing employment. Zhang [21] found that the significant reduction of industrial land in the central urban area of Hangzhou and the emergence of industrial clusters in the suburbs, active planning policies, controlled land prices, and market-based mechanisms have a great impact on the spatial restructuring and spatial pattern reconstruction of manufacturing industries. Ustaoglu [22] analyzed the lag effects between changes in industrial land and regional economic development in Dutch cities and found that GDP, employed population, and real estate prices are key influencing factors. Villarroya [23] and Pugh [24] analyzed the industrial land change in regions along highways in Spain and the United Kingdom, revealing the impact of industrial land change on spatial development, economic development, and job creation.

2.2. *Socioeconomic Value of Land Use and Resource Management*

Most papers are committed to the study of development intensity, input and output efficiency, land price management, functional validity, pollution and degradation degree, property ownership, operation mode, and the spatial effect of industrial land. Zhou [15] analyzed the characteristics of industrial land supply and its contribution to economic growth and found that over time, the supply of basic industrial land becomes increasingly concentrated while the supply of technology-intensive industrial land is gradually dispersed. Tonin [25] analyzed the effect of contamination and remediation schemes on industrial real estate properties and concluded that the effect of the size, location, accessibility, and other relevant economic indicators on the price differences of industrial land cannot be ignored. Chen [26,27] examined the spatial effect of industrial land diffusion on land price using a geographically weighted regression model and suggested that the government should guide the optimization of the manufacturing spatial layout through reasonable control of land prices. Lin [28] and Wu [29] found that the land price mechanism is an important way to control the expansion of urban industrial land through an empirical study in China, and Yang [30] further pointed out that the implementation of local industrial land price policies significantly improves the efficiency of industrial land use, and it plays a great role in promoting sustainable development of the regional economy. Li [31] proposed a model to measure the relationship between manufacturing input and output. Kumpula [32] analyzed the positive and negative impacts of industrial land changes on the social a ecological environment in the Arctic Russia region. Zhang [33] found that the productiv of urban industrial land has steadily increased and that capital density, labor density, urb population, economic growth, and industrial structure are key influencing factors.

2.3. Land Use Policy and Territorial Spatial Planning

To promote the rational and orderly development of industrial land, governme around the world have strengthened the management of industrial land developm planning and policy design, prompting the study of industrial land planning techniques a policy effects to attract the attention of government policy makers and planners. Lee [analyzed the location patterns of knowledge-intensive industries and their determinants the Seoul metropolitan area of Korea, with a focus on industrial land planning. Aktas [constructed a technical framework for industrial land planning suitability evaluation usi the weighted linear combination technique and the analytic hierarchy process meth and GIS tools to classify industrial land in Kemalpaa into five levels. Cheng [36] analyz the decision-making process and mechanism of bottom-up industrial land redevelopm planning in China based on a case study of Shenzhen, explaining the roles, conflicti interests, and different stakeholders, such as landowners, developers, and the governm at different stages of industrial land redevelopment. Danilo [37] analyzed the evoluti of industrial land policy in Chicago and explained the process by which industrial la planning and manufacturing reshape the urban space, society, and economy. Based on combination of the multi-criteria decision-making method with the geographic informati system, Salari [38] proposed an industrial land capacity and policy evaluation method the three dimensions of the economy, society, and environment. Silva [39] and Ustaoglu [constructed a model for predicting industrial land area according to the level of urb economic development and applied it to industrial land policy design. Sun [41] fou that the development zoning affects the use efficiency of urban industrial land throu the selection effect, policy effect, cluster effect, and location effect, and suggested a pol design for high-quality development based on this. Galarza [42] analyzed the industr land policy in Alava, summarized the theory and methods of incentivizing industrial la policy design, and proposed the establishment and effects of industrial parks and industr land management agencies.

2.4. Literature Limitations and New Breakthrough Directions

In summary, as an emerging research hotspot, urban industrial land planning a policy research has achieved good results in a variety of fields, including urban industr land planning methods and technical methods, industrial land development and re velopment management policies, and research on industrial land supply and dema patterns and their impacts. Meanwhile, studies on the current conditions, change tren problem diagnosis, spatio-temporal evolution of urban industrial land, and its influenci factors started early, with considerable literature and high maturity. These studies ha better revealed the spatio-temporal evolution of industrial land and its driving mechanis providing a basis for planners to carry out industrial policy design and urban developm planning, and offering a reference for the government to promote industrial transformati and upgrading and urban function optimization.

However, the existing studies also have obvious shortcomings. As mentioned abo urban industrial land is generally increasing in scale, with great differences between diff ent cities and regions. Such differences represent multi-dimensional spatial inequalities economic, social, political, and ecological dimensions in different cities and regions, a ignorance of them in urban industrial land planning and management policy design w lead to a serious dilution of planning and policy implementation performance. The curre papers are mainly based on regression analysis of time series and simulation, and the is insufficient analysis of spatial differentiation and correlation effects. In the new era increasing spatio-temporal uncertainty and interaction, some studies have no integrated specialized analysis of spatio-temporal coupling and fail to attract extensive attention from scholars and governments, although they have noticed the spatial differences in industrial land change and its utilization efficiency [43,44].

In addition, the research on the change dynamics, influencing factors, social and economic benefits, resource management policies, and spatial planning of urban industrial land is separated, lacking a technical framework to integrate different studies. At the same time, the urban industrial land use differentiation management method is still key and a difficult field in research and practice. Scholars such as Li [45], Meng [46], and Zirlewagen [47] proposed the concept of differentiated land management but did not address the workflow, technical methods, decision-making basis, or realization mechanism of differentiated urban industrial land management policies.

3. Materials and Methods

3.1. Study Area

The study area is the Yangtze River Delta of China, covering all administrative regions of Shanghai, Jiangsu, Zhejiang, and Anhui, including 41 cities such as Shanghai, Suzhou-JS, Changzhou, Nanjing, Hefei, Wuhu, Hangzhou, and Jiaxing (Figure 1). In this paper, Taizhou in Zhejiang province is abbreviated as Taizhou-ZJ, Taizhou in Jiangsu province is Taizhou-JS, Suzhou in Jiangsu province is Suzhou-JS, and Suzhou in Anhui province is Suzhou-AH to avoid confusion. The Yangtze River Delta is the region with the highest level of industrialization development in China, the core carrier used to implement China's manufacturing power strategy, and one of six world-class urban agglomerations identified by French geographer Jean Gottmann. Its urban industrial land changes are typical in China and the world.

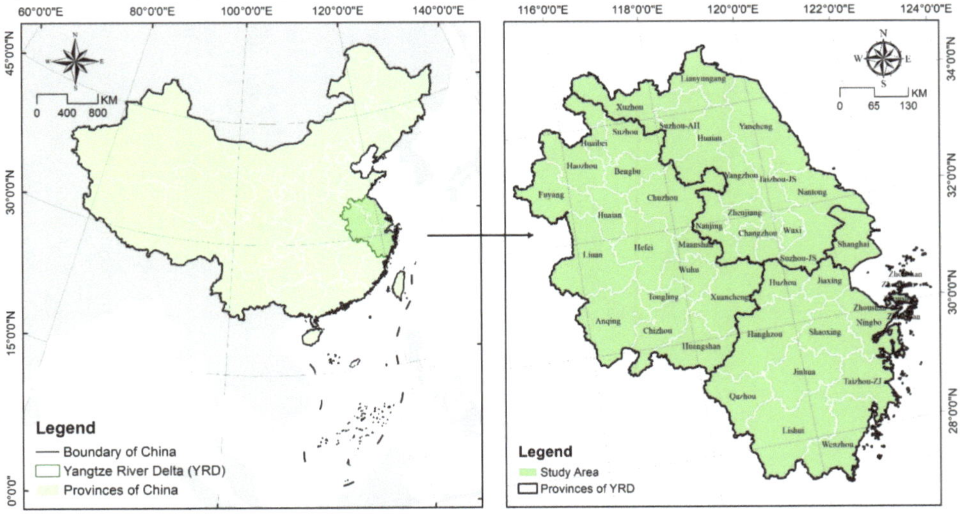

Figure 1. Study area.

Due to long-term high-intensity development, urban industrial land change in Yangtze River Delta is becoming increasingly complicated and the management needs are more diversified. In 2019, the urban industrial land in Yangtze River Delta covered an area of 2618.21 km^2, accounting for 22.62% of urban construction land; it created an industrial added value of 8,179.138 billion yuan, 34.47% of GDP. In the same year, the urban industrial land in Yangtze River Delta accounted for 22.81% of the total in China, and the industrial added value in Yangtze River Delta accounted for 25.87% of the total in China. An important goal of Yangtze River Delta spatial planning and urban planning is to enhance the protection and optimization of the use of urban industrial land in order to support and guarantee development of the real economy. Therefore, it is of great significance to analyze the urb industrial land change and its interaction with economic growth in Yangtze River Delta improve the industrial economic efficiency and effectively protect natural land resourc

3.2. Research Steps and Technical Route

The first step is to study the distribution and change spatial patterns of urban industr land in the three dimensions of relative share, change speed, and evolution mode. In t process, GIS tools and the Boston Consulting Group matrix are used to analyze the k characteristics of urban industrial land change in the Yangtze River Delta. The second step to analyze the main factors influencing the evolution of industrial land in the Yangtze Ri Delta cities and their intrinsic driving mechanisms using spatial econometric regressi methods. First, exploratory spatial data analysis and the Gini Index are introduced analyze the spatial correlation and heterogeneity of the industrial land distribution a changes in the Yangtze River Delta. Second, Geodetector is used to analyze the dir and interactive influences of different factors. The third step is to study the decoupli relationship between urban industrial land change and industrial economic growth fro two perspectives of added value and enterprise assets to analyze the external econon performance of land resource consumption management. The fourth step is to desi and study the zoning management policy. The technical method of planning zoning a supply governance is constructed, and specialized management policies for each zoni are proposed based on the overlay analysis of the research results reached in the previo steps, and the results are applied to spatial governance and urban planning (Figure 2).

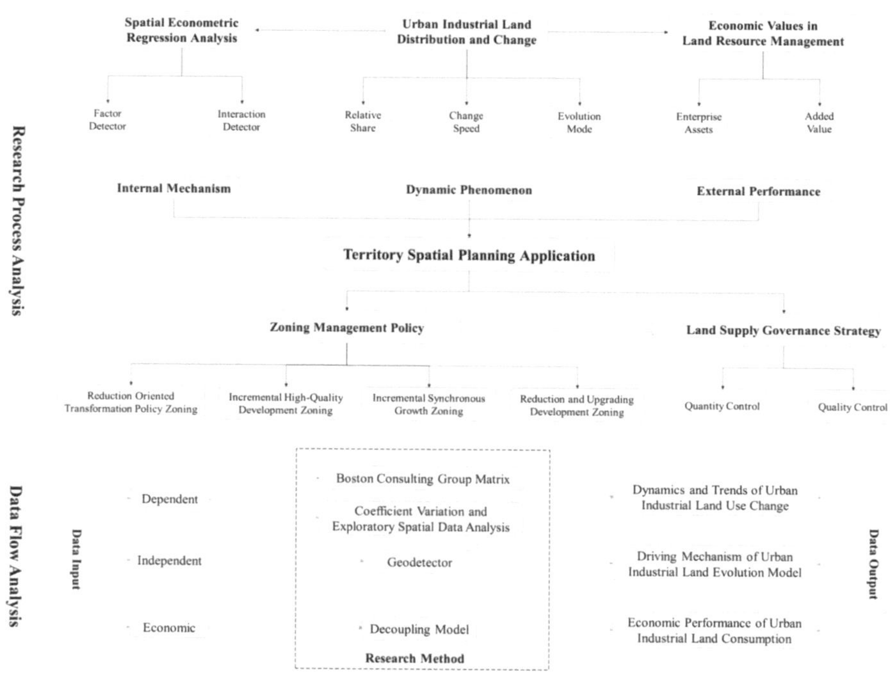

Figure 2. Research steps and technical route.

3.3. Variables and Data Sources

The study period of this paper is 2010–2019 and the variable types used include dependent, independent, and economic, with 15 indicators in total (Table 1). The urban industrial land in this paper coincides with the Code for Classification of Urban Land Use and Planning Standards of Development Land (GB50137-2011), referring to the land for independent factories, production workshops, handicraft workshops and their ancillary facilities. It includes production sites for construction and installation, slag (ash) discharge sites, and sites dedicated for railroads, docks, auxiliary roads, and parking lots. It is generally classified into three classes according to the degree of impact on the habitat. In Sections 4.1 and 4.3, the code of urban industrial land is L_i, which represents land area and is continuous data. In Section 4.2, the code is Y, which represents the evolution mode and is type data (integer with value range of 1–4).

Table 1. Description and analysis of the index system.

Variables	No.	Code	Indicators	Implication
Dependent	1	L_i/Y	Urban Industrial Land	Spatial Pattern/Evolution Model
Independent	2	X_1	Gross Domestic Product (GDP)	Government Demand
	3	X_2	Built-Up Area	
	4	X_3	Road Area	Environment
	5	X_4	Real Estate Investment	
	6	X_5	Per Capita GDP	Industrialization
	7	X_6	Tertiary Industry	
	8	X_7	Import	Globalization
	9	X_8	Export	
	10	X_9	Foreign Direct Investment	
	11	X_{10}	Patent Application Number	Innovation
	12	X_{11}	Higher Education Institution Number	
	13	X_{12}	Education investment	
Economic	14	Z_1	Secondary Industry Added Value	Government Value
	15	Z_2	Industrial Enterprise Assets	Market Value

The change in urban industrial land is a direct mapping of the synergistic development process of regional industrialization and urbanization, reflecting the real economic activities of cities and determining the main pattern of urban economic space. Industrial land is an important part of urban built-up area/construction land, and in the GDP-oriented performance appraisal system, the impact of a city government's pursuit of both economic growth and a larger city size on the change in industrial land must be considered first, which are represented by GDP and built-up area in this paper [48]. Industrial development is inseparable from the support of ancillary facilities, and their impact is measured by road area in view of the Chinese consensus that that transportation infrastructure is fundamental for a region's development [49,50]. As the industrial upgrading and urban renewal in the Yangtze River Delta has significantly accelerated in recent years, a lot of industrial land has been transformed into residential land, so the impact of real estate investment on the industrial business environment should not be ignored [51]. The supply and demand of urban industrial land are influenced by the industrialization stage and are also closely related to the economic structure. In this paper, per capita GDP and tertiary industry are adopted for measurement [52]. The Yangtze River Delta has been the frontier region of China's opening up to the outside world and has undertaken vast quantities of relocated international industries, so this paper chooses import, export, and foreign direct investment to analyze the impact of globalization on the evolution of urban industrial land [53]. The central government has put "innovation-driven" at the top of the "Five Development Ideas" and proposed in the *Development Plan for the Construction of Science and Technology*

Innovation Community in the Yangtze River Delta that the Yangtze River Delta should fully established as a leading global science and technology innovation community 2035. In this context, the influence of technology and education on land use and econo development in the Yangtze River Delta is becoming more important. In this paper, choose patent application number, higher education institution number, and educat investment to analyze the influence of innovation on industrial land in Yangtze River De cities [54].

The formation of production capacity in the manufacturing sector corresponding industrial land will lead to a continuous increase in industrial added value and employm thus promoting rapid economic and social development in the city. Therefore, the c government is strongly willing to keep expanding urban industrial land in order to attr manufacturing investment and boost the growth of the urban industrial economy. In t paper, we use the added value [55] of the secondary industry to represent governm actions and concerns, and to reflect the economic value created by urban industrial land. addition, the assets and operation status of enterprises, the direct users of urban indust land, are the key factors affecting and reflecting the level of land use. Accordingly, enterp assets [56,57] is used in this paper to represent the actions and concerns of enterprises, a reflect the economic benefits created by urban industrial land.

The data of urban industrial land are from the *China Urban Construction Statist Yearbook* released by the Ministry of Housing and Construction of China. The other d are from the *China City Statistical Yearbook* issued by China's National Bureau of Statist and some missing data come from provincial and urban statistical yearbooks, statisti bulletins, and government work reports in the study area, with attention paid to impact of adjustment of the administrative division in data collection and processi Tables A1 and A2 show the analysis results of the evolution mode of independent varia factors, which are calculated by the Boston Consulting Group matrix. Tables A3 and include the normalized data, and their source data are used by the decoupling model. T data of other years are integrated on the basis of the administrative divisions in 2019. example, as Gaochun County and Lishui County were merged into Nanjing City in 20 the data from 2010 to 2013 were directly included in those of Nanjing City; as anot example, as the prefecture-level city of Chaohu in Anhui province was split in 2011, and county-level cities of Chaohu and Lujiang under its jurisdiction were enclosed in Hefei C Wuwei County was enclosed in Wuhu City, and Huangshan County and Hexian Cou (excluding Shenxiang Town) were enclosed in Ma'anshan City, the data of 2010–2011 w processed at a ratio of 2/5 for Hefei, 2/5 for Ma'anshan, and 1/5 for Wuhu when the d were integrated.

3.4. Research Methods

3.4.1. Boston Consulting Group Matrix

For companies that provide more than one type of product or service, their susta ability should be evaluated in an integrated way through portfolio analysis among th businesses since each business has different market positions and value advantages. T Boston Consulting Group matrix is based on a methodology pioneered by the Bost Consulting Group in the 1970s to analyze and optimize a company's business portfolio the marketplace. In the Boston Consulting Group Matrix-based analysis, the horizo coordinate represents the relative market share of the company's revenue, and the verti coordinate represents the average annual growth rate of the revenue, with the average set value (e.g., 0.5 or 10%) as the threshold to classify the company's business into fo types: star, cow, question, and dog [58]. This division of business portfolio types fac tates the development of effective and appropriate strategies for different businesses a integration of corporate resources to improve competitiveness. Boston Consulting Gro Matrix-based analysis aims to differentiate business by dividing the products or servic into different quadrants to ensure that it continuously eliminates dog business with limit prospects and maintains a reasonable combination of star, cow, and question busine thus realizing a virtuous cycle of product, service, and resource allocation structure. The Boston Consulting Group Matrix has the advantage of improving the business analysis and strategic decision-making capabilities of corporate executives and providing a deeper understanding of the linkages between different business activities for them [59].

The Boston Consulting Group Matrix originated in the field of business management and is currently widely used in the fields of tourism management and foreign investment management and has also been tentatively applied in the field of land use change management [60]. In this paper, the Boston Consulting Group matrix is used to measure the spatial and temporal evolution trend of urban industrial land in Yangtze River Delta. It is also applied to analyze the spatial and temporal evolution trends of the independent variables, and the output results are then used as input variables for the Geodetector-based analysis. *RS* is defined as a relative share to reflect the position of urban industrial land stock in Yangtze River Delta. With L_i representing the area of urban industrial land in city i in a given year, L_{i-max} representing the maximum area of urban industrial land of 41 cities in Yangtze River Delta in a given year, L_{i-last} and L_{i-base} representing the urban industrial land area of city i in the base and end periods, t representing the time interval of the study period ($t = 4$ in 2010–2014, $t = 5$ in 2014–2019), and n being the number of cities, *RS* and *AV* are calculated as follows [61]:

$$RS = \frac{L_i}{L_{i-max}} \quad (1)$$

$$CS = \sqrt[t]{L_{i-last}/L_{i-base}} - 1 \quad (2)$$

According to Equations (1) and (2), the competition state of urban industrial land of the cities in Yangtze River Delta can be identified to reveal the spatial and temporal evolution trend of urban industrial land in Yangtze River Delta. With the average of *CS* and *RS* as a threshold, the 41 cities in Yangtze River Delta are divided into 4 types for analysis of the possible strategies for future management of urban industrial land in each city (Table 2). When both *CS* and *RS* are greater than or equal to the average value, it belongs to star-cities, and when both are less than the average value, it belongs to dog-cities. When *CS* is greater than or equal to the average and *RS* is less than the average, it belongs to cow-cities. When *CS* is less than the average and *RS* is not less than the average, it belongs to question-cities. At the same time, this paper also uses the method of the Boston Consulting Group matrix to deal with independent variables, which lays a foundation for the analysis of the driving mechanism.

3.4.2. Coefficient Variation and Exploratory Spatial Data Analysis

The spatial heterogeneity is analyzed by the coefficient of variation (CV), representing weak spatial heterogeneity when it is less than 0.15 or representing strong spatial heterogeneity if it is greater than 0.36 [62]. This paper uses the spatial autocorrelation analysis method to explore the correlation of the spatial distribution of urban industrial land. By calculating the global Moran's-I index, the overall spatial correlation and agglomeration degree of urban industrial land are measured [63]. The global Moran's I value ranges from −1 to 1, and a larger value indicates a higher degree of spatial correlation and agglomeration. When the global Moran's I index is greater than zero, it indicates a positive spatial correlation; on the contrary, if it is less than zero, it indicates a negative spatial correlation; if it is equal to zero, it indicates that the spatial distribution is random. The local Moran's I can reveal the similarity or correlation between spatial units and their neighbors, and further divide the spatial correlation mode into four types, including HH (High-High), HL (High-Low), LH (Low-High), and LL (Low-Low) [64]. This paper conducts the spatial autocorrelation analysis by Arcgis10.2 and GeoDa1.18 at the significance level of 0.1, with the spatial weight matrix based on the adjacent boundaries and all default parameters of the software. n represents the number of cities; M_i and M_j represent the observed values of cities i and j, respectively; $\overline{M}$ represents the average of the observed values; W_{ij} represents the spatial weight matrix in global spatial autocorrelation, or the row normalized value of spatial weight in local spatial autocorrelation; S_0 represents the sum of spatial wei matrices; and N_i and N_j represent the normalized values of the observations of cities i a j. The equations for global and local Moran's I are as follows [65]:

$$\text{Global Moran's } I = \frac{n}{S_0} \times \frac{\sum_{i=1}^{n}\sum_{j=1}^{n} W_{ij}(M_i - \overline{M})(M_j - \overline{M})}{\sum_{i=1}^{n}(M_i - \overline{M})^2},\ S_0 = \sum_{i=1}^{n}\sum_{i=1}^{n} = W_{ij}$$

$$\text{Local Moran's } I_i = N_i \sum_{i=1}^{n} W_{ij} N Y_j$$

Table 2. Decoupling type and decoupling indicator range.

State	*RS*	*CS*	Characteristic	Future Alternative Strategies
Star-cities	$\geq \overline{RS}$	$\geq \overline{CS}$	High growth rate and relative share of urban industrial land, with great development potential and good opportunities.	They are at the stage of rapid growth and priority should be given to expansionary strategies and greater investment in urban industrial land to promote urban economic and social development.
Cow-cities	$\geq \overline{RS}$	$< \overline{CS}$	High relative share of urban industrial land but low growth rate, high regional status but low development potential.	They are at the stage of maturity, and priori should be given to harvesting strategies to control or even reduce investment in urban industrial land to maximize the return on investment in land resources.
Question-cities	$< \overline{RS}$	$\geq \overline{CS}$	Low relative share of urban industrial land, high growth rate, with possibility to become a new spatial growth pole for industrial economic development.	They are at the take-off stage and priority should be given to selective strategies. Due the high uncertainty, it is necessary to be cautious and carefully analyze the real reasons for the increase in the growth rate, with a focus on cultivating cities that have th potential to become stars; otherwise, give u investment.
Dog-cities	$< \overline{RS}$	$< \overline{CS}$	Low relative share and growth rate of urban industrial land, and low regional status and development potential.	They are at the stage of recession and the priority should be given to withdrawal strategies. It is necessary to reduce the scale of land input to mitigate risks and avoid waste of resources due to blind investment.

3.4.3. Geodetector

When the geographical distribution of the research objects has spatial heterogene correlation, and agglomeration, research on the driving mechanism needs to use the spat econometric analysis method instead of the traditional regression model. Geodetector i research-oriented (and still non-commercial) geostatistical analysis application develop by Prof. Wang Jinfeng in the Chinese Academy of Sciences, including both excel a R language versions, which can be downloaded for free at http://www.geodetector.c (accessed on 20 July 2022). Geodetector software provides four functions: the first is factor detection, applied to quantitatively measure the degree of spatial heterogeneity the dependent variable and the influence of a single independent variable on the depend variable; the second is for interaction analysis, applied to identify the compound influen on the dependent variable when different independent variable factors act together; t third is for risk analysis, applied to determine whether the differences between differe classes of the independent variable factors are significant; and the fourth is for ecologi analysis, applied to compare whether there are significant differences in the influence different independent variable factors on the dependent variable [66,67]. The first a second functions are used in this paper.

In this paper, we use the *Factor−detector* function of Geodetector to analyze the fluence of different factors on the evolution pattern of urban industrial land, and t *Interaction−detector* function to analyze the synergy between different factors. For t independent variable factors (e.g., X_i, X_j) and the dependent variable (Y), if they both have a similar evolution mode, Geodetector will determine that the factor has a greater influence on urban industrial land change [68]. With the input patterns of change in the independent and dependent variables, i.e., the results of the analysis using the Boston Consulting Group Matrix, Geodetector analyzes the influence of different factors on land use change by comparing the similarity of the patterns of change in the independent and dependent variables based on an internal algorithm. The influence size is represented by the index q, where q (X_i) and q (X_j) represent the direct influence of the two factors i and j in the independent stage, and q ($X_i \cap X_j$) represents the interactive influence of the two factors i and j in the joint case. Their value range is [0,1], and a larger value implies a greater influence. In the Geodetector software, the independent needs to use type data instead of continuous data. With h representing the number of classifications of the independent variables (h = 4 in this paper), N_h and N representing the number of cities in stratum h and the study area, σ_h^2 and σ^2 representing the variance of the dependent variable in stratum h and the study area, respectively, SSW representing the within sum of squares, and SST representing the total sum of squares in the study area, the calculation equation for q is [69]:

$$q = 1 - \frac{\sum_{h=1}^{l} N_h \sigma_h^2}{N\sigma^2} = 1 - \frac{SSW}{SST},\ SSW = \sum_{h=1}^{l} N_h \sigma_h^2,\ SST = N\sigma^2 \quad (5)$$

The interactive influence is classified into five types based on the relationship between q ($X_i \cap X_j$) and the minimum (Min (q (X_i), q (X_j))), maximum (Max (q (X_i))), q (X_j))), and summation (q (X_i) + q (X_j)) values of direct influence [70,71]. The nonlinear weaken (q ($X_i \cap X_j$) < Min(q (X_i), q (X_j))) and single nonlinear weaken (Min (q (X_i), q (X_j)) < q ($X_i \cap X_j$) < Max(q (X_i)), q (X_j))) represent the antagonistic effect between different factors, indicating that the driving forces of i and j cancel each other when they act together on Y, and that their influence is weakened or even disappears, so the pairing of the two factors should be avoided in policy design when possible. The bifactor enhancement (q (X_i) + q (X_j) > q ($X_i \cap X_j$) > Max (q (X_i), q (X_j))) and nonlinear enhancement (q ($X_i \cap X_j$) > q (X_i) + q (X_j)) represent a synergy effect between different factors, indicating that the driving forces of i and j are mutually reinforcing when they act together on Y, and that the influence is enhanced or even significantly amplified, so the two factors should be paired in policy design when possible. Notably, when q ($X_i \cap X_j$) = q (X_i) + q (X_j), it represents that the processes between the different factors are independent, and they do not interfere or relate to each other, a rare and special phenomenon requiring no consideration of the interactive influence (but the direct driving force) in policy design (Figure 3) [72].

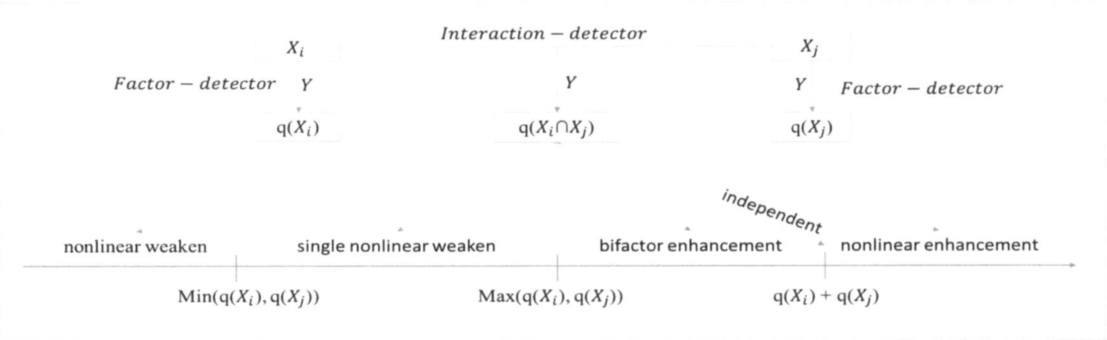

Figure 3. Analysis on the interaction between factors.

3.4.4. Decoupling Model

The decoupling model, proposed by Tapio [73], was originally used to analyze the connection of GDP with transport volume and carbon emissions in EU countries, and is now widely used in economics and ecology research. Analysis of the relationship betw land change and economic growth based on decoupling models has emerged, mainly the fields of urban construction land [74] and agricultural land [75], but it is still rare urban industrial land studies. This paper analyzes the relationship between the urb industrial land change and the growth of the industrial economy (real economy) in Yang River Delta using the decoupling model, essentially analyzing the process of de-landi and materialization in the development of the real economy in Yangtze River Delta cit i.e., the process of reducing the consumption of land resources by industrial econor activities [76]. With γ representing the decoupling index, $\Delta\alpha$ representing the growth rate urban industrial land in Yangtze River Delta, $\Delta\beta$ representing the growth rate of industr economic development-related indicators (including added value, total assets, gross pro and employed population), X_i and X_{i+n} representing the annual values of econor development-related indicators in years i and $i + n$, respectively, and k representing study period, the decoupling index is calculated as follows [77]:

$$\gamma = \frac{\Delta\alpha}{\Delta\beta}$$

$$\Delta\alpha = \frac{L_{i-last} - L_{i-base}}{L_{i-base}}$$

$$\Delta\beta = \frac{Z_{i-last} - Z_{i-base}}{Z_{i-base}}$$

The concept of "decoupling" emphasizes the long-term trending process. Based the relevant research experience [78,79], the study period is divided into three stages 2010–2014, 2015–2019, and 2010–2019, corresponding to short-term and long-term stud in accordance with the length of the research data time series in this paper. The decoupli is classified into 3 types and 8 sub-types based on whether $\Delta\alpha$ and $\Delta\beta$ are positive negative, with 0.8 and 1.2 as the thresholds for γ (Table 3) [80]. A remarkable fact is th strong or weak decoupling is ideal, which shows that the urban industrial land char is in high coordination with industrial economic growth, and the transformation of la investment into economic benefits has a nonlinear amplification effect. Cities in expans coupling or expansive negative decoupling should improve their land use efficiency a intensity early while those in strong or weak negative decoupling or in recessive coupli or decoupling are unhealthy and they are question-cities of regional development, wh should develop targeted policies and plans to reverse the development direction soon achieve sustainable development.

Table 3. Decoupling type and decoupling indicator range.

Type	$\Delta\alpha$	$\Delta\beta$	γ	Implication
SD	≤0	≥0	≤0	It indicates the first best state, where the industrial growth is accompanied by a steady decline in urban industrial land; it has been a benchmark for regional high-quality development since the development of the real economy got rid of the expansion of urban industrial land.
WD	>0	>0	(0,0.8]	It indicates the second-best state, where the growth of the industrial economy is faster than that of urban industrial land with efficient use and intensive development of land resources.
EC	>0	>0	(0.8,1.2]	It indicates the state of steady incremental expansion, with the growth of industrial economy largely synchronized with that of urban industrial land, and the development still heavily depending on land resources.
END	>0	>0	(1.2, +∞)	It indicates the state of incremental and extensive development, with the growth of the industrial economy being slower than that of urban industrial land, low utilization efficiency of land resources, and insufficient transformation of land investment into economic returns.
RD	<0	<0	(1.2, +∞)	It indicates that the cities are in contraction, with both the industrial economy and urban industrial land in negative growth, land resources decreasing faster than the economy, and a high level of land use efficiency and intensity.
RC	<0	<0	(0.8,1.2]	It indicates the stage of steady reduction and contraction, where the industrial economy and urban industrial land are largely declining in a synchronous manner and development is still dependent on land resources.
WND	<0	<0	(0,0.8]	It indicates the second-worst state, with the industrial economy reduction occurring faster than that of urban industrial land, the added value of land output gradually decreasing, and the land reduction having an unhealthy effect of nonlinear amplification on economic development.
SND	>0	<0	<0	It indicates the worst state, where the urban industrial land continues to grow, but the industrial economy is declining gradually, the land investment has not transformed into economic returns, and there is a waste of resources, leading to unsustainable development.

Note: SD stands for strong decoupling, WD stands for weak decoupling, EC stands for expansive coupling, END stands for expansive negative decoupling, RD stands for recessive decoupling, RC stands for recessive coupling, WND stands for weak negative decoupling, and SND stands for strong negative decoupling.

4. Results

4.1. Dynamics and Trends of Urban Industrial Land Use Change

4.1.1. Relative Share

The average of the relative share of urban industrial land is 0.07. Shanghai is in first place and Lishui is at the bottom. The coefficient of variation has decreased from 2.19 to 2.09, indicating high-level but stable spatial heterogeneity. In 2014, high and higher cities were clustered in three agglomeration zones of Shanghai-Nanjing-Hefei, Shanghai-Huzhou-Hangzhou, and Ningbo-Wenzhou in a "finger-shaped" distribution, with Shanghai as the core while low and lower cities were mostly clustered in Anhui and western Zhejiang. In 2019, high and higher cities formed two "arc-shaped" agglomeration belts in Jiangsu and Zhejiang, and the agglomeration center of low and lower cities shifted further to the west of the study area, especially to western Anhui (Figures 4 and 5). In summary, the relative share of industrial land in the Yangtze River Delta cities is distributed in an "east-west" gradient and shows the characteristics of a belt-like agglomeration, with the status of cities in Jiangsu rising significantly and the cities in western Anhui remaining marginalized for a long time.

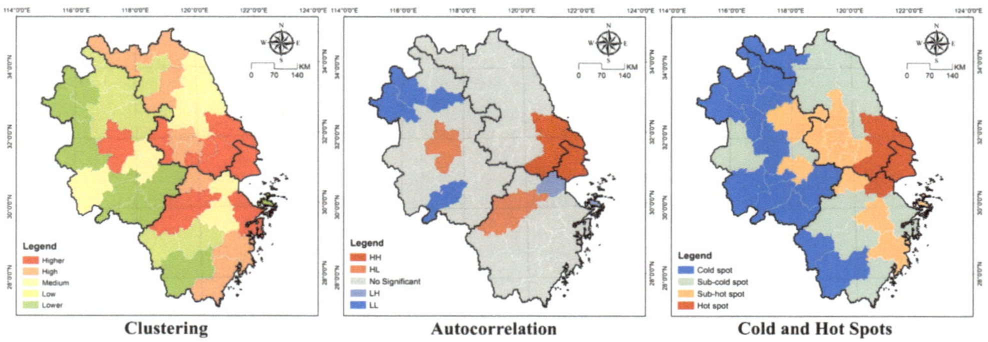

Figure 4. Spatial analysis of the relative share in 2010–2014.

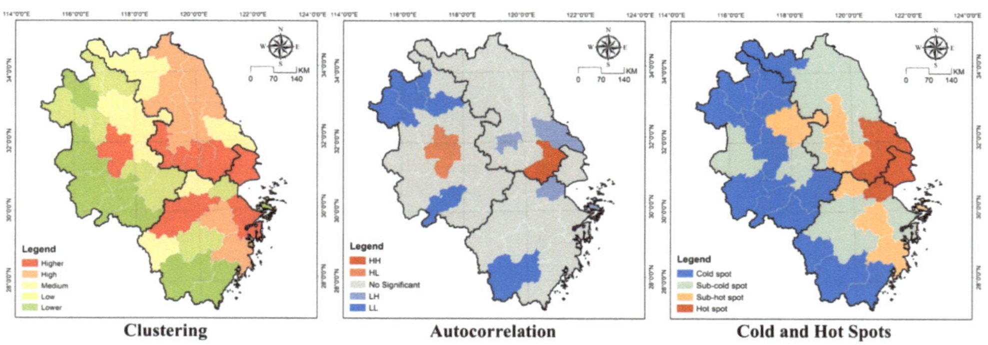

Figure 5. Spatial analysis of the relative share in 2015–2019.

The global Moran's I in 2014 was 0.072 ($p < 0.05$, Z = 2.28), indicating a spatia positive correlation between the relative shares of urban industrial land. HH cities w clustered in the Shanghai metropolitan area, including Shanghai, Nantong, and Suzhou while LL cities were clustered in Anhui, including Chizhou, Bozhou, Fuyang, and Beng Hot spot cities were clustered in the Shanghai metropolitan area, with secondary hot sp cities extending to the Nanjing metropolitan area, while cold spot cities were clustered i band in Anhui and western Zhejiang. Global Moran's I index in 2019 was 0.023 ($p > 0.$ Z = 1.12), indicating that the spatial autocorrelation was significantly reduced and statistically significant. Suzhou-JS was the only HH city, and there was one more LL c Lishui than in 2014. The cold hotspot cities remained largely unchanged in both geograp pattern and spatial structure.

4.1.2. Change Speed

The average change range of urban industrial land in Yangtze River Delta from 2(to 2014 was 1.81 km^2, with the largest increase in Shaoxing of up to 35.71 km^2, and largest decrease in Nantong of −37.81 km^2. The average change range of urban industr land in Yangtze River Delta from 2015 to 2019 was −1.55 km^2, with the largest increase Hangzhou of 44.76 km^2, and the largest decrease in Shanghai of −159.37 km^2. From results of the quantile spatial clustering analysis, most of the highest and high cities w clustered in the Hangzhou-Ningbo development belt, and a few were concentrated in junction between Jiangsu and Anhui; the lowest and low cities were mainly concentra in the Shanghai metropolitan area, southern Zhejiang, and western Anhui, where urb industrial land has been experiencing active (e.g., Shanghai and Suzhou-JS, etc., making development strategic plans for active transformation) or passive (e.g., Wenzhou, Zhoushan, Huai'an, etc., subjected to the siphon effect of big cities) reduction (Figures 6 and 7). To sum up, urban industrial land in Yangtze River Delta experienced a small change on the whole, and the spatial pattern remained stable in general. However, it showed a tendency to shift from increment-led development to reduction-led development over time.

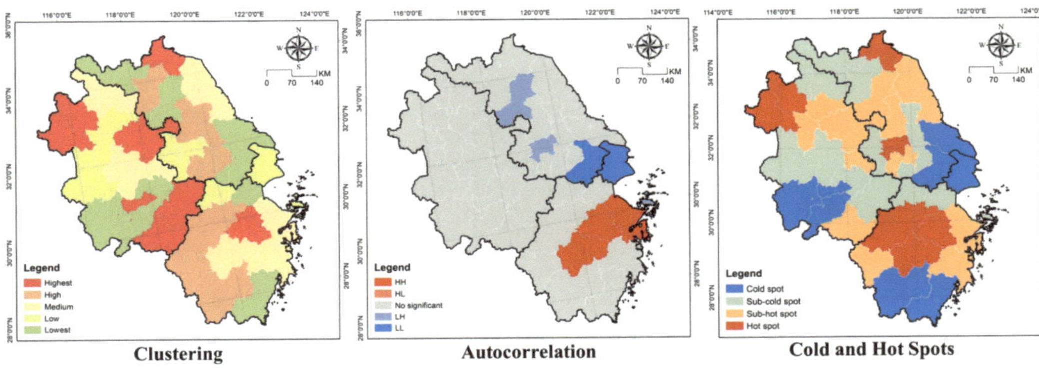

Figure 6. Spatial analysis of the change speed in 2010–2014.

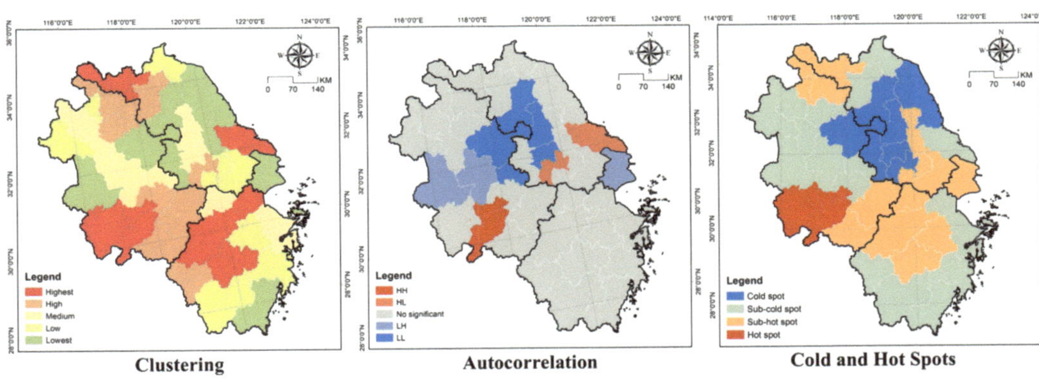

Figure 7. Spatial analysis of the change speed in 2015–2019.

The coefficient of variation decreased from 9.26 to 2.62, indicating that the level of spatial heterogeneity is decreasing but still high. It is worth noting that the global Moran's I index changed from 0.113 to −0.123, indicating that the change range of urban industrial land changed from positive to negative spatial correlation. In 2010–2014, HH cities clustered in Zhejiang province in a band shape but scattered in Zhejiang and Anhui provinces in 2015–2019. From 2010 to 2014, LL cities clustered in the Shanghai metropolitan area, and then all disappeared. There were no HL cities in 2010–2014, and most of them were concentrated in Jiangsu province in 2015–2019, especially in the peripheral areas of the Shanghai metropolitan area. The number of LH cities has been sparse for a long time and distributed randomly. From 2010 to 2014, hotspot cities were clustered in the Hangzhou metropolitan area and northern Anhui while cold spot cities were clustered in the Shanghai metropolitan area, Anhui, and southern Zhejiang. From 2015 to 2019, hotspot cities were clustered in southern Anhui, with secondary hotspot cities mostly clustered in the Anhui-

Zhejiang-Jiangsu junction area and Huaihai Economic Zone, and cold spots in cen
Jiangsu.

4.1.3. Evolution Mode

From the perspective of statistical distribution: The cities of the different types 2010–2014 are ranked by number as stars = cows < dogs < question, and as stars < cow question < dogs for the cities in 2015–2019. In the urban evolution mode, star-cities are the best state and dog-cities are in the worst state while cow-cities and question-cities in the middle between the two. The cities in different states evolve into three distribut structures by the quantity statistics: olive, pyramid, and dumbbell. The olive struct is the most stable, representing a small number of star-cities and dog-cities and a la number of cow-cities and question-cities, which is the best combination to adapt to sustainable development of the region. In complete contrast to olive, dumbbell ha large number of star-cities and dog-cities and a small number of cow-cities and questi cities, indicating that the regional development is polarized at both ends, which is least desirable combination. Pyramid is a structure between dumbbell and olive, with number of star-cities, cow-cities, question-cities, and dog-cities increasing in that or The distribution structure of the four types of patterns has changed from an "olive" sha to a "pyramid" shape, which indicates that the urban system in the region is becomi increasingly better.

From the perspective of geographical pattern: Except for question-cities, all types cities were distributed randomly from 2010 to 2014, with star-cities being Wuxi, Hangzh Ningbo, and Hefei; cow-cities being Nanjing, Suzhou-JS, and Nantong; question-cit including Changzhou, Lianyungang, Yangzhou, Huzhou, Shaoxing, and Huangshan; a dog-cities including Xuzhou, Huai'an, Wenzhou, Jiaxing, Wuhu, and Bengbu. From 2(to 2019, star-cities and cow-cities formed clustering zonings along the Shanghai-Suzh JS-Nanjing and Hangzhou-Ningbo axes while question-cities were mostly clustered southwestern Anhui, and dog-cities were concentrated in a contiguous distribution central Yangtze River Delta (central-northern Anhui and Jiangsu) and southern Zhejia (Figures 8 and 9 and Table 4).

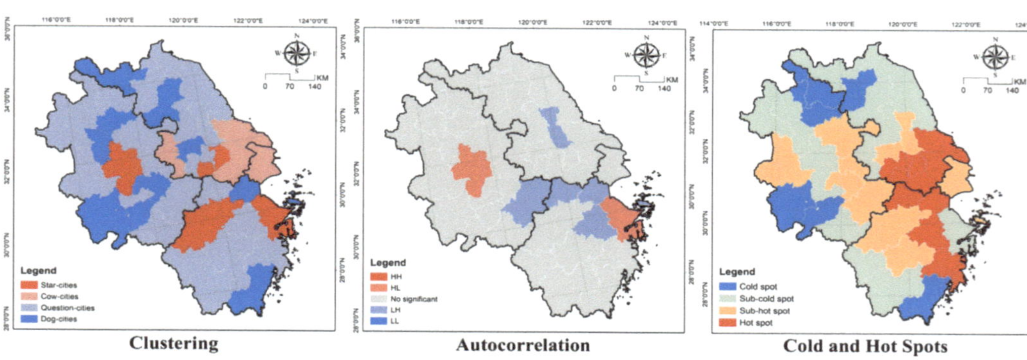

Figure 8. Spatial analysis of the evolution mode in 2010–2014.

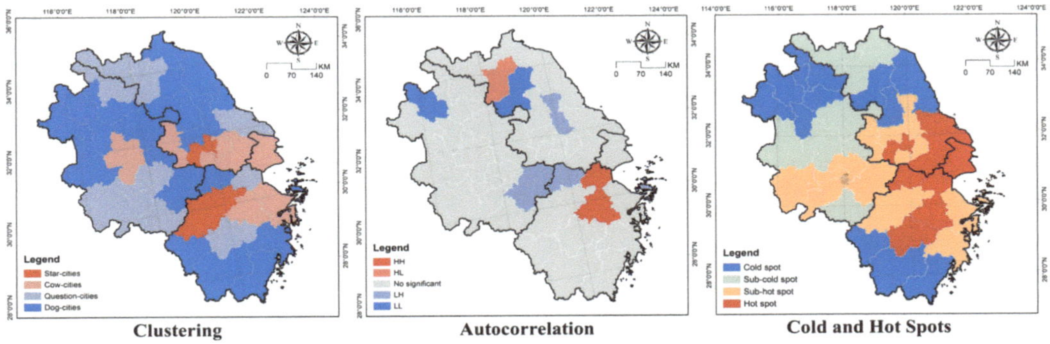

Figure 9. Spatial analysis of the evolution mode in 2015–2019.

Table 4. Analysis of the spatial heterogeneity parameters of urban industrial land in Yangtze River Delta.

	2010–2014	2015–2019
RS (%)	0.07	0.07
CS (%)	1.24	3.82
Star-cities	4	2
Cow-cities	4	7
Question-cities	22	11
Dog-cities	11	21

From the perspective of spatial effect: The coefficient of variation increased from 0.43 to 0.52, indicating that the spatial heterogeneity increased slightly and remained at a high level for a long time. The global Moran's I index changed from −0.129 to 0.080, indicating that the change trend of urban industrial land changed from a negative to positive spatial correlation. No cities belonged to HH and LL in 2010–2014, and HH cities were clustered in northern Zhejiang and LL cities were scattered in Jiangsu and Anhui provinces in 2015–2019. The number of HL cities has been sparse for a long time and distributed randomly. The LH cities were concentrated in northern Zhejiang in 2010–2014, and the spatial scope shrank significantly in 2015–2019. From 2010 to 2014, hotspot cities were clustered in the south of Jiangsu and extended to Zhejiang, with secondary hotspot cities in their periphery. Cold spot cities were clustered in Huaihai Economic Zone and southwestern Anhui. From 2015 to 2019, hotspot cities were clustered in the Shanghai metropolitan area and central Zhejiang, with secondary hotspot cities distributed in their periphery, while cold spot cities formed three clusters in northern Anhui, central Jiangsu, and southern Zhejiang.

4.2. *Driving Mechanism of the Urban Industrial Land Evolution Model*

4.2.1. Direct Influence

The results of the analysis of *Factor−detector* are shown in Table 5. Per capita GDP, export, and education investment had a low direct influence among the 12 factors, and they were not statistically significant from 2010 to 2014. The average direct influence of the remaining nine factors was 0.37, which was used as a threshold to classify the influence factors. The direct influence of real estate investment, tertiary industry, higher education institution number, and road area was greater than the average and they were important factors; import, foreign direct investment, built-up area, gross domestic product, and patent application number had a direct influence that was less than the average as auxiliary factors. Out of the 12 factors, only import had a low and non-statistically significant direct influence from 2015 to 2019. The average direct influence of the rest of the factors was 0.52. Export, built-up area, tertiary industry, patent application number, road area, real estate investment, education investment, and gross domestic product (GDP) had a direct influence that w greater than the average and they were important factors; while foreign direct investme higher education institution number, and per capita GDP had a direct influence that w less than the average as auxiliary factors (Table 5).

Table 5. Analysis of the evolution mode of independent variables in 2015–2019.

Indicators	Code	2010–2014		2015–2019		Change
		q	p	q	p	q
Gross Domestic Product (GDP)	X_1	0.34	0.03	0.54	0.01	0.2
Built-Up Area	X_2	0.34	0.04	0.6	0	0.26
Road Area	X_3	0.38	0.01	0.57	0	0.19
Real Estate Investment	X_4	0.44	0.01	0.55	0.01	0.11
Per Capita GDP	X_5	0.09	0.98	0.31	0.02	#
Tertiary Industry	X_6	0.43	0.01	0.58	0	0.15
Import	X_7	0.36	0.01	0.34	0.26	#
Export	X_8	0.29	0.22	0.64	0	#
Foreign Direct Investment	X_9	0.35	0.01	0.44	0.02	0.09
Patent Application Number	X_{10}	0.28	0.04	0.57	0	0.3
Higher Education Institution Number	X_{11}	0.42	0.01	0.41	0.04	−0.01
Education investment	X_{12}	0.22	0.12	0.54	0	#

Notes: "#" represents that there were non-statistically significant phenomena in 2010–2014 or 2015–2019.

The comparison between 2010–2014 and 2015–2019 shows that the average grow of the factor influence exceeded 40%, and the factors insignificantly decreased from th to one, indicating a significant increase in their power to explain the evolution patte of urban industrial land. Export and education investment showed the most significa change, transforming from non-statistically significant factors to important factors, a per capita GDP, despite a significant increase in the direct influence, still remained auxiliary factor. Patent application number, gross domestic product, and built-up a experienced a great increase in direct influence and changed from auxiliary factors important factors, playing a key role in the evolution of urban industrial land in t long term. Factors such as foreign direct investment, real estate investment, and tertia industry experienced a small range of growth in their direct influence and their drivi force remained stable over time. The influence of higher education institution numb decreased but to a lesser extent and its driving force remained stable. It is notable th import shifted from an auxiliary factor to a non-statistically significant factor, with t steepest decline. The influence of industrial structure serviceization increased while t industrialization stage saw a significant decrease; the influence of foreign investme remained stable in the long term while the role of import and export was reversed; t driving force of applied innovation (patent) and education investment soared rapidly wh the role of higher education institution number generally remained stable. The governme demand, in general, is the key driving force of the evolution of urban industrial land, t influence of supporting facilities and business environment has long remained stable, a the mechanism of action of industrialization, globalization, and innovation is becomi more complex.

4.2.2. Interactive Influence

The results of the analysis of *Interaction−detector* are shown in Tables 6 and 7. T interaction of factor pairs is dominated by bifactor enhancement, with only a few showi nonlinear enhancement effects found in 2010–2014, including $X_1 \cap X_9$, $X_1 \cap X_{10}$, $X_3 \cap X$ $X_9 \cap X_{12}$, and $X_{12} \cap X_{10}$. The difference between the interactive influence and the dire influence is calculated and defined as the enhancement range as the synergy effect factor interaction. For example, the direct influence of X_1 was 0.34, and the interacti influence of X_1 with X_2, X_3 ......X_{11}, and X_{12} was 0.40, 0.69 ......0.54, and 0.51, respective with an enhancement range of 0.06, 0.35, and ......0.19, 0.17 and an average of 0.23 (Table 8). Education investment, per capita GDP, patent application number, road area, export, gross domestic product, and foreign direct investment experienced a large enhancement range from 2010 to 2014, and from 2015 to 2019, per capita GDP, import, higher education institution number, foreign direct investment, and patent application number experienced a large enhancement range, and they can be considered as super interaction factors. The maximum interaction forces were 0.76 ($X_3 \cap X_{10}$) and 0.84 ($X_1 \cap X_8$, $X_3 \cap X_{10}$, $X_8 \cap X_9$, $X_8 \cap X_{12}$, $X_{10} \cap X_9$), and the minimum values were 0.26 ($X_5 \cap X_{12}$) and 0.49 ($X_7 \cap X_{11}$) for 2010–2014 and 2015–2019, respectively. Notably, the interactive influence of $X_1 \cap X_3$, $X_1 \cap X_4$, $X_1 \cap X_9$, $X_1 \cap X_{10}$, $X_2 \cap X_3$, $X_3 \cap X_4$, $X_3 \cap X_6$, and $X_4 \cap X_{12}$ was close to or more than 0.7 from 2010 to 2014, and more than half of the factor pairs were close to or more than 0.7 from 2015 to 2019, including $X_1 \cap X_2$, $X_1 \cap X_9$, $X_2 \cap X_3$, $X_2 \cap X_4$, $X_2 \cap X_5$, $X_2 \cap X_6$, $X_2 \cap X_9$, $X_2 \cap X_{12}$, $X_3 \cap X_8$, $X_4 \cap X_8$, $X_4 \cap X_{10}$, $X_6 \cap X_8$, $X_6 \cap X_{10}$, and $X_{12} \cap X_{10}$, with an interactive influence of even more than 0.8, playing a pivotal role in the evolution of urban industrial land. Generally, the interaction of per capita GDP, foreign direct investment, and patent application number with other factors remained at a high level for a long time, and they must be given priority attention in urban industrial land management policy design and spatial planning.

Table 6. Analysis of the evolution mode of independent variables in 2015–2019.

	X_1	X_2	X_3	X_4	X_5	X_6	X_7	X_8	X_9	X_{10}	X_{11}	X_{12}
X_1	0.34											
X_2	0.40	0.34										
X_3	0.69	0.69	0.38									
X_4	0.74	0.53	0.71	0.44								
X_5	0.38	0.38	0.41	0.47	0.09							
X_6	0.49	0.47	0.74	0.66	0.45	0.43						
X_7	0.54	0.45	0.63	0.50	0.40	0.52	0.36					
X_8	0.53	0.55	0.64	0.53	0.34	0.56	0.50	0.29				
X_9	0.75	0.54	0.46	0.50	0.38	0.65	0.56	0.58	0.35			
X_{10}	0.74	0.51	**0.76**	0.50	0.32	0.64	0.42	0.53	0.49	0.28		
X_{11}	0.54	0.47	0.54	0.62	0.44	0.51	0.61	0.50	0.52	0.49	0.42	
X_{12}	0.51	0.48	0.63	0.69	**0.26**	0.58	0.53	0.52	0.67	0.54	0.51	0.22

Notes: Red represents the nonlinear enhancement relationship, bold represents the maximum and minimum values, and underline represents the significant high values.

Table 7. Analysis of the evolution mode of independent variables in 2015–2019.

	X_1	X_2	X_3	X_4	X_5	X_6	X_7	X_8	X_9	X_{10}	X_{11}	X_{12}
X_1	0.54											
X_2	0.82	0.60										
X_3	0.73	0.81	0.57									
X_4	0.57	0.81	0.74	0.55								
X_5	0.64	0.81	0.73	0.63	0.31							
X_6	0.64	0.80	0.74	0.60	0.73	0.58						
X_7	0.61	0.72	0.72	0.59	0.51	0.62	0.34					
X_8	**0.84**	0.77	0.81	0.83	0.79	0.81	0.75	0.64				
X_9	0.59	0.81	0.78	0.61	0.65	0.65	0.56	**0.84**	0.44			
X_{10}	0.82	0.74	**0.84**	0.81	0.71	0.80	0.78	0.69	**0.84**	0.57		
X_{11}	0.66	0.72	0.75	0.68	0.60	0.63	**0.49**	0.78	0.56	0.77	0.41	
X_{12}	0.56	0.81	0.73	0.56	0.62	0.60	0.61	**0.84**	0.60	0.81	0.66	0.54

Notes: Same as Table 6.

Table 8. Analysis of the evolution mode of independent variables in 2015–2019.

	X_1	X_2	X_3	X_4	X_5	X_6	X_7	X_8	X_9	X_{10}	X_{11}	X_{12}	Ave
2010–2014	0.23	0.16	0.25	0.14	0.30	0.14	0.16	0.24	0.21	0.26	0.10	0.32	0.2
2015–2019	0.14	0.18	0.19	0.12	0.37	0.11	0.29	0.16	0.24	0.21	0.26	0.13	0.20

4.3. Economic Performance of Urban Industrial Land Consumption

4.3.1. Government: Secondary Industry Added Value

Changes in urban industrial land and industrial value-added growth are well coo nation on the whole, with a stable spatial pattern in general, but there are more diversif decoupling types. More than 85% of the cities were in strong or weak decoupling fr 2010 to 2014, with land consumption in good coordination with economic developm The cities of the former type were mainly found in the Shanghai metropolitan area a southwestern Anhui while those of the latter type were widely distributed in Yang River Delta. Other cities were decoupling, scattered in a random distribution. From 2 to 2019, the cities in strong or weak decoupling decreased to about 70%, with the forn clustered in the junction of three provinces and southeastern Zhejiang, including Sha hai, Suzhou-JS, Huzhou, Nanjing, Suqian, Wenzhou, Taizhou-ZJ, and the latter wid distributed in Yangtze River Delta. Zhenjiang, Jiaxing, Jinhua, Wuhu, and Bengbu w in expansive coupling. Cities in strong or expansive negative decoupling emerged, w the former being Huaibei and Tongling, and the latter being Xuzhou, Nantong, Hangzh and Anqing, all in the problem-space of Yangtze River Delta (Figure 10). The decoupli types were changing with increasing complexity. A total of 31.71% of the cities were progressive decoupling, distributed in clusters in southwest Anhui and northwest Z jiang; 36.59% remained unchanged in the decoupling state, clustered in northwest Jiang southeast Zhejiang, and central Yangtze River Delta (in the junction of many provin along Shanghai-Huangshan). Notably, 31.71% of the cities were in regressive decoupli concentrated in Jiangsu and Anhui in a band. Measures should be taken in the future prevent them from degenerating from decoupling to coupling or even negative decoupli

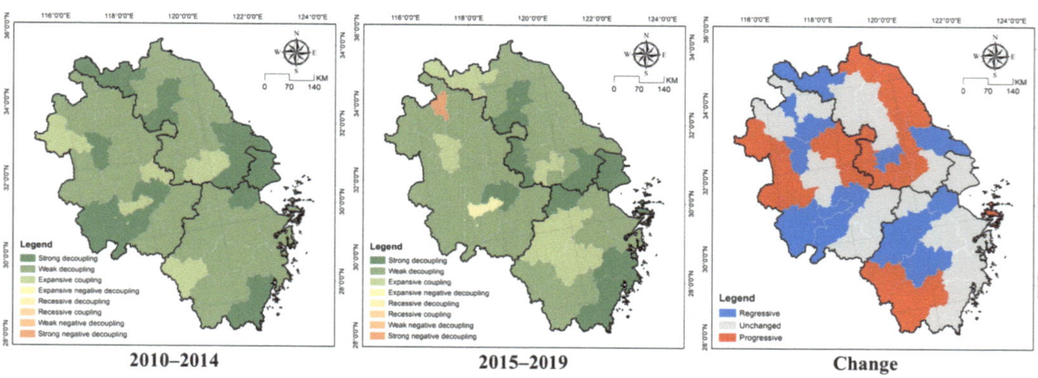

Figure 10. Spatial analysis of the decoupling relationship between urban industrial land and add value in Yangtze River Delta.

4.3.2. Market: Secondary Industry Enterprise Assets

The coordination between the urban industrial land change and industrial enterpr asset growth is acceptable on the whole and the decoupling types and spatial patterns stable in general, with large changes in some areas of Anhui and Zhejiang. About 80% of t cities were in strong or weak decoupling from 2010 to 2014, including Changzhou, Ningbo, Fuyang, and Xuancheng in expansive coupling; Wuxi, Jinhua, and Taizhou-ZJ in expansive negative decoupling; and Wenzhou in recessive decoupling. The cities in strong or weak decoupling decreased to 56.10% from 2015 to 2019 while the cities in strong or expansive negative decoupling expanded to about 20%, with Huai'an and Zhoushan in recessive decoupling. From the perspective of de-coupling changes, 31.71% of the cities reached progressive decoupling, distributed in three clusters in southwestern Anhui, central Jiangsu, and northern Zhejiang; 24.39% remained unchanged, mostly concentrated in the Shanghai metropolitan area and central Zhejiang; Regressive decoupling was found in 43.90%, which were concentrated in southeastern Zhejiang, northeastern Jiangsu, western Anhui, Nanjing metropolitan area, and its sur-rounding areas (Figure 11).

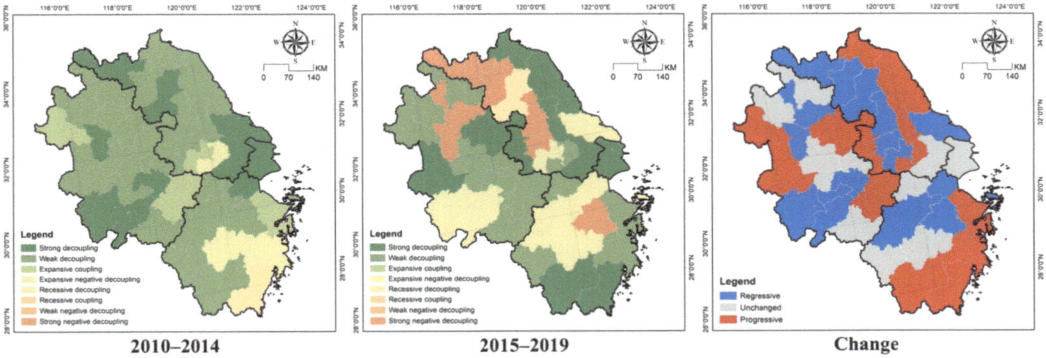

Figure 11. Spatial analysis of the decoupling relationship between urban industrial land and enterprise assets in Yangtze River Delta.

5. Discussion

The evolution model and spatial pattern of urban industrial land in the Yangtze River Delta are becoming increasingly more complicated, and the level of spatial agglomeration, heterogeneity, and relevance is decreasing. This complexity is shown in many areas. At the micro level, the evolution patterns, economic values, and driving mechanisms of different cities changed significantly in 2010–2014 and 2015–2019; at the macro level, the evolution pattern of urban industrial land changed from a "pyramid-" to an "olive"-shaped structure in the quantitative combination, with a shift from a random geographical distribution to geographical agglomeration, and from positive spatial correlation to negative spatial correlation. This conclusion confirms the views of other scholars. Louw [81] found significant spatial heterogeneity in the productivity of industrial areas in the Netherlands. Wang [82] and Cui [83] found that industrial land in China has in uneven spatial distribution with spatial convergence. A notable fact is that the viewpoints of some papers are not exactly the same or even opposite to the findings in this paper. Zhu [84] found, using a centrifugal model and contribution degree, that urban agglomerations in the middle reaches of the Yangtze River are significantly uneven in terms of the spatial distribution of industrial land, and that the spatial heterogeneity is increasing rather than decreasing over time, which is different from the conclusion reached in this paper. The possible reason for this discrepancy is that the study areas are different. Yangtze River Delta is located in the lower reaches of the Yangtze River as a developed region while the urban agglomerations in the middle reaches of the Yangtze River are located in the middle reaches and lag behind Yangtze River Delta in terms of development [85]. He found that the scale of urban industrial land in 38 counties of Chongqing changed repeatedly between regional imbalance and balance from 2009 to 2018, and there was no continuous decrease in spatial heterogeneity [86]. The scale may be a key factor leading to this discrepancy. T paper analyzed cities at the mesoscopic scale, compared to the microscopic scale at wh he analyzed counties. These differences indicate that there are still controversies in research on urban industrial land and its changing characteristics. More empirical stud and case studies are needed in the future to refine common patterns and identify individ characteristics through comparative analysis of new findings and the contributions different papers to lay the foundation for establishing a characteristic industrial la management theory.

Due to the special background of land system and land management policies in Chi where the primary market of industrial land supply is monopolized by the government, city government tends to use land allocation as a policy tool to attract industrial investm and manufacturing enterprises [87]. Many city governments have adopted the developm mode of boosting industrialization and urbanization by land supply, which, together w open-up policies and demographic dividends, did have a positive effect on urban industr economic development. As for the value added, most cities remained in weak decoupli and strong decoupling, indicating that "seek-development-with-land" is still valid for c governments. However, from the perspective of enterprise assets, an increasing number cities degenerated into the states of strong and expansive negative decoupling, with gr variation in the type of decoupling between the two stages, indicating an unstable mar performance and increasing challenges. This view is also supported by scholars in the fie Qi [88] found that the relationship between industrial land and economic growth is in we decoupling in most Chinese cities, and there are a growing number of cities in negati decoupling. Against the background of increasingly strict land resource management a constraints, promotion of the transformation of the urban development mode from rou expansion to refined utilization through diversified land supply means such as expansi contraction, and dynamic balance is becoming a new development trend. The impact temporal and spatial evolution complexity and its driving factors must be considered changes in the management system of urban industrial land supply, the development spatial allocation schemes, and the design of planning and governance policies [89].

In the new era of high-quality development, it has become an important issue for government and scholars to work together to improve the coordination between the char in industrial land and industrial economic growth, suppress the negative effect of la change, and reactivate the positive effect by reasonably controlling the scale of industr land and optimizing the quantity, type structure, spatial layout, and development mode industrial land supply through land spatial planning and land management reform [9 There is a need to implement zoning management of urban industrial land changes Yangtze River Delta, and to clarify the specialization and targeted strategies for differe zonings, which are of great value to government decision makers. Based on the decoupli relationship between urban industrial land change and industrial economic growth a taking the land evolution model and its driving mechanism as constraints, the Yangtze Riv Delta is divided into four management zonings (Figure 2). In other words, the decoupli relationship between urban industrial land change and manufacturing economic grow should be the center, and cities in an ideal state should try to maintain the current evoluti pattern while those in an unhealthy state should use a transformation strategy to chan the unsustainable evolution pattern or trend by controlling the quantity and quality urban industrial land supply.

Shanghai, Nanjing, Suzhou-JS, Lianyungang, Huai'an, Yancheng, Taizhou-JS, W zhou, Zhoushan, Lishui, Chuzhou, and Lu'an are in the best state of strong decoupli In the future, a path-dependent strategy should be adopted to maintain land reducti development plans and policies, with these cities included in the reduction-oriented tra formation policy zoning. Wuxi, Changzhou, Yangzhou, Suqian, Ningbo, Huzhou, Sha ing, Quzhou, Taizhou-ZJ, Hefei, Huainan, Maanshan, Huangshan, Fuyang, Suzhou-A Bozhou, Chizhou, and Xuancheng are in weak decoupling, characterized by synergistic velopment between land and economy. The current land supply and management polic should be maintained in the future, with such cities included in the incremental high-quality development zoning. Moreover, Zhenjiang, Jiaxing, Jinhua, Wuhu, and Bengbu are in expansive coupling with low land use intensification, and the focus should be placed on improving land use efficiency in the future and these cities should be included in the incremental high-quality development zoning as well. Xuzhou, Nantong, Hangzhou, and Anqing are in expansive negative decoupling with extensive land use. In the future, strict control should be exercised over the land supply, with a focus on improving the quality of land use to ensure land consumption keeps pace with economic growth at least, and these cities are included in the incremental synchronous growth zoning. Huaibei and Tongling are in strong negative decoupling with serious land wastage. In the future, the focus should be on promoting industrial upgrading and reducing the land supply appropriately, with the inclusion of these cities in the reduction and upgrading development zoning (Figure 12).

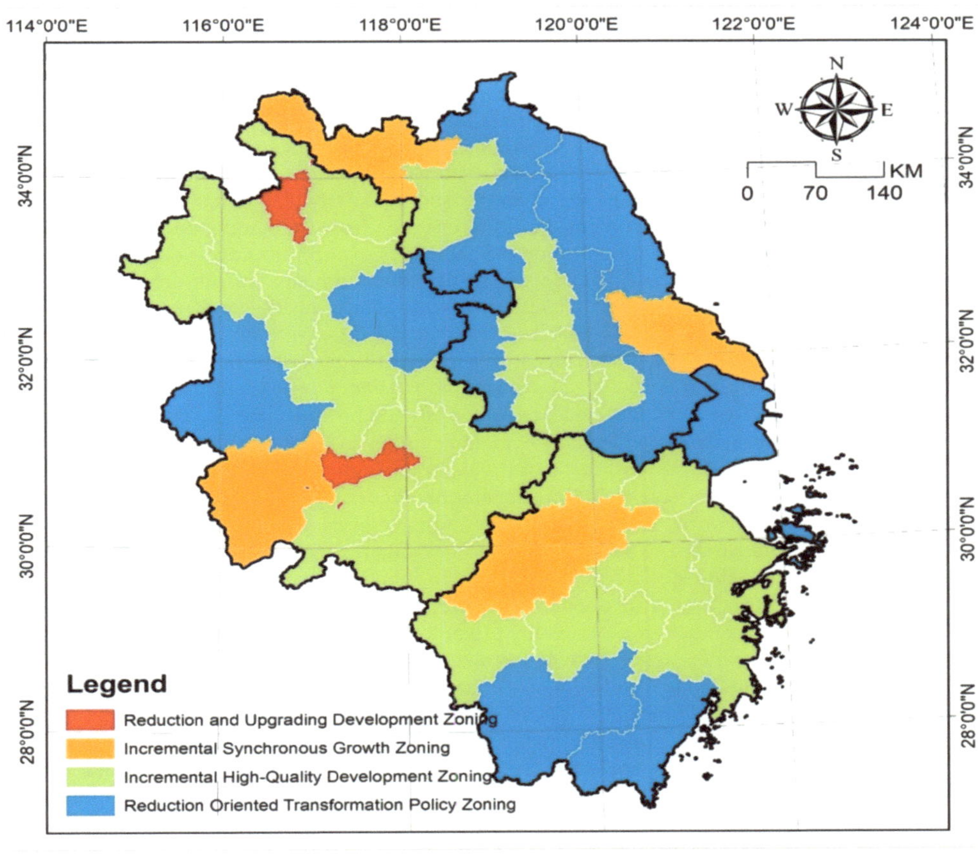

Figure 12. Analysis on the zoning management in Yangtze River Delta.

Additionally, differentiated governance strategies should be adopted for land supply in policy areas, with a combination of both quantitative and qualitative control methods. The government requires the coordination of territorial spatial planning and development planning and has announced the target of industrial added value (average annual growth rate) by 2025 in the 14th Five-Year Industrial Development Plan. The added value of the secondary industry is the core of the performance appraisal of city governments, and its decoupling from industrial land has long remained stable. Therefore, this paper predicts the industrial land area in 2025 based on the decoupling relationship between the char in industrial land and added value in urban areas, coupled with the objectives of the 1 Five-Year Plan and policy zoning, to provide a basis for controlling the quantity of la supply (Table 9). The cities in reduction-oriented transformation policy zoning shou increase the input of capital, technology, talents, and other innovative factors per unit a of construction land; innovate the use model of industrial land; and promote industr development transformation. These cities are in the stage of industrialization transiti in essence a process of optimizing and reconfiguring the relationship and combination production factors such as land, natural resources, labor, capital, and technology. M efforts should be carried out to promote industrial parks [91], guide enterprises to en industrial parks for development, and set up higher standards of investment access manufacturing, trying to force the improvement of industrial land efficiency by means the system in incremental high-quality development zoning [92].

For the cities in incremental synchronous growth zoning, relying on industrial la change to drive industrial economic growth is still an effective development mode. Hc ever, due to their low land use efficiency and unsatisfactory conversion of economic retur measures should be taken to improve the quality of land use. Future work is increasi the supply of Class I and Class II industrial land, encouraging and supporting industr upgrading and the development of new business models, and promoting the optimizati of the urban industrial land structure in urban industrial land increment according to development stage and characteristics of the city [93]. In the process of reusing industr land, the government should be changed from the leading party to the guiding party have a direct influence on psychological expectations and policy factors. Governme driven industrial land development and investment attraction are often disconnected fr the market, leading to inefficient use of some industrial land, or even to it lying idle unc the background of economic downturn and fierce competition between industrial par Therefore, reasonable incentive and penalty policies for the renewal of inefficient industr land, such as floor area ratio awards, transfer of development rights, relaxation of planni controls, and land price reductions, should be formulated according to the land use nee and selection preferences of manufacturing and high-technology enterprises [94] to indu land-using enterprises to form their own willingness to renew and allow market forces play a decisive role in the transformation of urban industrial land stock.

For the cities in reduction and upgrading development zoning, the scale of urb industrial land should be strictly controlled in the future, and the redevelopment of ind trial land stock should be pushed hard in the process of reduction. First, it is necessary carry out an evaluation of the suitability of urban industrial land reduction, identify spa with reduction, and steadily promote and gradually explore the mode of withdrawal a redevelopment [95], renewal mechanisms [96], and reclamation programs [97] of urb industrial land stock. Second, the redevelopment of industrial land stock should be co bined with the development of urban public space to promote the improvement of urb habitats and raise the quality of life of citizens. From the perspective of land planni it is necessary to respect the change rules of urban industrial land, take into account common and individual needs of different types of manufacturing enterprises [98], gi priority to ensuring land for the development of new industries and new forms of bu ness, support high-tech industries and environment-friendly enterprises, and drive upgrading of industrial structures.

Table 9. Prediction of urban industrial land area based on the decoupling model.

City	Decoupling Index	Secondary Industry Added Value Growth Rate	Urban Industrial Land Growth Rate	Urban Industrial Land Area
Shanghai	−1.09	5.00	−5.46	383.95
Nanjing	−1.91	6.73	−12.87	41.08
Wuxi	0.11	6.79	0.74	74.67
Xuzhou	1.00	7.12	7.12	75.51
Changzhou	0.71	5.88	4.15	114.27
Suzhou-JS	−0.03	4.86	−0.15	127.66
Nantong	1.00	6.81	6.81	94.83
Lianyungang	0.00	10.00	−0.05	51.95
Huai'an	−0.92	10.42	−9.55	19.71
Yancheng	−1.40	7.79	−10.89	19.55
Yangzhou	0.66	5.94	3.90	48.86
Zhenjiang	0.40	2.52	1.01	42.70
Taizhou-JS	−0.80	7.78	−6.25	23.06
Suqian	0.59	10.00	5.92	41.67
Hangzhou	1.00	8.90	8.90	207.50
Ningbo	0.22	9.93	2.21	150.56
Wenzhou	−0.26	7.79	−1.99	5.96
Jiaxing	0.40	7.06	2.82	40.92
Huzhou	0.12	7.09	0.82	33.00
Shaoxing	0.05	9.90	0.51	67.52
Jinhua	0.40	7.72	3.09	30.02
Quzhou	0.43	9.47	4.07	35.49
Zhoushan	−3.71	10.00	−37.12	0.39
Taizhou-ZJ	0.08	5.83	0.44	35.42
Lishui	0.00	12.66	0.00	4.36
Hefei	0.55	10.38	5.66	119.35
Wuhu	0.40	6.48	2.59	19.22
Bengbu	0.40	6.36	2.54	30.52
Huainan	0.68	8.45	5.74	26.08
Ma'anshan	0.21	7.46	1.57	38.64
Huaibei	−0.15	13.40	−2.01	17.17
Tongling	−0.50	13.08	−6.54	15.93
Anqing	1.00	13.32	13.32	58.88
Huangshan	0.33	9.41	3.07	12.17
Chuzhou	−0.20	5.08	−1.03	24.17
Fuyang	0.08	7.62	0.58	21.45
Suzhou-AH	0.72	7.24	5.24	24.68
Lu'an	−0.01	9.36	−0.13	13.38
Bozhou	0.13	8.19	1.07	15.24
Chizhou	0.77	9.52	7.34	9.56
Xuancheng	0.34	6.49	2.24	20.18

6. Conclusions

(1) Behind the increasingly complex spatial and temporal evolution of urban industrial land dynamics, regular features of the process and spatial patterns of urban industrial land change are hidden. According to the Boston Consulting Group matrix, the spatio-temporal evolution model of urban industrial land can be divided into four types of stars, cows, dogs, and question, and the spatial agglomeration, heterogeneity, and correlation of different patterns have gradually decreased.

(2) The forces of different factors on the evolution of urban industrial land are increasingly differentiated, and their direct and interactive influences are significantly enhanced, with bifactor enhancement dominating the interaction of factor pairs. It should be noted that the government demand is the key driving force of the evolution of urban industrial land, the influence of supporting facilities and the business environment has

long remained stable, and the mechanism of action of industrialization, globalizati and innovation is becoming more complex.

(3) The match and synergy between changes in urban industrial land and industr economic growth are fine in general, and the land resource management policy "seek-development-with-land" is still effective for the government, but its effecti ness for enterprises (market) is declining rapidly. The progressive, unchanged, a regressive types of decoupling exist side by side, and the much higher long-te stability than that of assets makes the added value more suitable for future urb industrial land-scale projections.

(4) Based on the decoupling relationship, a technical framework for zoning managem and classification governance of urban industrial land is constructed in this pap with the land evolution pattern and its influencing factors taken into account. T Yangtze River Delta is divided into reduction-oriented transformation policy zoni incremental high-quality development zoning, incremental synchronous growth z ing, and reduction and upgrading development zoning, and adaptive land quant and quality control strategies are proposed for zonings.

The biggest innovation in this paper is the construction of a technical framework ir grating "model evolution + driving mechanism + performance evaluation + policy desig which can be used for urban industrial land resources management and territory spat planning based on the combination of GIS tools, Geodetector software, Boston Consulti Group matrix, and decoupling model. The technical framework and findings presented this paper are not only applicable to China but also provide valuable references for decisi making in countries undergoing rapid industrialization and re-industrialization. For co tries in the early and middle stages of industrialization such as Vietnam, Indonesia, Inc Malaysia, Iran, Uzbekistan, Brazil, and Egypt, and post-industrialized countries such as United States, Japan, Russia, Germany, and France, which put forward re-industrializati strategies after the financial crisis, how to use land resources to attract large-scale inve ment in manufacturing, promote the development of the real economy or the return manufacturing industries, and achieve scientific management of urban industrial land becoming a new challenge for land management and spatial planning. The change in urb industrial land is a key manifestation of this urban industrial economic evolution. T two are showing an increasingly obvious non-linear, dynamic, and phased characteris Empirical studies and case studies of these countries based on decoupling models help accurately capture and quantify the non-linear relationship and asymmetric effect betwe them, providing a scientific basis for government policy makers and planning designer

Due to the limitation of data and information availability, this paper inevitably h some shortcomings in the selection of indicators and policy recommendations. For examp China's industrial land supply is monopolized by the government, and the system a policy have an important impact. However, because it is very difficult to obtain syste and policy data, this paper did not include them in the analysis framework. If they can incorporated into the analysis process in the future, it will help to improve the accuracy the results. As another example, urban industrial land can be further subdivided int variety of types, with some differences in the evolution mode and decoupling relationshi between different types of industrial land, but there no detailed study of them was carri out in this paper. In addition, industrial land change creates economic benefits and soc values while having some impact on both ecology and the environment. To gain a mc comprehensive understanding of the combined effects of industrial land change, we c for future research to further focus on the social and ecological values of industrial la consumption.

Author Contributions: Conceptualization, F.Q. and K.Z.; methodology, F.X. and K.Z.; software, F and S.Z.; validation, F.Q., F.X. and S.Z.; formal analysis, F.Q.; investigation, F.X. and S.Z.; resourc F.X.; data curation, F.X. and S.Z.; writing—original draft preparation, F.X. and S.Z.; writing—revi and editing, F.Q. and K.Z.; visualization, F.X. and S.Z.; supervision, F.Q.; project administration, F and F.Q.; funding acquisition, F.X. and F.Q. All authors have read and agreed to the published version of the manuscript.

Funding: This research was funded by Teaching Reform Research of Education Department of Guangdong Province, grant number 2019yjjg011.

Data Availability Statement: https://www.mohurd.gov.cn/index.html (accessed on 17 May 2022) and http://www.stats.gov.cn/(accessed on 21 May 2022).

Conflicts of Interest: The authors declare no conflict of interest.

Appendix A

Table A1. Analysis of the evolution mode of independent variables in 2010–2014.

City	X_1	X_2	X_3	X_4	X_5	X_6	X_7	X_8	X_9	X_{10}	X_{11}	X_{12}
Shanghai	3	3	3	3	1	3	3	4	3	3	3	3
Nanjing	3	3	4	3	1	3	3	3	3	3	3	3
Wuxi	3	4	3	3	1	3	3	3	3	3	3	3
Xuzhou	3	3	4	1	1	3	1	1	1	1	1	3
Changzhou	3	3	3	3	1	3	3	1	3	3	1	1
Suzhou-JS	4	3	3	4	1	4	3	3	3	3	3	3
Nantong	4	4	4	3	1	4	3	1	3	3	2	4
Lianyungang	2	2	1	1	1	2	1	2	1	2	1	2
Huai'an	1	2	4	1	1	1	2	1	1	2	1	1
Yancheng	4	1	2	2	1	2	1	2	1	2	1	4
Yangzhou	2	1	1	1	1	2	1	1	1	3	2	1
Zhenjiang	2	2	1	2	1	2	1	1	1	3	1	1
Taizhou-JS	2	1	1	2	1	2	1	1	1	4	1	2
Suqian	2	1	2	2	1	2	2	1	2	2	2	2
Hangzhou	3	3	3	3	4	3	3	4	3	3	3	3
Ningbo	3	3	3	3	1	3	3	4	3	3	3	3
Wenzhou	3	4	1	4	1	4	1	4	2	4	2	3
Jiaxing	2	1	1	2	1	2	2	4	4	4	1	2
Huzhou	1	2	1	2	1	2	1	1	1	1	1	1
Shaoxing	4	4	2	4	1	4	1	4	2	4	2	2
Jinhua	2	1	1	2	1	2	1	4	1	1	1	2
Quzhou	1	1	2	1	1	1	1	1	1	1	1	1
Zhoushan	1	1	2	2	1	1	2	2	2	1	2	1
Taizhou-ZJ	2	1	1	2	1	2	1	4	2	1	1	1
Lishui	2	1	1	2	1	2	1	2	2	1	1	2
Hefei	3	3	3	3	1	3	1	2	3	3	3	3
Wuhu	1	1	3	1	1	1	2	1	3	1	1	2
Bengbu	1	2	2	2	1	1	2	1	2	1	1	2
Huainan	1	1	2	1	1	1	2	2	1	1	1	1
Ma'anshan	1	1	1	1	1	1	1	2	1	2	1	1
Huaibei	1	2	1	1	1	1	2	1	2	2	1	1
Tongling	1	2	1	1	1	1	1	2	1	1	1	1
Anqing	2	1	1	1	1	2	2	1	1	2	2	2
Huangshan	1	2	1	1	1	1	2	1	1	1	2	1
Chuzhou	2	2	2	2	1	2	2	1	2	2	1	2
Fuyang	2	2	2	2	1	2	2	1	1	2	1	2
Suzhou-AH	2	2	2	2	1	2	2	1	2	2	2	2
Lu'an	2	1	2	2	2	2	2	1	2	2	1	2
Bozhou	2	2	2	2	1	2	2	1	2	1	1	2
Chizhou	1	1	1	1	1	1	1	2	1	2	1	1
Xuancheng	2	1	2	2	2	2	2	1	2	2	2	2

Table A2. Analysis of the evolution mode of independent variables in 2015–2019.

City	X_1	X_2	X_3	X_4	X_5	X_6	X_7	X_8	X_9	X_{10}	X_{11}	X_{12}
Shanghai	4	4	3	4	3	4	4	3	4	4	3	4
Nanjing	4	3	3	4	3	4	4	3	4	4	4	4
Wuxi	3	3	3	4	3	3	4	3	3	3	3	4
Xuzhou	1	3	3	2	4	1	1	2	2	4	1	1
Changzhou	4	3	3	4	3	4	1	4	4	3	1	4
Suzhou-JS	3	3	4	4	3	3	3	3	3	4	4	4
Nantong	1	4	3	1	4	1	1	4	1	3	1	1
Lianyungang	2	3	1	2	2	2	2	1	1	1	2	2
Huai'an	2	2	4	1	1	2	1	2	1	1	1	2
Yancheng	1	2	2	1	1	1	1	1	1	1	2	1
Yangzhou	2	2	1	2	3	2	1	2	2	4	2	2
Zhenjiang	1	1	1	1	3	1	1	1	1	1	2	1
Taizhou-JS	1	2	2	1	3	1	2	1	2	1	2	1
Suqian	1	2	1	1	1	1	1	2	1	1	1	1
Hangzhou	4	4	4	4	3	4	4	3	4	4	3	4
Ningbo	4	3	4	4	3	4	4	3	3	3	3	4
Wenzhou	1	3	1	1	1	1	2	3	2	4	4	1
Jiaxing	1	2	2	1	3	1	2	3	1	4	1	1
Huzhou	1	2	1	2	1	1	2	1	2	1	1	1
Shaoxing	2	4	2	2	3	1	2	3	1	3	4	2
Jinhua	1	2	2	1	1	1	2	3	1	4	2	1
Quzhou	1	1	1	2	1	1	2	1	2	2	1	1
Zhoushan	2	1	1	2	3	2	2	1	2	1	1	2
Taizhou-ZJ	1	1	1	2	1	1	2	1	2	4	1	1
Lishui	1	2	2	1	1	1	2	1	1	2	1	1
Hefei	4	3	4	3	4	4	2	2	4	4	3	4
Wuhu	2	1	3	1	4	2	1	2	3	1	1	2
Bengbu	2	1	2	2	2	2	1	1	1	1	1	1
Huainan	2	1	2	2	1	2	1	2	2	2	2	2
Ma'anshan	2	1	1	1	4	2	2	1	4	2	1	1
Huaibei	2	1	2	2	1	2	1	2	1	2	1	2
Tongling	2	1	2	2	1	2	2	1	2	1	1	2
Anqing	1	2	2	2	2	1	1	2	2	1	1	1
Huangshan	2	1	2	2	2	2	1	2	1	2	1	1
Chuzhou	1	1	1	1	4	2	2	2	1	1	1	1
Fuyang	1	2	2	2	2	2	1	2	2	1	1	1
Suzhou-AH	1	2	1	1	2	1	1	2	1	2	1	1
Lu'an	2	1	1	2	2	2	1	2	1	2	1	1
Bozhou	1	2	1	2	2	1	1	2	1	2	1	2
Chizhou	2	1	1	1	2	1	2	1	1	1	1	1
Xuancheng	1	2	1	1	1	1	1	2	1	1	1	1

Table A3. Index data based on the max-min standardization method in 2010 and 2014.

City	Industrial Land		Added Value		Enterprise Assets	
	2010	2014	2010	2014	2010	2014
Shanghai	1.0000	1.0000	1.0000	1.0000	1.0000	1.0000
Nanjing	0.2151	0.2256	0.2803	0.4414	0.2447	0.3024
Wuxi	0.0721	0.0896	0.2161	0.2293	0.1859	0.1798
Xuzhou	0.0430	0.0310	0.1297	0.1683	0.0856	0.1289
Changzhou	0.0512	0.0719	0.1791	0.2219	0.1642	0.1961
Suzhou-JS	0.2083	0.1752	0.2687	0.4334	0.2356	0.3457
Nantong	0.0718	0.0195	0.0989	0.1152	0.0701	0.0879
Lianyungang	0.0415	0.0589	0.0240	0.0465	0.0378	0.0563

Table A3. *Cont.*

City	Industrial Land		Added Value		Enterprise Assets	
	2010	2014	2010	2014	2010	2014
Huai'an	0.0581	0.0503	0.0524	0.0726	0.0280	0.0415
Yancheng	0.0345	0.0404	0.0414	0.0635	0.0255	0.0367
Yangzhou	0.0329	0.0461	0.0723	0.1440	0.0539	0.0809
Zhenjiang	0.0437	0.0440	0.0616	0.0790	0.0473	0.0631
Taizhou-JS	0.0342	0.0466	0.0417	0.0840	0.0353	0.0549
Suqian	0.0175	0.0244	0.0199	0.0304	0.0141	0.0313
Hangzhou	0.0816	0.1027	0.2966	0.3885	0.2885	0.3430
Ningbo	0.1645	0.1992	0.2300	0.2805	0.2173	0.2199
Wenzhou	0.0423	0.0005	0.0764	0.0888	0.0537	0.0372
Jiaxing	0.0327	0.0201	0.0351	0.0367	0.0398	0.0463
Huzhou	0.0431	0.0453	0.0390	0.0404	0.0313	0.0342
Shaoxing	0.0332	0.0814	0.0234	0.1495	0.0344	0.1623
Jinhua	0.0170	0.0192	0.0184	0.0172	0.0220	0.0177
Quzhou	0.0203	0.0257	0.0168	0.0147	0.0146	0.0198
Zhoushan	0.0053	0.0045	0.0208	0.0264	0.0265	0.0293
Taizhou-ZJ	0.0581	0.0701	0.0514	0.0581	0.0445	0.0403
Lishui	0.0000	0.0000	0.0031	0.0002	0.0073	0.0077
Hefei	0.0843	0.0968	0.1387	0.2177	0.0854	0.1184
Wuhu	0.0330	0.0125	0.0711	0.1004	0.0519	0.0855
Bengbu	0.0240	0.0236	0.0177	0.0363	0.0119	0.0208
Huainan	0.0163	0.0144	0.0307	0.0257	0.0609	0.0633
Ma'anshan	0.0385	0.0442	0.0571	0.0514	0.0412	0.0445
Huaibei	0.0211	0.0235	0.0272	0.0337	0.0368	0.0428
Tongling	0.0078	0.0121	0.0333	0.0399	0.0281	0.0360
Anqing	0.0278	0.0018	0.0140	0.0110	0.0070	0.0111
Huangshan	0.0033	0.0059	0.0014	0.0000	0.0000	0.0000
Chuzhou	0.0214	0.0353	0.0073	0.0122	0.0060	0.0094
Fuyang	0.0075	0.0166	0.0056	0.0086	0.0048	0.0082
Suzhou-AH	0.0131	0.0141	0.0068	0.0152	0.0050	0.0060
Lu'an	0.0116	0.0121	0.0000	0.0080	0.0028	0.0070
Bozhou	0.0060	0.0108	0.0028	0.0029	0.0007	0.0036
Chizhou	0.0024	0.0001	0.0035	0.0042	0.0031	0.0062
Xuancheng	0.0093	0.0148	0.0006	0.0009	0.0027	0.0023

Table A4. Index data based on the max-min standardization method in 2015 and 2020.

City	Industrial Land		Added Value		Enterprise Assets	
	2015	2019	2015	2019	2015	2019
Shanghai	1.0000	1.0000	1.0000	1.0000	1.0000	1.0000
Nanjing	0.2133	0.1678	0.4899	0.4833	0.2950	0.2981
Wuxi	0.0887	0.1258	0.2391	0.2495	0.1766	0.2088
Xuzhou	0.0319	0.0855	0.1728	0.1344	0.1236	0.0718
Changzhou	0.0905	0.1597	0.2640	0.2842	0.2196	0.2101
Suzhou-JS	0.1716	0.2334	0.4455	0.3947	0.3402	0.3634
Nantong	0.0266	0.1116	0.1271	0.1445	0.0876	0.0772
Lianyungang	0.0657	0.0895	0.0552	0.0712	0.0592	0.0630
Huai'an	0.0681	0.0593	0.0836	0.0959	0.0466	0.0291
Yancheng	0.0578	0.0650	0.1026	0.0860	0.0618	0.0571
Yangzhou	0.0412	0.0646	0.1583	0.1440	0.0803	0.0631
Zhenjiang	0.0455	0.0672	0.0889	0.0699	0.0633	0.0411
Taizhou-JS	0.0544	0.0555	0.0917	0.0942	0.0576	0.0539
Suqian	0.0273	0.0472	0.0380	0.0396	0.0334	0.0216

Table A4. *Cont.*

City	Industrial Land		Added Value		Enterprise Assets	
	2015	2019	2015	2019	2015	2019
Hangzhou	0.1033	0.2250	0.4034	0.4238	0.3437	0.3779
Ningbo	0.1619	0.2394	0.2988	0.3192	0.2209	0.2616
Wenzhou	0.0035	0.0044	0.1009	0.0844	0.0331	0.0358
Jiaxing	0.0215	0.0568	0.0392	0.0546	0.0447	0.0538
Huzhou	0.0346	0.0507	0.0455	0.0584	0.0359	0.0425
Shaoxing	0.0829	0.1146	0.1574	0.1466	0.1498	0.1121
Jinhua	0.0198	0.0387	0.0207	0.0188	0.0187	0.0186
Quzhou	0.0272	0.0442	0.0169	0.0173	0.0222	0.0239
Zhoushan	0.0045	0.0036	0.0317	0.0220	0.0280	0.0207
Taizhou-ZJ	0.0400	0.0565	0.0615	0.0691	0.0413	0.0419
Lishui	0.0000	0.0000	0.0029	0.0000	0.0072	0.0059
Hefei	0.1020	0.1526	0.2357	0.2096	0.1432	0.1887
Wuhu	0.0132	0.0227	0.1067	0.0916	0.0893	0.0947
Bengbu	0.0258	0.0410	0.0430	0.0354	0.0257	0.0192
Huainan	0.0167	0.0268	0.0232	0.0189	0.0616	0.0252
Ma'anshan	0.0392	0.0578	0.0487	0.0500	0.0441	0.0498
Huaibei	0.0194	0.0282	0.0304	0.0102	0.0474	0.0133
Tongling	0.0122	0.0366	0.0477	0.0278	0.0387	0.0321
Anqing	0.0020	0.0439	0.0134	0.0194	0.0134	0.0108
Huangshan	0.0060	0.0109	0.0000	0.0006	0.0000	0.0000
Chuzhou	0.0364	0.0400	0.0177	0.0358	0.0127	0.0216
Fuyang	0.0212	0.0307	0.0109	0.0166	0.0102	0.0094
Suzhou-AH	0.0143	0.0259	0.0205	0.0195	0.0065	0.0115
Lu'an	0.0126	0.0171	0.0117	0.0123	0.0067	0.0048
Bozhou	0.0118	0.0186	0.0074	0.0157	0.0049	0.0122
Chizhou	0.0003	0.0035	0.0078	0.0091	0.0060	0.0062
Xuancheng	0.0151	0.0250	0.0036	0.0041	0.0002	0.0057

References

1. Louw, E.; Bontekoning, Y. Planning of industrial land in the Netherlands: Its rationales and consequences. *Tijdschr. V Econo-Mische En Soc. Geogr.* **2007**, *98*, 121–129. [CrossRef]
2. Roost, F.; Jeckel, E. Post-Fordist Production and Urban Industrial Land Use Patterns. *Urban Plan.* **2021**, *6*, 321–333. [CrossRef
3. Funderburg, R.G.; Zhou, X.X. Trading industry clusters amid the legacy of industrial land-use planning in southern Califor *Environ. Plan. A—Econ. Space* **2014**, *45*, 2752–2770. [CrossRef]
4. Mavrakis, A.; Papavasileiou, C.; Salvati, L. Towards (Un)sustainable urban growth? Industrial development, land-use, s depletion and climate aridity in a Greek agro-forest area. *J. Arid. Environ.* **2015**, *121*, 1–6. [CrossRef]
5. Huang, Z.J.; He, C.F.; Li, H. Local government intervention, firm-government connection, and industrial land expansion in Chi *J. Urban Aff.* **2019**, *41*, 206–222. [CrossRef]
6. Tu, F.; Yu, X.F.; Ruan, J.Q. Industrial land use efficiency under government intervention: Evidence from Hangzhou, China. *Hab. Int.* **2014**, *43*, 1–10. [CrossRef]
7. Zhou, L.; Tian, L.; Gao, Y.; Ling, Y.K.; Fan, C.J.; Hou, D.Y.; Shen, T.Y.; Zhou, W.T. How did industrial land supply respond transitions in state strategy? An analysis of prefecture-level cities in China from 2007 to 2016. *Land Use Policy* **2019**, *87*, 104C [CrossRef]
8. Song, M.L.; Wang, S.H.; Wu, K.Y. Environment-biased technological progress and industrial land-use efficiency in China's n normal. *Ann. Oper. Res.* **2018**, *268*, 425–440. [CrossRef]
9. Busscher, N.; Parra, C.; Vanclay, F. Environmental justice implications of land grabbing for industrial agriculture and forestry Argentina. *J. Environ. Plan. Manag.* **2019**, *63*, 500–522. [CrossRef]
10. Abel, T.D.; White, J.; Clauson, S. Risky Business: Sustainability and Industrial Land Use across Seattle's Gentrifying Risksca *Sustainability* **2015**, *7*, 15718–15753. [CrossRef]
11. Egidi, G.; Cividino, S.; Quaranta, G.; Alhuseen, A.; Salvati, L. Land mismatches, urban growth and spatial planning: contribution to metropolitan sustainability. *Environ. Impact Assess. Rev.* **2020**, *84*, 106439. [CrossRef]
12. Wang, Z.S.; Zhao, J.; Lin, S.J.; Liu, Y. Identification of Industrial Land Parcels and Its Implications for Environmental R Management in the Beijing-Tianjin-Hebei Urban Agglomeration. *Sustainability* **2020**, *12*, 174. [CrossRef]
13. Kapungwe, E.M. Industrial land use and heavy metal contaminated wastewater used for irrigation in periurban Zambia. *Sing J. Trop. Geogr.* **2011**, *32*, 71–84. [CrossRef]

Tang, B.; Winky, K. Land-use planning and market adjustment under de-industrialization: Restructuring of industrial space in Hong Kong. *Land Use Policy* **2015**, *43*, 28–36. [CrossRef]
Zhou, L.; Tian, L.; Cao, Y.D.; Yang, L.C. Industrial land supply at different technological intensities and its contribution to economic growth in China: A case study of the Beijing-Tianjin-Hebei region. *Land Use Policy* **2021**, *101*, 105087. [CrossRef]
Gao, J.B.; Qiao, W.F.; Ji, Q.Q.; Yu, C.; Sun, J.W.; Ma, Z.F. Intensive-use-oriented identification and optimization of industrial land readjustment during transformation and development: A case study of Huai'an, China. *Habitat Int.* **2021**, *118*, 102451. [CrossRef]
Zambon, I.; Cerda, A.; Gambella, F.; Egidi, G.; Salvati, L. Industrial Sprawl and Residential Housing: Exploring the Interplay between Local Development and Land-Use Change in the Valencian Community, Spain. *Land* **2019**, *8*, 143. [CrossRef]
Debela, D.D.; Stellmacher, T.; Azadi, H.; Kelboro, G.; Lebailly, P.; Ghorbani, M. The Impact of Industrial Investments on Land Use and Smallholder Farmers' Livelihoods in Ethiopia. *Land Use Policy* **2021**, *99*, 105091. [CrossRef]
Xiong, C.; Lu, J.Y.; Niu, F.Q. Urban Industrial Land Expansion and Its Influencing Factors in Shunde: 1995–2017. *Complexity* **2020**, *2020*, 6769176. [CrossRef]
Park, J.I.; Leigh, N.G. Urban Industrial Land Loss and Foreign Direct Investment-Related Manufacturing Job Sprawl: An Atlanta, Georgia MSA Case Study. *J. Urban Technol.* **2017**, *24*, 95–113. [CrossRef]
Zhang, L.L.; Yue, W.Z.; Liu, Y.; Fan, P.L.; Wei, Y.H.D. Suburban industrial land development in transitional China: Spatial restructuring and determinants. *Cities* **2018**, *78*, 96–107. [CrossRef]
Ustaoglu, E.; Lavalle, C. Examining lag effects between industrial land development and regional economic changes: The Netherlands experience. *PLoS ONE* **2017**, *12*, e0183285. [CrossRef]
Villarroya, A.; Puig, J. Urban and Industrial Land-Use Changes Alongside Motorways within the Pyrenean Area of Navarre, Spain. *Environ. Eng. Manag. J.* **2012**, *11*, 1213–1220.
Pugh, G.; Fairburn, J. Evaluating the effects of the M6 toll road on industrial land development and employment. *Reg. Stud.* **2008**, *42*, 977–990. [CrossRef]
Tonin, S.; Turvani, M. Redeveloping industrial land: Are contamination and remediation schemes affecting the price of industrial real estate properties? *J. Environ. Plan. Manag.* **2014**, *57*, 1043–1065. [CrossRef]
Chen, W.; Shen, Y.; Wang, Y.A. Does industrial land price lead to industrial diffusion in China? An empirical study from a spatial perspective. *Sustain. Cities Soc.* **2018**, *40*, 307–316. [CrossRef]
Chen, W.; Shen, Y.; Wang, Y.A.; Wu, Q. How do industrial land price variations affect industrial diffusion? Evidence from a spatial analysis of China. *Land Use Policy* **2018**, *71*, 384–394. [CrossRef]
Lin, Y.T.; Qin, Y.; Yang, Y.; Zhu, H.J. Can price regulation increase land-use intensity? Evidence from China's industrial land market. *Reg. Sci. Urban Econ.* **2020**, *81*, 103501. [CrossRef]
Wu, Y.; Zhang, X.; Skitmore, M.; Song, Y.; Hui, E. Industrial land price and its impact on urban growth: A Chinese case study. *Land Use Policy* **2014**, *36*, 199–209. [CrossRef]
Yang, S.F.; Hu, S.G.; Li, W.D.; Zhang, C.R.; Song, D.D. Spatio-Temporal Nonstationary Effects of Impact Factors on Industrial Land Price in Industrializing Cities of China. *Sustainability* **2020**, *12*, 2792. [CrossRef]
Li, C.X.; Wu, K.N. An input-output analysis of transportation equipment manufacturing industrial transfer: Evidence from Beijing-Tianjin-Hebei region, China. *Growth Change* **2022**, *53*, 91–111. [CrossRef]
Kumpula, T.; Pajunen, A.; Kaarlejarvi, E.; Forbes, B.C.; Stammler, F. Land use and land cover change in Arctic Russia: Ecological and social implications of industrial development. *Glob. Environ. Chang.—Hum. Policy Dimens.* **2011**, *21*, 550–562. [CrossRef]
Zhang, J.F.; Zhang, D.X.; Huang, L.Y.; Wen, H.Z.; Zhao, G.C.; Zhan, D.S. Spatial distribution and influential factors of industrial land productivity in China's rapid urbanization. *J. Clean. Prod.* **2019**, *234*, 1287–1295. [CrossRef]
Lee, J.; Jung, S. Industrial land use planning and the growth of knowledge industry: Location pattern of knowledge-intensive services and their determinants in the Seoul metropolitan area. *Land Use Policy* **2020**, *95*, 104632. [CrossRef]
Aktas, E.; Tahta, B.T. Investigation of suitable land use potential for industrial sites: The case of Kemalpaa. *Environ. Monit. Assess.* **2018**, *190*, 654. [CrossRef] [PubMed]
Cheng, H.; Lai, Y.N.; Tong, D. Decoding the decision-making in the new wave of urban redevelopment in China: A case study of a bottom-up industrial land redevelopment in Shenzhen. *Land Use Policy* **2021**, *111*, 105774. [CrossRef]
Danilo, V.D. The evolution process of Chicago's industrial land-use policy. *Urban Res. Pract.* **2018**, *11*, 478–491. [CrossRef]
Salari, M.; Shariat, S.M.; Rahimi, R.; Dashti, S. Land capability evaluation for identifying industrial zones: Combination multi-criteria decision-making method with geographic information system. *Int. J. Environ. Sci. Technol.* **2019**, *16*, 5501–5512. [CrossRef]
Silva, F.B.E.; Koomen, E.; Diogo, V.; Lavalle, C. Estimating Demand for Industrial and Commercial Land Use Given Economic Forecasts. *PLoS ONE* **2014**, *9*, e91991. [CrossRef]
Ustaoglu, E.; Silva, F.B.E.; Lavalle, C. Quantifying and modelling industrial and commercial land-use demand in France. *Environ. Dev. Sustain.* **2020**, *22*, 519–549. [CrossRef]
Sun, Y.F.; Ma, A.B.; Su, H.R.; Su, S.L.; Chen, F.; Wang, W.; Weng, M. Does the establishment of development zones really improve industrial land use efficiency? Implications for China's high-quality development policy. *Land Use Policy* **2020**, *90*, 104265. [CrossRef]
Galarza, M.E.; Gonzalez, I.A.; Enjuto, M.C.T. Industrial Land Policies in Alava. *Bol. De La Asoc. De Geogr. Esp.* **2012**, *58*, 459–462. [CrossRef]

43. Kuang, W.H.; Liu, J.Y.; Dong, J.W.; Chi, W.F.; Zhang, C. The rapid and massive urban and industrial land expansions in Ch between 1990 and 2010: A CLUD-based analysis of their trajectories, patterns, and drivers. *Landsc. Urban Plan.* **2016**, *145*, 21- [CrossRef]
44. Li, L.; Pan, C.Z.; Ling, S.; Li, M.Q. Ecological Efficiency of Urban Industrial Land in Metropolitan Areas: Evidence from Ch *Land* **2022**, *11*, 104. [CrossRef]
45. Li, X.; Yu, X.; Wu, K.N.; Feng, Z.; Liu, Y.N.; Li, X.L. Land-use zoning management to protecting the Regional Key Ecosyst Services: A case study in the city belt along the Chaobai River, China. *Sci. Total Environ.* **2021**, *762*, 143167. [CrossRef]
46. Zhang, J.; Zhang, A. Theoretical Explanation and Policy System Construction of Differentiated Land Management. *Chin. J. Ag Resour. Reg. Plan.* **2018**, *39*, 10–16. [CrossRef]
47. Zirlewagen, D.; von Wilpert, K. Upscaling of Environmental Information: Support of Land-Use Management Decisions Spatio-Temporal Regionalization Approaches. *Environ. Manag.* **2010**, *46*, 878–893. [CrossRef]
48. Jiang, G.; Ma, W.; Qu, Y.; Zhang, R.; Zhou, D. How does sprawl differ across urban built-up land types in China? A spat temporal analysis of the Beijing metropolitan area using granted land parcel data. *Cities.* **2016**, *58*, 1–9. [CrossRef]
49. Dai, P.C.; Sheng, R.X.; Miao, Z.Z.; Chen, Z.X.; Zhou, Y. Analysis of Spatial-Temporal Characteristics of Industrial Land Sup Scale in Relation to Industrial Structure in China. *Land* **2021**, *10*, 1272. [CrossRef]
50. Zeng, C.; Zhao, Z.; Wen, C.; Yang, J.; Lv, T. Effect of complex road networks on intensive land use in China's Beijing-Tianjin-He urban agglomeration. *Land.* **2020**, *9*, 532. [CrossRef]
51. Li, W.; Weng, L.; Zhao, K.; Zhao, S.; Zhang, P. Research on the Evaluation of Real Estate Inventory Management in China. *L* **2021**, *10*, 1283. [CrossRef]
52. Liu, S.C.; Lin, Y.B.; Ye, Y.M.; Xiao, W. Spatial-temporal characteristics of industrial land use efficiency in provincial China ba on a stochastic frontier production function approach. *J. Clean. Prod.* **2021**, *295*, 126432. [CrossRef]
53. Zhang, L.X.; Bi, X.; Huang, Z.J. Urban industrial land use efficiency under the background of economic transformation in Yangtze River Economic Belt. *Resour. Sci.* **2020**, *42*, 1728–1738. [CrossRef]
54. Huang, Z.J.; Zhu, S.J.; Shi, T. Industrial Land Transfer, Technological Relatedness and Industrial Entry Dynamics. *Econ. Ge* **2022**, *42*, 144–155. [CrossRef]
55. Huang, Z.H.; Du, X.J. Strategic interaction in local governments' industrial land supply: Evidence from China. *Urban Stud.* **20** *54*, 1328–1346. [CrossRef]
56. Rahman, A.H.; Gandhi, D.K.; Smith, J.B. Industrial Land-Value and Asset Size—Theory and Evidence. *Revue Canadienne Sciences de l'Administration/Can. J. Adm. Sci.* **1994**, *11*, 177–180. [CrossRef]
57. Lou, J.; Zhao, X.; Wang, L.; Zhu, F.; Jin, P. Industrial Land Provision Based on Enterprise Life Cycle: A Case Study of Jiangsu. *M Urban Res.* **2017**, *10*, 18–24. [CrossRef]
58. Mai, P.T.Q.; Kyoung-Hoon, C.; Gyei-Kark, P. Efficiency Analysis for Major Ports in Korea and China using Boston Consult Group and Data Envelopment Analysis Model. *J. Navig. Port Res.* **2018**, *42*, 107–116. [CrossRef]
59. Myllyla, Y.; Kaivo-oja, J. Integrating Delphi methodology to some classical concepts of the Boston consulting group framew Arctic maritime technology BCG Delphi foresight-a pilot study from Finland. *Eur. J. Futures Res.* **2015**, *3*, 2. [CrossRef]
60. Li, W.; Yang, Y.; Zhao, S. Analyses on the Pattern of Chinese FDI Utilization and Strategic Choice Based on the Competition St *Econ. Geogr.* **2012**, *32*, 23–29. [CrossRef]
61. Zhao, S.; Zhang, P.; Li, W. A Study on Evaluation of Influencing Factors for Sustainable Development of Smart Construct En-terprises: Case Study from China. *Buildings* **2021**, *11*, 221. [CrossRef]
62. Guan, X.Y.; Wang, S.L.; Gao, Z.Y.; Lv, Y.; Fu, X.J. Spatio-temporal variability of soil salinity and its relationship with the deptl groundwater in salinization irrigation district. *Acta Ecol. Sin.* **2012**, *32*, 198–206.
63. Salvati, L.; Venanzoni, G.; Serra, P.; Carlucci, M. Scattered or polycentric? Untangling urban growth in three southern Europ metropolitan regions through exploratory spatial data analysis. *Ann. Reg. Sci.* **2016**, *57*, 1–29. [CrossRef]
64. Anselin, L. *Spatial Econometrics: Methods and Models*, 1st ed.; Springer: Dordrecht, The Netherlands, 1988; pp. 119–136.
65. Zhang, P.; Li, W.; Zhao, K.; Zhao, S. Spatial Pattern and Driving Mechanism of Urban–Rural Income Gap in Gansu Provinc China. *Land* **2021**, *10*, 1002. [CrossRef]
66. Shrestha, A.; Luo, W. Analysis of Groundwater Nitrate Contamination in the Central Valley: Comparison of the Geodetec Method, Principal Component Analysis and Geographically Weighted Regression. *ISPRS Int. J. Geo-Inf.* **2017**, *6*, 297. [CrossR
67. Polykretis, C.; Grillakis, M.G.; Argyriou, A.V.; Papadopoulos, N.; Alexakis, D.D. Integrating Multivariate (GeoDetector) a Bivariate (IV) Statistics for Hybrid Landslide Susceptibility Modeling: A Case of the Vicinity of Pinios Artificial Lake, Ilia, Gre *Land* **2021**, *10*, 973. [CrossRef]
68. Wang, J.F.; Li, X.H.; Christakos, G.; Liao, Y.L.; Zhang, T.; Gu, X.; Zheng, X.Y. Geographical detectors-based health risk assessm and its application in the neural tube defects study of the Heshun region, China. *Int. J. Geogr. Inf. Sci.* **2010**, *24*, 107–127. [CrossR
69. Wang, J.F.; Xu, C.D. Geodetector: Principle and prospective. *Acta Geogr. Sin.* **2017**, *72*, 116–134.
70. Zhao, S.; Yan, Y.; Han, J. Industrial Land Change in Chinese Silk Road Cities and Its Influence on Environments. *Land* **2(** *10*, 806. [CrossRef]
71. Ma, Y.; Zhang, P.; Zhao, K.; Zhou, Y.; Zhao, S. A Dynamic Performance and Differentiation Management Policy for Urb Construction Land Use Change in Gansu, China. *Land* **2022**, *11*, 942. [CrossRef]

Zhao, S.; Zhang, C.; Qi, J. The Key Factors Driving the Development of New Towns by Mother Cities and Regions: Evidence from China. *ISPRS Int. J. Geo-Inf.* **2021**, *10*, 223. [CrossRef]
Tapio, P. Towards a theory of decoupling: Degrees of decoupling in the EU and the case of road traffic in Finland between 1970 and 2001. *Transp. Policy* **2005**, *12*, 137–151. [CrossRef]
Li, M.; Shi, Y.Y.; Duan, W.K.; Chen, A.Q.; Wang, N.; Hao, J.M. Spatiotemporal Decoupling of Population, Economy and Construction Land Changes in Hebei Province. *Sustainability* **2019**, *11*, 6794. [CrossRef]
Zhou, M.; Hu, B.X. Decoupling of carbon emissions from agricultural land utilisation from economic growth in China. *Agric. Econ.-Zemed. Ekon.* **2020**, *66*, 510–518. [CrossRef]
Zhang, P.; Hu, J.; Zhao, K.; Chen, H.; Zhao, S.; Li, W. Dynamics and Decoupling Analysis of Carbon Emissions from Construction Industry in China. *Buildings* **2022**, *12*, 257. [CrossRef]
Casadio Tarabusi, E.; Guarini, G. An axiomatic approach to decoupling indicators for green growth. *Ecol. Indic.* **2018**, *84*, 515–524. [CrossRef]
Gupta, S. Decoupling: A step toward sustainable development with reference to OECD countries. *Int. J. Sustain. Dev. World Ecol.* **2015**, *22*, 510–519. [CrossRef]
Zhao, S.; Zhao, K.; Yan, Y.; Zhu, K.; Guan, C. Spatio-Temporal Evolution Characteristics and Influencing Factors of Urban Service-Industry Land in China. *Land* **2022**, *11*, 13. [CrossRef]
Longhofer, W.; Jorgenson, A. Decoupling reconsidered: Does world society integration influence the relationship between the environment and economic development? *Soc. Sci. Res.* **2017**, *65*, 17–29. [CrossRef]
Louw, E.; van der Krabben, E.; van Amsterdam, H. The Spatial Productivity of Industrial Land. *Reg. Studies* **2012**, *46*, 137–147. [CrossRef]
Wang, Y.; Gu, H.; Zhou, L.; Shen, T. Spatial Patterns and Evolving Characteristics of Industrial Land Conveyance in China from 2007 to 2016. *Areal Res. Dev.* **2018**, *37*, 148–154. [CrossRef]
Cui, X.; Meng, X.; Wang, D. Regional Differences and Convergence Characteristics of Industrial Land Marketization in Urban Agglomerations of China from the Spatial Perspective. *China Land Sci.* **2020**, *34*, 34–43. [CrossRef]
Zhu, Z.; He, Q.; Qu, X. Evolution Characteristics of Industrial Land of Urban Agglomerations in the Middle Reaches of the Yangtze River. *Econ. Geogr.* **2021**, *41*, 106–114. [CrossRef]
Xu, J.; He, Q.; Zhu, Z. The Evolution Characteristics of the Industrial Lands of the Changsha-Zhuzhou-Xiangtan Urban Agglomeration. *Econ. Geogr.* **2018**, *38*, 89–97. [CrossRef]
He, H.; Yang, Q.; Bi, G.; Liang, W. Spatio-temporal Characteristics of Industrial Land Supply in Chongqing from 2009 to 2018. *Resour. Environ. Yangtze Basin* **2021**, *30*, 808–817. [CrossRef]
Dong, Z.Y.Z.; Li, Y.C.; Balland, P.A.; Zheng, S.Q. Industrial land policy and economic complexity of Chinese Cities. *Ind. Innov.* **2021**, *29*, 367–395. [CrossRef]
Qi, J.; Hu, M.; Han, B.; Zheng, J.; Wang, H. Decoupling Relationship between Industrial Land Expansion and Economic Development in China. *Land* **2022**, *11*, 1209. [CrossRef]
Mei, L.; Wei, X. Industrial Land Allocation, Spatial Heterogeneity and Industrial Efficiency. *Res. Econ. Manag.* **2022**, *43*, 78–96. [CrossRef]
Peng, S.; Wang, Y.; Chen, C.; Wang, J.; Lei, G.; Wu, X. Rationality evaluation and driving force of industrial land expansion in Shandong Province. *Resour. Sci.* **2016**, *38*, 228–238. [CrossRef]
Del Pozo, P.B. Promotion and evaluation of industrial land in Asturias. *Bol. De La Asoc. De Geogr. Esp.* **2006**, *42*, 99–119.
Zhao, X.F.; Zhang, L.; Huang, X.J.; Zhao, Y.T.; Zhang, Y.P. Evolution of the Spatiotemporal Pattern of Urban Industrial Land Use Efficiency in China. *Sustainability.* **2018**, *10*, 2174. [CrossRef]
Qian, L.; Ma, X.; Shi, X.; Zou, X.; Lan, J. Does the supply of industrial land affect industial energy carbon emissions? *China Popul. Resour. Environ.* **2019**, *29*, 57–67. [CrossRef]
Xie, W.; Qu, W. Influencing Factors of Enterprises' Land Use Preferences in the Reform of Flexible Leasing of Industrial Land in Chengdu City. *China Land Sci.* **2021**, *35*, 39–46. [CrossRef]
Lai, Y.N.; Peng, Y.; Li, B.; Lin, Y.L. Industrial land development in urban villages in China: A property rights perspective. *Habitat Int.* **2014**, *41*, 185–194. [CrossRef]
Zeng, P.; Li, J. Spatiotemporal Responses Between Urban Regeneration and Policy Change of Stock Industrial Land: A Case Study of Tianjin Downtown Area. *City Plan. Rev.* **2020**, *29*, 36–46. [CrossRef]
Powter, C.; Chymko, N.; Dinwoodie, G.; Howat, D.; Janz, A.; Puhlmann, R.; Richens, T.; Watson, D.; Sinton, H.; Ball, K. Regulatory history of Alberta's industrial land conservation and reclamation program. *Can. J. Soil Sci.* **2012**, *92*, 39–51. [CrossRef]
Needham, B.; Louw, E.; Metzemakers, P. An economic theory for industrial land policy. *Land Use Policy* **2013**, *33*, 227–234. [CrossRef]

land

Article

The Value of Farmland and Its Determinants—The Current State of the Art

Vasco Capela Tavares [1], Fernando Tavares [2,*] and Eulália Santos [3]

1. ISEG-UL—Lisbon School of Economics and Management, University of Lisbon, Rua do Quelhas, 6, 1200-781 Lisbon, Portugal
2. ISCET—Higher Institute of Business and Tourism, Rua de Cedofeita, 285, 4050-180 Porto, Portugal
3. Higher School of Education and Social Sciences, CI&DEI, Polytechnic of Leiria, Campus 1, Rua Dr. João Soares, Apartado 4045, 2411-901 Leiria, Portugal

* Correspondence: ftavares@iscet.pt

Abstract: This article aims to perform a literature review on the topic of farmland valuation, cover the determinants of farmland value and the models that are used to price land. To do so, rec literature on the topic was combined with classical and well-known papers. All the factors conside in these papers to explain farmland prices and/or to model them were retrieved, presented, a compared. Then, the main models proposed in the literature are presented and their suitability a goals are explained. This study can help academics as it gives an overview of the current state of art, summarizes the main factors proposed by researchers to explain farmland prices, and sheds li on new lines of research. Besides that, it is also relevant for policymakers because farmland valuat and its use have implications on society and on urban planning, which is a hot topic under discuss

Keywords: farmland valuation; urban pressure; farmland price determinants; hedonic valuat models; econometric models

Citation: Tavares, V.C.; Tavares, F.; Santos, E. The Value of Farmland and Its Determinants—The Current State of the Art. *Land* **2022**, *11*, 1908. https://doi.org/10.3390/land11111908

Academic Editor: Matjaž Glavan

Received: 25 September 2022
Accepted: 25 October 2022
Published: 27 October 2022

1. Introduction

The existence of arable land has been galvanizing the development of mankind generations. It is no coincidence that throughout time, major cities were formed in prosp ous and fertile land, following the well-known Von Thünen model [1]. Indeed, in Ref. it is argued that for a long time the main sources of wealth in the western world w property and agriculture. Presently, the agricultural land market is key for the sustainal development of rural areas [3]. In Ref. [4], it was observed that the agro-industry was h by COVID-19 because agricultural producers took losses due to their strong dependency imported phytopharmaceutical products and equipment. Moreover, mankind's producti of food and model of consumption needs areas available for agriculture and livestock a nationwide policies to manage agricultural soils [5]. In that study, it is also mentioned t it is important that human activity of hunting in wildlife areas and the wildlife comme are controlled and have proper areas for their practice. Thus, it is also necessary to ha public policies to manage soils.

Farmland is the livelihood of many people across the globe, such as agricultural a livestock producers, developers, and investors. These, and in due course, governme officials, would all benefit from understanding what triggers farmland price volatility a knowing the factors that influence farmland prices [6]. Farmland accounts for 85% of asset base of the USA's farm sector; consequently, any changes in farmland values ha enormous implications for the financial health of this sector, in the biggest economy the world [7]. Therefore, it becomes crucial in the modern world to appraise farmla accurately.

Traditional economic theory states that farmland values are determined by the d counted stream of future rents. So, in rural areas where agricultural land is only used agricultural production, land prices are not influenced by the demand for its use in urban activities [6]. However, if there is the possibility of future development driven by urban necessities, then one needs to account for the expected return related to those changes in the current land value. Those necessities do not only include urban sprawl and expansion (e.g., [6]), but also the purchase of land for portfolio diversification (see [8]) and the push by urbanizations for developments capable of creating food and energy (see [9]). The market value of properties in areas where land-use changes are likely to occur intrinsically has an element of hope value [10]. Besides that, the amenity value of farmland is relevant. Not only are there natural amenities that influence the value of farmland, but also a land parcel itself may constitute a natural amenity for the surrounding properties. Furthermore, there is not enough liquidity in the real estate market to keep its players updated, and information asymmetry enhances the differences between the property's value and its sale price [11]. This fact threatens the concept of fair value. Hence, traditional economic theory is only able to partially value farmland, because agricultural land values are driven by a complex set of factors [12].

Although agriculture plays a big role in economic development and sustainability, agricultural producers are less keen to remain in the sector because of its increasingly challenging environment [13]. The continuity of the sector depends on the funding of their projects, preferably by using agricultural land as collateral [13]. Thus, farmland valuation becomes essential for all the people involved in the process: not only the farmers, but also financiers, investors, and ultimately consumers.

Farm real estate is a significant source of value in the farming sector and in the typical investment portfolios of families that live off agriculture [14]. Also, changes in the values of farm real estate have significant implications on the sector's health and the households' well-being because nearly a quarter of agricultural lands in the USA are subject to urban influences and consequently to the changes in the demand of residential markets [14]. Furthermore, real estate is often used in investment portfolios because it lowers risk while offering returns. In a study for New Zealand, it is confirmed that the risk-return trade-off of portfolios of financial assets can be improved if farm real estate is included in them since the diversification benefits are robust under high and low inflationary periods [8]. Moreover, farm real estate is a consistent part of risk efficient portfolios, as the risk reduction benefits of diversifying with farm real estate outweigh the risk enhancement benefits. The results show that farm real estate is more of a risk-reducer rather than a return-enhancer in a globally diversified portfolio [8]. However, there are some factors brought in by other researchers that are contrary to this view. For example, in a mixed-asset portfolio, the autocorrelation in the returns of agricultural land increases risk in the long run, which results in a lower diversification effect when compared to other traditional investment assets [15]. For the authors, given farmland's illiquidity, indivisibility, and high transaction costs, investors could select it for a longer retention period, but it is precisely during that period that autocorrelation and risk become significant.

Despite its relevance, in a study for Portugal, it is stated that the valuation of farmland is not popular among professionals and it is difficult to justify its different prices [16]. Since individuals rely heavily on experts to make decisions in the real estate market [11] and farmland valuation is difficult, the pathway for information asymmetry is formed. The lack of information, together with the differences in the search for land and bargaining costs in agricultural land impact price dispersion [17]. In the same study, it is claimed that information asymmetry causes an uneven playing field because when it is present institutional sellers can sell with markups whereas other sellers incur losses. Besides that, the authors also state that local farmers and buyers have an edge in the level of information they have on a property and take advantage of it, obtaining lower prices and search costs. So, it is suggested that other sellers could eventually halve their costs due to the lack of information if they were more professional and that tenant-buyers can benefit from informational advantages during the harvest season [17]. The significance of the characteristics of buyers and sellers in information asymmetry is also examined in [18], who determined that social and economic factors impact information asymmetry beca realtors avoid making deals with clients who resort to credit, live far from the prope or look for cheap assets. Moreover, blindly trusting expert opinions to track the farm estate conditions induces measurement errors labeled appraisal smoothing, which n negatively impact investment decisions and policy in agriculture [19].

Furthermore, at the time of the study in [20], the valuation of farmland was regain interest because of tax adjustments, the purchase of land for public building and so purposes, and because it is necessary when the owners wish to claim EU grants. This factor becomes even more pertinent today, as the EU's financial stimulus in reaction to COVID-19 pandemic and subsequent crisis is the biggest ever in Europe [21]. Additiona the growing demand for land for urban use impacts society as a whole. An example that is the region of Flanders, where the remaining open spaces are being pressured urbanization to deliver food and biomass, but not always with success for the society The societal impacts also include health. Residents of more compact urban counties sh better health indicators because the urban environment encourages health-related go behaviors, while urban sprawl is often associated with obesity, hypertension, diabetes, a the low number of minutes walked [22]. Control in generalized urban sprawl is necess as it lacks accessibility and open spaces, and conveys exaggerated public spending a loss of resource lands [23]. However, there are ways to preserve farmland in the fa of urbanization, according to [24]. In his article, the author provides a representat example of Oregon's policies, where there are exclusive farm use zones, boundaries urban and exurban sprawl, and tax incentives for farmers to keep their activity. In a m recent article ([9]), it is mentioned that the optimal land use depends on the location a socio-economic context of the land and denotes the difference in regions with high a low population densities: in the former, cultural values are becoming more import for post-production rural development, while for the latter an exclusively producti oriented philosophy is preferred. For the authors, this is one of the aspects that shi the preference toward more unconventional land-use alternatives, which integrate ag ecological production with nature development. These shifts in the preference for la use and the aforementioned tax incentives are relevant for farmland valuation since pol changes impact land price because the prospects for that land are changed [25].

There is a gap in the literature on farmland valuation, according to [26]. The auth suggest that the current state of the art lacks sufficient detail on farm management a agronomic conditions, and the heterogeneity in behavioral constraints, and suggest inc porating learning and collective structures, and modeling complex adaptive systems. T literature, according to the authors, also needs articles that work on farm interaction a incorporate spatial dimensions: interactions should be modeled directly and establish on empirical data, and researchers must use statistically sound methods to initialize t population, including its positioning in space. Moreover, land valuation can be inaccur because of factors of a methodological, technical, and legal nature.

2. Materials and Methods

Considering the relevance of the topic, the aim of this literature review is to prese the current state of the art on the topic of the value of farmland and its determinants. To so, a literature review was conducted based on more than 50 research items, which are cit in this article. Following the lines of the present introduction, in Section 2, we delve into t factors presented in the literature that are utilized to explain the value of farmland, wh are in line with the factors that researchers use in models they construct to value farmla These factors will start by being exhibited, alongside the authors that mention them or u them in their valuation models. Then, the researchers' conclusions about those factors w be compared. This section is also meant to name the models used in farmland valuati and to see their limitations. Then, Section 3 is dedicated to the results and discussion. He the main results of the methodologies presented in Section 2 are presented and compar and some implications are discussed. Last but not least, in Section 4, the main conclusions on the topic of farmland valuation will be drawn.

It is known that valuing land can be a difficult task because many variables may influence farmland value. As previously mentioned, the value of land in rural areas goes beyond its discounted stream of agricultural rents or net returns. The literature on the topic of farmland valuation puts forward many factors which impact the value of farmland. While some researchers mention their effect, others build models to evaluate it. It is also possible to, for instance, measure the amenity benefits of farmland (e.g., [27]) and study the transfer of development rights (e.g., [28]).

However, in Table 1, our goal is to present the factors mentioned in the covered literature that are utilized to explain the value of the land. Therefore, no distinction is made between the methodology of the studies, as it is intended to go through the main factors in the literature first.

Table 1. Factors that impact farmland value and supporting literature.

Variables	Sources
A—Seller/Owner characteristics	[3,17,29–32].
B—Buyer characteristics	[17,30,31,33].
C—Location (including the characteristics of the neighborhood—industrial, commercial, agricultural, …)	[2,3,12,16,17,29,30,32–37].
D—Distance	[3,12,16,20,28,30,33,37–41].
E—Travel time	[12,33,35,42].
F—Farm income	[12,31,33,38,42–46]
G—Cultivated crops	[3,6,7,20,28,29,33,45].
H—Good access	[3,6,16,17,20,28,30,34–36,39,41,47].
I—Macroeconomic conditions	[7,32,38,42–44,48].
L—Land characteristics	
L1—Water rights	[7,16,30,34,36,45,47,49–51].
L2—Soil productivity	[2,3,6,17,20,28–30,33–37,41,45,50,52].
L3—Climate	[2,17,20,29,32,39,43,49].
L4—Tree cover	[12].
O—Other	Percentage of non-farm employment in manufacturing [6]; Percentages of people below the poverty line, agriculture in the state's GDP, and non-agricultural workers in the state [47]; Hunting licenses per square mile [12]; Debt to asset ratio and consumer distress [43]; Per capita income in each county [2,42]; Tenant farmer legislation [33].
R—Environment	[2,3,29,45].
S—Size	[6,12,16,17,28,33,34,37,41,45,49,50].
T—Date or time	[28,42].
U—Population	[2,6,12,33,38,45,47].
U′—Population growth	[6,42,49].
V—Property characteristics	
V1—Fence	[16,20].

Table 1. *Cont.*

Variables	Sources
V2—Infrastructure	[16,20,33,49].
V3—Geometry of the plot	[3,20,39,41].
V4—Unevenness of the plot	[3,20,39,50,52].
V5—Immediate and future development potential	[3,12,28,33,35,40,44,50].
W—Loss of farmland to sprawl	[6].
X—Government/Other subsidies	[17,29,32,38,43,44,53].
Y—Farm yields or production	[6,20,29,36,42,44,47,48].

Source: Own elaboration.

As for Table 1, it should be noticed that macroeconomic conditions include mentic of interest rates, annual returns on investment in the S&P 500 index as a proxy of t economy, and commodity price volatility. According to [43], the farmland option values p acre are positively related to inflation, commodity price volatility, weather, S&P500 annu return, and GDP; and are negatively related to the home price index and consumer distre index. On the other hand, although there is a consensus on the impact of inflation on t value of farmland, it was found to be irrelevant in the study [44]. Besides that, there a superscripts in the sources that correspond to [47] because this paper analyzes the Indi territory on two different levels: (1) corresponds to the study conducted for the district a village levels, and (2) to the all-India level. The authors discovered that the most importa factors that influenced land prices on a village level were population density, road dens distance to the nearest town, percentage of people living below the poverty line, cr yield, non-agriculture share in the state's GDP, and workers in the non-agricultural sec as a ratio of agricultural workers. On a district level, the demand for more land frc urban areas plays a big role in farmland prices, as industrialization and tourism drive t expansion of cities [47]. Interest rates are another macroeconomic factor that should considered to explain real estate, and in particular, farmland prices [48]. Low interest ra allow investors to finance themselves at a cheaper price. Therefore, there is more dema and prices increase. On the other hand, with high interest rates, money becomes mc expensive to borrow and demand decreases. Moreover, the increase in marked-to-mark mortgages may increase the supply in the real estate market. So, in such a scenario, pric are expected to decrease.

An example of a paper that includes both the seller's and the buyer's characteristi is [17], as it was aforementioned. This subject is also tackled by [30], where the author stat that farmland is exchanged at a lower price when both market participants are individu (ceteris paribus) because when one of the parties is a corporation, due to its liquidi the transaction occurs closer to the time of conversion to urban use. The authors of [3 introduce the characteristics of buyers and sellers in the model's equilibrium, in which t increase in a farmer's bid given a marginal increase in one of the characteristics must equal to the increase in the market of the parcel's rent with a marginal improvement that characteristic, or the farmer would profit by using land with different characteristi Another condition is added in equilibrium: the farmer's total bid must equal the parce rental price. Since different farmers have different skills, the bid function is not equal to t hedonic price and locational preferences emerge [31].

Observing Table 1, it is important to note the characteristic called "T—date or tim The sale date was included, for example, in Ref. [28], as additional data that could impa selling prices. However, the authors do not mention any direct relation between it ar selling prices. Furthermore, the discount stream of agricultural rents, which according traditional economic theory determines farmland values, is mostly dependent on intere rates and/or inflation, so the impact of the transaction date on farmland prices is expecte to be captured by other macroeconomic factors.

The location also impacts farmland value in different ways depending on its price range [29]. It is explained in Ref. [29] that at a higher price range of farmland, natural amenity heavily impacts the value because it is more "luxury" than "necessity". Attributes such as water area as a proportion of the total county area and high-value crop farms are also "luxury". At lower quantiles, land retirement programs positively impact farmland values [29].

The factor named distance generally comprises the inclusion of the distance to the nearest town, city, or metropolitan area in the study. However, in Ref. [12], the distance to services and recreational areas is also included. Curiously, the authors find that proximity to urban areas on land values is not statistically significant, as measured by the driving time to small and large town centers. Therefore, population interaction and development pressure can be more useful in explaining agricultural land values than simply the distance to an urban area [12]. This disagrees with the findings in [33], who conclude that farmland prices decrease steeply with distance when they are close to the city and then gently further away using an econometric approach. The authors were even able to study the gradient at which agricultural markets respond to distance from the city. The distance to technological centers and sales markets is also included in a study that uses fuzzy logic theory [3]. Furthermore, a peculiar finding in Ref. [42] is also included in the category of distance. The authors conduct a study where they include not only the distance to the nearest metropolitan area but also the second nearest metropolitan area, in the New York region. As expected, an increase in the distance to the closest metropolitan area is correlated with a decrease in land value, but land values increase when the distance to the second nearest metropolitan area increases, as is also the case for travel times [42].

Whether land has good access or not also impacts its value, according to the literature. For example, in Ref. [30] it is argued that the effect of distance to the nearest town on land price may be reduced in part by the location of parcels relative to the boundaries of incorporated towns, as closeness to such areas increases farmland prices by 40%. The author presented an example that considers the parcels located on non-township roads and with good access to motorways are also more expensive.

Logically, the value of farmland in the urban fringe is impacted by immediate and future development potential. The urban fringe is defined in Ref. [30] as the areas that border central cities and that consist of the surrounding close-in suburbs and non-contiguous nearby towns, extending into the adjacent, open countryside. However, apart from distance, location, and travel times, other characteristics measure immediate and future development potential, which is the reason why there is also a category named after that factor. For example, the per-acre premium related to entitlements: one-home-site entitlement increases the value of an agricultural parcel by 130,000$ [28]. The expectation of future development can go beyond the distance to the closest city or metropolitan area, and that expectation influences the farmland price [40,50].

The main task when valuating land is to identify the variables, which affect its value [20]. So far, some variables which are present and common in the literature are identified alongside their sources. Having checked those variables, some studies build models to value land. The following subsections are intended to ascertain the models that are utilized in the literature and their applicability, performance, and shortcomings.

2.1. Hedonic Valuation Models

The hedonic methodology can be traced back to 1939 and became widely used in the 1960s, according to [31]. The author also states that the hedonic methodology received a theoretical model that consolidates the empirical techniques in 1974. Economists started using the hedonic price method to value farmland amenities in the late 1990s [54]. Hedonic pricing can be used to estimate the values of individual farmland characteristics [6] and allows for a more explicit model of the complex set of attributes that form the basis for farmland valuation [12]. Therefore, hedonic valuation models are widely used in the literature, due to the fact that they are well-established and easily applied.

An example of such a model is the one developed in Ref. [16] for farmland valuati in Portugal. The hedonic valuation model shows that there are factors that impact farmland's value, besides the factors which include the quality of the soil and other char teristics that generate a certain cash flow for that land [16]. The authors explicitly includ those factors, such as proximity to highly populated areas, population density, and possi land use, in the hedonic pricing model, and they could explain why asking square me prices change so much, for all the studied land sizes (up to 100.000 square meters).

In Ref. [6], farmland values are also modeled as a hedonic function of factors t include the location (proximity to metropolitan areas), physical land characteristics, and potential of converting farmland to alternative uses. The authors included dummy variab to measure the influence of urban access on farmland prices for sales in non-metropoli areas and the natural logarithm of the land's price per square meter as the target.

In Ref. [12], the authors decided to take a slightly different approach to the hedo price model. Since the authors intended to model both the agricultural and non-agricultu determinants of farmland values, they think of the portion of farmland's market va not attributed to agricultural use as a "residual" value. This regression shows the deg to which non-agricultural factors explain the portion of land values not attributed agricultural returns [12]. To account for spatial dependence and spatial autocorrelati the authors use spatial fixed effects and spatial error clustering. Another study [34] us the same methodology, achieving results in agreement with the aforementioned study; hedonic price model is also utilized to estimate the impact on the price of Chilean farmla of the following variables: size, soil quality, water rights, connectivity (distance to nearest paved road) and location.

The hedonic price model is also used in [30] to ascertain the natural and man-ma factors that impact farmland prices in an urban fringe market. This model is adequ because it identifies the specific characteristics of different parcels, accommodating heterogeneous qualities of land [30].

Although hedonic price models generally include a simple linear regression in th methodology, a different approach is carried out in Ref. [31]. In the article, the aut starts with a hedonic price model for land rents and mentions that it does not need be linear. Then, the bid function (payment the farmer is willing to pay for the use o parcel) is derived and it depends on the characteristics of the parcel, the prices of outp and other inputs, the desired profit level, and the farmer's production skills. The auth then proceed to derive conditions for equilibrium: the increase in a farmer's bid give marginal increase in one of the characteristics must be equal to the increase in the mar of the parcel's rent with a marginal improvement in that characteristic, and the farme total bid must equal the parcel's rental price. Since different farmers have different ski the bid function is not equal to the hedonic price and locational preferences emerge [31

2.2. Econometric Models

Econometric models, on the other hand, have been widely used in this subject, nam regarding the incidence of USA agricultural subsidies on farmland rental rates, accordi to [46]. In their study, the authors apply a general spatial model that combines a spat lag and a spatial error model and justify that it is necessary to obtain consistent resu An econometric model is also built in [42], now for the price of farmland, using average per-acre net return from agricultural land, the population change in the t nearest metropolitan areas, the travel times to those areas, and the per-capita income of county. Essentially, what the authors were able to grasp was the impact of urban press on farmland values. An interesting approach to ascertain the urban influence on farmla prices is also undertaken in [33]. The utilized methodology is also an econometric mo which was revealed to be suitable for this task.

In a study for Midwest [48], the authors use an econometric approach and state t farmland has grown in value due to low interest rates, an increase in the demand for gra and lower stocks. These authors conclude that there is some speculation, especially for high and medium-quality land, which is in agreement with the warnings made in [33].

2.3. Other Alternative Models

So far, the literature on farmland valuation using hedonic pricing models and econometric models has been addressed. However, other valid and adequate models should also be presented. For instance, multiple correspondence analysis (MCA) is widely used in social sciences, and the authors of [20] state that it is an adequate method for farmland valuation in Spain due to the possibility of multicollinearity and redundant information. Using this method, followed by a linear model with the variables that resulted from MCA, the authors managed to build a simple model with R-square values greater than 0.92.

A semi-parametric quantile regression was used in [29], which according to the authors has advantages over the conventional ordinary least squares method. Namely, it provides a more complete picture of the conditional distributions of a dependent variable given a set of regressors, it is more robust to outliers, it can be more efficient when the error is non-normal, and heteroskedasticity can be conveniently analyzed by estimating quantile regressions. This method allowed the authors to group explanatory variables into "luxury" or "necessity" factors and to verify spatial differences in farmland values.

A Samuelson-McKean model is proposed in [10], which is an option pricing model, to determine the hope value of farmland. In this article, the authors prove its efficacy in the valuation of land plots with development potential. The application of an option pricing model is also seen in [43], a study that uses a real options approach for the farmland market of Illinois.

Furthermore, the authors of [3] propose the application of fuzzy logic theory to valuate agricultural lands. This method is cheap, efficient, and transparent throughout the entire valuation procedure; thus it is useful in a limited market with an insufficient number of sales transactions or in underdeveloped regions.

In Ref. [54], a literature review work on amenity values generated by farmland, the authors indicate that contingent valuation studies were widely used in the 1980s to estimate willingness to pay to protect farmland. In Ref. [27], the authors estimated the amenity value of horse farmland in Kentucky using the contingent valuation and the hedonic pricing model methods. A different approach is taken in [38], using a so-called gravity model, based on the idea from Marketing research that the attraction of retail between two cities is proportional to their populations and inversely proportional to the square of the distance between both cities. The authors successfully grasp the urban impact on land prices of farmland located in the rural-urban fringe, as their model explains more than 95% of the variations in farmland prices. A classic and simple methodology is proposed by the authors of [6], who state that according to economic theory, farmland values are based on the expected economic returns. Thus, net present value models are usually used to evaluate farmland values. In these models, the net present value of farmland represents the sum of all future income streams from farming, accounting for the difference between the value of money today and at some future date [6]. The authors of [6] also note that farmland values are based on expected returns, not necessarily historical revenues. These observations are backed by another study [42], which states that in a competitive land market, the price for land equals the present discounted value of the stream of future rent.

3. Results and Discussion

3.1. Results of the Methodologies

In this subsection, we list the main conclusions and results of farmland valuation using the aforementioned methodologies.

Something that is a result of many studies is the impact of urban necessities on farmland value. For instance, using a hedonic valuation model, the authors of [12] state that development potential and population interaction are significant determinants of market value, as a 1% increase in the land area subject to immediate development potential has a

0.43% increase in cropland values and a 0.74% increase in pastureland values. Moreov future development potential shows similar premiums [12]. In fact, with a similar meth ology, in Ref. [34] it is seen that the most important attribute is the location of farmland a the second most important attribute is the quality of the land, and that both have a doub digit percent impact on the price of farmland. The importance of the location of farmla reflects the market's preference for places where non-farm profits can be obtained from future real estate development, while the remaining variables are of little impact, accordi to the authors. The distance to paved roads and the parcel's size slightly decrease the val of land, while the impact of water rights on farmland prices is only marginally positive [3 However, according to a previous study [35], farmland values tend to be higher near urb areas for more reasons, apart from susceptibility to development pressures. Those reaso according to the authors, include the soil's productivity, as many urban centers initia grew among particularly fertile soils, greater access to markets and ports and therefo lower transportation costs, and recreational opportunities and lifestyle amenities offered the urban fringe to the nearby population. As a consequence, farmland values are high in urban-influenced areas with a premium at the median of approximately USD 2000 p acre [35]. A hedonic valuation model [30] also allows concluding that neighboring wat impact farmland prices negatively because they increase the risk of floods or present un ceptable levels of water quality, and the uses of neighboring land significantly impact urb fringe farmland prices. The insertion of land in an industrial/commercial zone increases value by 28% over agriculturally zoned land [30].

Econometric approaches validate the previous methodology's conclusions on t impact of urban pressure on farmland value. For instance, in a study for New York [4 the authors found that since land development is perceived as imminent due to high rates of population change, land values increase. An increase of 1000 people in the rate population change in the nearest metropolitan area increases the land value by USD 97 p acre; similarly, the corresponding number for the second-nearest metropolitan area USD 101 per acre [42]. The conclusions in Ref. [42] are backed by another study whe an econometric model is used [33]. According to the authors, in peri-urban areas, sin farmland is expected to be converted to urban uses which are translated into potent capital gains from such development, farmland prices decrease with distance to citi More specifically, in Ref. [33] it is found that farmland prices are only determined agricultural factors far from urban influence, approximately 35–40 km away from t closest city. It is also stated that for land for residential use, the land price decrease p kilometer is −12.8% at the urban pole boundary, −3.3% at 20 km from the boundary, a almost flat at 35 km. Land rents were also studied by the authors, who concluded th the main determinant of land rent is the trade-off between commuting costs and la rent and that its impact is bigger than that of development expectations because on there is residential use, we are dealing with actual conversion instead of expectations. T limitation of urban sprawl caused by the ideas and policies of urban planners is the ma driver of the farmland prices bubble, according to [33]. Thus, cities must be dense a surrounding rural areas must be preserved, and consequently, expectations of dema for new residential land will disappear [33]. Rental prices are also an object of study Ref. [46], where it is seen that farmland rental prices are mainly influenced by the ren prices in the parcel's neighborhood and by regional livestock density, which are heav influenced by agricultural policies.

Moreover, an option pricing method in Ref. [43] led to the conclusion that risi farmland values are primarily dependent on agricultural commodity prices and inter rates and that the farmland market value is also influenced by uncertainty about futu growth and capital gains. Furthermore, the farmland option values per acre are positiv related to the inflation rate, commodity price volatility, weather, S&P500 annual return, a GDP; and are negatively related to the home price index and consumer distress index [4

A combination of hedonic pricing models and contingent valuation in Ref. [27] allow concluding that Kentucky's residents feel that horse farms are a positive amenity, becau of their cultural heritage, prettiness, and their service as an impediment to gentrification. Farmland itself generates positive externalities, and that is shown in both methods [27].

Like the authors of [33], who took on an econometric approach, the authors of [38] found that expectation adds more value to farmland than actual present returns, with a gravity model: a one-dollar increase of the net real returns to land from farming adds USD 0.74 to the value of farmland, while the expectation of a one-dollar growth in real capital gains increases farmland value by USD 1.27.

3.2. Comparison of the Methodologies

Hedonic pricing methods are arguably the most straightforward models to conceive and implement. In most standard cases, the researcher collects in the literature a set of impactful characteristics in farmland valuation and builds a simple regression model.

An example of a successful application of such a model is found in [6]. The model was significant at the 0.001 level, with an R-square value equal to 0.46, condition indices below 30, and no variable had two or more variance proportion values greater than 0.5 [6]. Therefore, multicollinearity was assumed not to be a problem and the methodology was appropriate, according to the authors. Hedonic pricing methods also allow for the usage of supplementary statistical treatment of the data. As was mentioned before, an example of that is found in Ref. [12], where the authors use cluster analysis to account for spatial autocorrelation.

A good alternative when there is a possibility of multicollinearity and redundant information is multiple correspondence analysis [20]. As seen in the previous section, in Ref. [29] a semi-parametric quantile regression is used, which is useful when the error is non-normal and to analyze any potential case of heteroskedasticity.

Spatial econometrics is a very robust way of seeing the impact of proximity to urban centers, for example. This variable is no ordinary example; across all the studies reviewed that include it, its impact was always measured to be positive. Since spatial econometrics is capable of grasping the spatial correlations and including them in the model, it is a strong and classical candidate to measure the impact of these factors, in particular.

3.3. Discussion

As for the determinants of farmland value, there is an agreement in some of the determinants, although the literature covered in this article often disagrees about the effects of some factors on farmland value.

Soil productivity is an example of a factor in which there is agreement about its positive impact on the farmland's value (see, for example, [6,29,41]). However, there is some disagreement about to which extent it is significant. On one hand, the types of soil and their diversity is the most important characteristic when valuing farmland, according to the authors of [41], and soil productivity's double-digit percent impact on farmland price is outlined in [34]. On the other hand, the authors of [30] observed that the soil productivity index is not statistically relevant because farmland transactions are mainly for speculative and urban purposes. Therefore, it is concluded that when valuing farmland, it is important to keep in mind its future use. The productivity of the soil is important if the land is used for agricultural purposes; on the contrary, if there is a potential for urban development, the impact of soil productivity on the farmland's value will be diluted.

Urban development pressure and potential are also driving factors of farmland prices, according to the literature. In this article, it was shown that, given a land parcel, researchers propose different measures of urban influence on it [20]; in Ref. [47], for example, the authors utilize the distance to the nearest urban center and observe its significant and positive impact on land prices. Distance to the urban fringe is also investigated by the authors of [33], who put a timespan for the conversion of farmland to urban use of roughly 30–50 years for parcels 5–40 km away from the urban fringe, in a non-linear form. An urban influence index that relates the size of the nearest urban pole and the distance to it was proposed by the authors of [38], who found out that a one-unit increase in the numerical value of that index will increase per acre real land value by $132.60. Howe the authors of [12] warn that measuring population interaction can be more effective explaining land prices than simply the distance to an urban area, and the authors of [state that there is no evidence to attest to the impact of distance on farmland amenity valu Furthermore, the proximity of farmland to urban centers affects both the developm component of farmland values and agricultural rents, because places with better access population centers have a larger share of high-value crops, which is consistent with the v Thünen formulation [55]. Moreover, when there are policies that prevent the conversion farmland from agricultural to urban uses, farmland prices are not influenced by distan to the city [33]. This evidence indicates that each land plot must be attentively conside and looked into along with the existing legislation. Thus, researchers must be cautic when using distance when trying to measure the urban influence in a plot, as there c potentially exist multicollinearity issues, as the aforementioned observation in Ref. [points out.

The literature review conducted in this work shows the difficulty in developin method for valuing and determining the factors that impact farmland value that can sta across time and different regions. Another example of such a factor is water rights or wa irrigation. It is stated in a study [50] that the prices of irrigated lands, regardless of th productive orientation, are approximately twice the prices of dry lands. The authors of [have a different view, stating that the water area as a proportion of the total county a positively impacts farmland value because it is viewed as a "luxury" factor. On the otl hand, it is observed in [30] that neighboring waters impact farmland prices negativ because they increase the risk of floods or present unacceptable levels of water qual This example illustrates the necessity of integrating a factor with a variety of others tl influence it. It is expected that irrigated lands are more valuable if they are located in area where the climate is usually propitious to draughts, for example, so the climate sho be taken into account.

Due to some characteristics of the farmland market, such as high transaction co and low turnover, they have characteristics of inefficient markets [43]. That explains w plattage value (the value of breaking one larger parcel into several smaller parcels) exi in the market because of subdivision costs, the liquidity of the buyers, and the lack market information held by the sellers [30]. The real estate market is not efficient a information asymmetry is a particular problem that buyers must face, as it is crucial to fi both the positive and the negative aspects of a property [56]. If information asymmetr are correlated with hedonic characteristics, not acknowledging their implications can le to inefficient estimation and biased coefficients for the hedonic function [17]. For examp heterogeneity is extremely common in Spanish orography, as geographically close distri do not share their main land features [20], which reveals how spatial autocorrelation mig not always be as strong as expected. For buyers and sellers, this means that knowi the prices of properties sold in geographically close areas may not be a reliable source information about the fair value of a given plot.

Therefore, appropriate and fair farmland valuation serves individuals and soci as a whole. The urban fringe is a delicate problem because the remaining space must used innovatively [9]. Moreover, with the urban development sprawl, conflict regardi land use in rural communities has grown [6]. The grassland ecosystems are in danger d to the expansion of their conversion to row crops, driven by commodity price increas technological improvements, and agricultural policy [57]. Today, land use is a cent concern in evaluating agri-environmental policies and the role of the government in ag culture is a major source of uncertainty affecting agricultural production and farmla markets [43]. This often leads to contradictory views by urban planners and policym ers. The process of the demand for the expansion of urban land increases the interest single-family detached homes, which also require land for household support services (e schools and shopping centers) [28]. Private investments must be channeled transparen and with a long-run vision for the implementation of sustainable urban land developme according to [58]. The author also reinforces the idea that the government must be involved to support this objective and achieve economic sustainability, providing funds to socially and environmentally sustainable realms.

4. Conclusions

The objective of this work was to highlight the importance of farmland valuation, break down the determinants of farmland prices and valuation, and describe some of the models and methodology used in the literature to value farmland.

Furthermore, this work is relevant for policymakers. At this point, it is important to remember that there is more to the value of farmland than simply the value of the land itself. From a social point of view, farmland is crucial to the people who live off it and to the industries that depend on it. Public policymakers and stakeholders must weigh both potential uses and returns: urban development and rural use. Therefore, in a certain region, it becomes essential for policymakers to know the value of farmland if it is exclusively used for urban development and if it is used to create jobs and value via farming. The topic of farmland valuation and the relationship between urban and rural poles must be studied together by academia and decision-makers because of the reasons stated above. As for academia, this topic of research is still open because this study must often be conducted on a micro-level for there to be an agreement about the determinants of farmland prices. Any future lines of work can use this article to ascertain which land features to use, as well as to see the models covered and their suitability to the work that is intended to be carried out. It is our belief that this article can be a good starting point for such studies due to the extensive amount of articles reviewed, the explanatory factors presented (which are extracted from those articles), and the introduction to several valuation models and their suitability for the desired research direction.

Starting from this literature review, in future work we intend to develop a multiple linear regression model through which we can test the dependent variables that impact the value of farmland. That model will allow the verification, in practical terms, of the value of the different agricultural properties, which, by comparing to the market prices, will allow inferring if the price is below or above the model's reference value.

Author Contributions: Conceptualization, V.C.T., F.T. and E.S.; investigation V.C.T., F.T. and E.S.; writing—original draft preparation, V.C.T. and F.T.; writing—review and editing, V.C.T., F.T. and E.S.; visualization, V.C.T.; supervision, V.C.T.; project administration, V.C.T. All authors have read and agreed to the published version of the manuscript.

Funding: This research received no external funding.

Data Availability Statement: Not applicable.

Conflicts of Interest: The authors declare no conflict of interest.

References

Silva, Z.A.G.P. Economic aspects of the demand for commercial Forest land in the state of Acre, Brazil. *For. Policy Econ.* **2022**, *138*, 102704. [CrossRef]
Caballer, V. Nuevas tendencias en la valoración territorial. *CT Catastro* **2002**, *45*, 135–146.
Źróbek, S.; Kovalyshyn, O.; Renigier-Biłozor, M.; Kovalyshyn, S.; Kovalyshyn, O. Fuzzy logic method of valuation supporting sustainable development of the agricultural land market. *Sustain. Dev.* **2020**, *28*, 1094–1105. [CrossRef]
Manachynska, Y.; Moshkovska, O.; Yevdoshchak, V.; Svitlana Luchyk, S.; Luchyk, V.; Luchyk, M. Actuarial Multidimensional Model of Ukrainian Agricultural Companies' Valuation. In Proceedings of the 2021 11th International Conference on Advanced Computer Information Technologies (ACIT), Online, 15–17 September 2021; pp. 455–459, ISBN 978-1-6654-1854-6. [CrossRef]
Kumar, R. Valuing variability: Agriculture, ecology and COVID-19. *Soc. Cult. South Asia* **2021**, *7*, 105–111. [CrossRef]
Drescher, K.; Henderson, J.R.; McNamara, K.T. Farmland Price Determinants. In Proceedings of the American Agricultural Economics Association (New Name 2008: Agricultural and Applied Economics Association), Chicago, IL, USA, 5–8 August 2001; Available online: https://EconPapers.repec.org/RePEc:ags:aaea01:20685 (accessed on 1 July 2022).
Henderson, J. Will farmland values keep booming? *Econ. Rev.* **2008**, *93*, 81–104.
Nartea, G.; Eves, C. Role of farm real estate in a globally diversified asset portfolio. *J. Prop. Invest. Financ.* **2010**, *28*, 198–220. [CrossRef]

9. Lerouge, F.; Sannen, K.; Gulinck, H.; Vranken, L. Revisiting production and ecosystem services on the farm scale for evaluat land use alternatives. *Environ. Sci. Policy* **2016**, *57*, 50–59. [CrossRef]
10. Drapikovskyi, O.; Ivanova, I.; Renigier-Biłozor, M.; Źróbek, S. How to assess the impact of hope on a change in the use of land market value? *Land Use Policy* **2020**, *97*, 104746. [CrossRef]
11. Tavares, F.; Moreira, A.; Pereira, E. Assimetria de Informação no Mercado Imobiliário: Uma Revisão da Literatura. *Rev. Unive Contábil* **2012**, *8*, 146–164. [CrossRef]
12. Borchers, A.; Ifft, J.; Kuethe, T. Linking the Price of Agricultural Land to Use Values and Amenities. *Am. J. Agric. Econ.* **2014**, 1307–1320. [CrossRef]
13. Middelberg, S.L. Agricultural land valuation methods used by financiers: The case of South Africa. *Agrekon* **2014**, *53*, 101–1 [CrossRef]
14. Zhang, W.; Nickerson, C.J. Housing market bust and farmland values: Identifying the changing influence of proximity to urb centers. *Land Econ.* **2015**, *91*, 605–626. [CrossRef]
15. Feng, X.; Hayes, D.J. Farmland Investment Characteristics from a Forward-Looking Perspective: An Explanation for the "Hi Return/Low Risk" Paradox. *Land Econ.* **2020**, *96*, 291–303. [CrossRef]
16. Lima, A.; Soares, V.S. Modelo de avaliação hedónico de terrenos rústicos e seus desafios: O estudo de caso da realidade da reg do Porto, norte de Portugal. *Popul. E Soc.* **2015**, *23*, 145–159.
17. Seifert, S.; Kahle, C.; Hüttel, S. Price Dispersion in Farmland Markets: What Is the Role of Asymmetric Information? *Am. J. Ag Econ.* **2020**, *103*, 1545–1568. [CrossRef]
18. Tavares, F.; Moreira, A.; Pereira, E. Assimetria de Informação no Mercado Imobiliário em Portugal. *Rev. De Adm. Unimep* **2013**, 196–220. [CrossRef]
19. Kuethe, T. The risk and return of farm real estate and the "bad data problem". *Agric. Financ. Rev.* **2016**, *76*, 140–150. [CrossRe
20. Garcia, T.; Grande, I. A model for the valuation of farmland in Spain: The case for the use of multivariate analysis. *J. Prop. Inv Financ.* **2003**, *21*, 136–153. [CrossRef]
21. Fuest, C. The NGEU Economic Recovery Fund. *CESifo Forum* **2021**, *22*, 3–8. Available online: http://hdl.handle.net/10419/232 (accessed on 1 July 2022).
22. Ewing, R.; Schmid, T.; Killingsworth, R.; Zlot, A.; Raudenbush, S. Relationship between Urban Sprawl and Physical Activ Obesity, and Morbidity. *Am. J. Health Promot.* **2003**, *18*, 47–57. [CrossRef]
23. Ewing, R. Is Los Angeles-Style Sprawl Desirable? *J. Am. Plan. Assoc.* **1997**, *63*, 107–126. [CrossRef]
24. Nelson, A.C. Preserving Prime Farmland in the Face of Urbanization: Lessons from Oregon. *J. Am. Plan. Assoc.* **1992**, *58*, 467–4 [CrossRef]
25. Libby, L.W.; Irwin, E.G. Rural Amenities and Farmland Values. In *Government Policy and Farmland Markets: The Maintenance Farmland Wealth*; Moss, C.B., Schmitz, A., Eds.; Iowa State Press: Ames, IA, USA; John Wiley & Sons, Ltd.: Hoboken, NJ, US 2003; pp. 343–364, Chapter 19.
26. Kremmydas, D.; Athanasiadis, I.N.; Rozakis, S. A review of Agent Based Modeling for agricultural policy evaluation. *Agric. S* **2018**, *164*, 95–106. [CrossRef]
27. Ready, R.C.; Berger, M.C.; Blomquist, G.C. Measuring amenity benefits from farmland: Hedonic pricing vs. contingent valuati *Growth Chang.* **1997**, *28*, 438–458. [CrossRef]
28. Mundy, B.; Lane, T. Preserving agricultural and forest land: A TDR approach. *J. Prop. Invest. Financ.* **2011**, *29*, 566–574. [CrossR
29. Uematsu, H.; Khanal, A.R.; Mishra, A.K. The impact of natural amenity on farmland values: A quantile regression approa *Land Use Policy* **2013**, *33*, 151–160. [CrossRef]
30. Chicoine, D.L. Farmland Values at the Urban Fringe: An Analysis of Sale Prices. *Land Econ.* **1981**, *57*, 353–362. [CrossRef]
31. Palmquist, R.B. Land as a Differentiated Factor of Production: A Hedonic Model and Its Implications for Welfare Measureme *Land Econ.* **1989**, *65*, 23–28. [CrossRef]
32. Wang, H. The Spatial Structure of Farmland Values: A Semiparametric Approach. *Agric. Resour. Econ. Rev.* **2018**, *47*, 568–5 [CrossRef]
33. Cavailhès, J.; Wavresky, P. Urban influences on periurban farmland prices. *Eur. Rev. Agric. Econ.* **2003**, *30*, 333–357. [CrossRef
34. Troncoso, J.L.; Aguirre, M.; Manriquez, P.; Labarra, V.; Ormazábal, Y. The influence of physical attributes on the price of land: T case of the province of Talca, Chile. *Cienc. E Investig. Agrar. Rev. Latinoam. Cienc. Agric.* **2010**, *37*, 105–112. [CrossRef]
35. Kuethe, T.H.; Ifft, J.; Morehart, M. The Influence of Urban Areas on Farmland Values. *Choices Mag. Food Farm Resour. Issues* **20** *26*. Available online: http://www.choicesmagazine.org/UserFiles/file/cmsarticle_30.pdf (accessed on 1 July 2022).
36. Gripp, J., Jr.; Marques, M.É.; Gonçalves, R.P.; Oliveira Andrade, R.J. Avaliação de imóveis rurais. In Proceedings of the Congres Brasileiro de Cadastro Técnico Multifinalitário-UFSC, Florianópolis, SC, Brazil, 15 October 2006; p. 15.
37. León, R.C.; Requena, J.C. Valoración de fincas olivareras de secano mediante métodos econométricos. Investigación agraria. *Pr Y Protección Veg.* **2000**, *15*, 91–104.
38. Shi, Y.J.; Phipps, T.T.; Colyer, D. Agricultural Land Values under Urbanizing Influences. *Land Econ.* **1997**, *73*, 90–100. [CrossR
39. Meneses, T.G. Un modelo analógico para la valoración catastral. *Rev. Española Estud. Agrosoc. Y Pesq.* **2000**, *186*, 105–127.
40. Isgin, T.; Forster, D.L. A hedonic price analysis of farmland option premiums under urban influences. *Can. J. Agric. Econ.* **2006**, 327–340. [CrossRef]

Kozioł-Kaczorek, D. Characteristics Determining a Market Value of an Agricultural Real Estate with Use of Multiple Correspondence Analysis. *Sci. J. Wars. Univ. Life Sci.—SGGW* **2014**, *14*, 101–107. Available online: http://yadda.icm.edu.pl/yadda/element/bwmeta1.element.cejsh-from-agro-02deadac-f26d-4bee-b5f1-203a1c982c0d (accessed on 1 July 2022).
Plantinga, A.J.; Miller, D.J. Agricultural Land Values and the Value of Rights to Future Land Development. *Land Econ.* **2001**, *77*, 56–67. [CrossRef]
Li, X. The Farmland Valuation Revisited. *Int. J. Food Agric. Econ.* **2016**, *4*, 11–24. [CrossRef]
Rios, M.S.; Solé, T.T. Factores no agronómicos: Análisis de su influencia en los precios de la tierra agraria. *Estud. De Econ. Apl.* **2007**, *25*, 145–164.
Guadalajara, N.; Caballer, M.-T.; Osca, J.M. Assessing localization impact on land values: A spatial hedonic study. *Span. J. Agric. Res.* **2019**, *17*, e0110. [CrossRef]
Breustedt, G.; Habermann, H. The incidence of EU per-hectare payments on farmland rental rates: A spatial econometric analysis of German farm-level data. *J. Agric. Econ.* **2011**, *62*, 225–243. [CrossRef]
Kumar, P.; Pradhan, B.K.; Subramanian, A. Farmland Prices in a Developing Economy: Some Stylised Facts and Determinants. *J. Int. Area Stud.* **2005**, *12*, 93–113. [CrossRef]
Stokes, J.; Cox, A. The Speculative Value of Farm Real Estate. *J. Real Estate Res.* **2014**, *36*, 169–186. [CrossRef]
Donoso, G.; Cancino, J.; Foster, W. Farmland values and agricultural growth: The case of Chile. *Econ. Agrar. Y Recur. Nat.* **2013**, *13*, 33–52.
Gracia, A.; Pérez, L.; Sanjuán, A.I.; Hurlé, J.B. Análisis hedónico de los precios de la tierra en la provincia de Zaragoza. *Estud. Agrosoc. Y Pesq.* **2004**, *202*, 51–69.
Meneses, T.G. Desarrollo de la valoración catastral de fincas rústicas. Aplicación a la Comunidad Foral de Navarra. *CT Catastro* **2003**, *47*, 63–74.
Fernandes, H.C.; Burla, E.R.; Leite, E.D.; Minette, L.J. Avaliação técnica e econômica de um "Harvester" em diferentes condições de terreno e produtividade da Floresta. *Sci. For.* **2013**, *41*, 141–151.
Zhou, T.; Koomen, E.; Ke, X. Determinants of farmland abandonment on the urban-rural fringe. *Environ. Manag.* **2020**, *65*, 369–384. [CrossRef]
Bergstrom, J.C.; Ready, R.C. What Have We Learned from Over 20 Years of Farmland Amenity Valuation Research in North America? *Appl. Econ. Perspect. Policy* **2009**, *31*, 21–49. [CrossRef]
Livanis, G.; Moss, C.B.; Breneman, V.E.; Nehring, R.F. Urban Sprawl and Farmland Prices. *Am. J. Agric. Econ.* **2006**, *88*, 915–929. [CrossRef]
Tavares, F.; Santos, E. Validation of an Information Asymmetry Scale in the Portuguese Real Estate Market. *Rev. Bus. Manag.* **2021**, *23*, 586–599. [CrossRef]
Sweikert, L.A.; Gigliotti, L.M. Evaluating the role of Farm Bill conservation program participation in conserving America's grasslands. *Land Use Policy* **2019**, *81*, 392–399. [CrossRef]
Kauko, T.; Siniak, N.; Źróbek, S. Sustainable Land Development in an Urban Context. *Real Estate Manag. Valuat.* **2015**, *23*, 110–119. [CrossRef]

land

MDPI

Article

Spatio-Temporal Evolution, Spillover Effects of Land Resource Use Efficiency in Urban Built-Up Area: A Further Analysis Based on Economic Agglomeration

Naifu Yu [1,2], Yingkai Tang [1] and Ying Ma [3,4,*]

1 Department of Accounting and Finance, School of Business, Sichuan University, Chengdu 610064, China
2 School of Economics, Northwest Minzu University, Lanzhou 730030, China
3 School of Economics, Lanzhou University, Lanzhou 730000, China
4 School of International Economy and Trade, Lanzhou University of Finance and Economics, Lanzhou 730101, China
* Correspondence: may2019@lzu.edu.cn

Abstract: The Chinese "New Normal" economic model is a national strategy for adapting to susta able development and offers important implications for the development of new economies. T "New Normal" economic model aims at improving the use efficiency of land resources in the fra work of sustainable development. A discussion of the spatio-temporal evolution of land resou use efficiency (LRUE) in urban built-up areas can help in better assessing LRUE. In this paper, super-efficiency slack-based measure (Super-SBM) method and spatial econometric models are us to study 281 prefecture-level cities in China between 2004 and 2020. Further, this paper explores relationship between economic agglomeration and LRUE, which is of great value in managing la resources. The results show that there is a spatial spillover in LRUE and a U-shaped relations between it and economic agglomeration.

Keywords: economic agglomeration; urban built-up area; urban land resource; spatial economet analysis

Citation: Yu, N.; Tang, Y.; Ma, Y. Spatio-Temporal Evolution, Spillover Effects of Land Resource Use Efficiency in Urban Built-Up Area: A Further Analysis Based on Economic Agglomeration. *Land* **2023**, *12*, 553. https://doi.org/10.3390/land12030553

Academic Editor: Matjaž Glavan

Received: 7 February 2023
Revised: 22 February 2023
Accepted: 22 February 2023
Published: 24 February 2023

1. Introduction

Since China's reform and opening-up policy was established in 1978, China's econor has been growing rapidly, the level of economic agglomeration has been continuou improving, and industries and population have shown a trend of concentration [1]. Desp the brilliant achievements in the economy, many problems such as resource depletion low land utilization rate [3], serious environmental pollution [4,5], and uneven regior development [6] have arisen. After the 18th National Congress of the Communist Pa of China, Chinese President Xi Jinping has brought forth a concept of the "New Norm on various occasions. As China's economy is gradually entering the new normal stage a medium to high rate of growth, sustainable development has become one of the cc policies of the Chinese government for the environment and the economy [7]. The extensi economic development model of the past has had difficulty promoting the sustained a healthy development of China's economy or has even caused huge losses [8]. Therefo solving the distribution issues and environmental problems in the process of econon development has become a major strategic goal of China's sustainable development.

The increase in land resource use efficiency (LRUE) should not solely be bound by improving the economic output of land resources, but it should include ecologi protection measures as part of the evaluation. The perspective of development at the cost land resources loss and urban ecological environment pollution has been facing difficult in meeting the requirements of high-quality economic development under the era of t "New Normal" [9]. Land resource use efficiency has become a key indicator for formulati economic and environmental development policies in many countries and regions, and it has begun to play an increasingly important role in measuring the efficiency of economic activities in relation to natural resources and the environment [10].

The rapid growth of China's economy in the past is closely related to economic agglomeration. According to the seventh national population census in 2020, compared with the sixth national population census in 2010, China's floating population has increased beyond expectations in the past ten years, and the population has continued to flood into big cities and metropolitan areas. However, while promoting economic growth, economic agglomeration, as a compact economic behavior, also has some negative externalities. For individuals at the micro level, the advantages brought by agglomeration such as productivity increase and economic aggregate growth can be barely felt in the short term, while it is very intuitive to feel the environmental pollution, traffic congestion, and other problems. Therefore, exploiting the relationship between economic agglomeration and urban land resource utilization not only is an action serving the needs of national development strategies, but also has important guiding significance for developing countries rationally allocating land resources and guiding the agglomeration of production factors and ecological construction [11].

In the past, a large number of studies have focused on the formation factors of economic agglomeration and the measure of land use efficiency, and there are few related studies for the relationship between both in China. On top of summarizing the existing economic agglomeration and land use efficiency-related theory and research, this thesis conducts an analysis of economic agglomeration under the influence of various scales for land use efficiency to establish a framework for theoretical connections. Based on the comprehensive use of the location theory, externality theory, new economic geography theory, and economic growth theory, through the study of the effect on post-economic agglomeration formation, this thesis analyzes the intrinsic mechanism between economic agglomeration and LRUE and tries to clarify the transmission pathway of economic agglomeration toward LRUE to provide important theoretical support for ecological environment facilitation, social equality, and economic development coordination.

The possible marginal contributions of this paper are as follows: (1) This study examines the spatial effects of economic agglomeration on LRUE in urban built-up areas at different scales, which enriches the relevant studies on LRUE. Moreover, the study passed the Moran's I test, robustness test, and IV method test, which made the results more reliable. (2) This study clarifies the transmission paths through which economic agglomeration improves LRUE, i.e., the scale effect, the knowledge spillover effect, and the crowding effect. In addition, this paper further examines the regional heterogeneity of economic agglomeration and LRUE.

In the next section, we review relevant studies and bring forward research ideas of our own. Section 3 describes the method, variable design, and data sources. Section 4 reports the regression results of the spatial econometric model and a series of tests. Section 5 discusses the spatio-temporal evolution characteristics of LRUE and the spatial spillover phenomenon and proposes corresponding policy recommendations.

2. Literature Review

2.1. Basic Concept

2.1.1. Economic Agglomeration

Scholars already have a relatively full understanding of the reasons for the formation of economic agglomeration. Early literature holds the view that agglomeration usually occurs in areas with superior geographical environments and abundant resource endowments [12–14]. Krugman took spatial factors into the category of economics and changed the assumption of a constant return to scale in traditional economics into that the return to scale is increasing instead [15]. The positive externalities brought by economies of scale are an important factor attracting similar enterprises to gather together in one place, tempting a greater labor force and more industries to enter [16–18]. In addition to the increasing return to scale, transportation cost is also an important factor affecting agglomeration in the new economic geography theory [19,20]. Increasing return to scale and transportation c saving are both from the perspective of the supplying side when it comes to the formati of economic agglomeration, as well as market demand [16,21]. Meanwhile, knowled spillover is also an endogenous and important factor [22–24].

In addition, some scholars mention that institutional factors also play certain ro in promoting the formation of economic agglomeration. After the emergence of the n economic geography, the study of the correlation between geography and economic fact continued for a long time, which lead to the ignorance of how the policy system may ha an impact on economic development. In addition, due to various national conditions, some countries (such as the United States), the impact of trade protectionism among regic on regional trader protection policy is not severe [25]. In China, however, the impact policies and regulations is quite obvious [26,27].

2.1.2. Land Resource Use Efficiency

Considerable research on LRUE has been conducted by scholars, mainly on the f lowing two aspects:

Measurement and evaluation of LRUE. Conventional measurement of LRUE is mai based on the single-factor evaluation method, which is simple and efficient, but it canr reflect the efficiency relationship among multiple inputs and multiple outputs in the proc of urban land resource use. The data envelopment analysis (DEA) model is a comm method that can better deal with the relationship between input–output factors. Based this, the DEA-SBM model [28,29] and the Super-SBM model [30,31] further remedy t problem that the efficiency of multiple DMUs had. At the level of research objectives, contains the usage of watershed [9], provincial [32], city [33], and county [34] objectiv The Super-SBM model can solve the comparison and sequencing problems that appear be in the forefront of the study of the efficiency of multiple land resource uses.

Driving-factor analysis of LRUE. Existing studies have focused more on the econon benefits generated by land resources, and less on their impact on ecosystems. This imp is worthy of attention [35], and the inclusion of undesired outputs in the LRUE evaluati system would lead to more valid conclusions [31]. In terms of research models, t traditional linear regression model was gradually shifted to the spatial measurement moc based on spatial factors, and the Tobit regression model and spatial lag model [36] we used to explore the driving factors of LRUE.

2.2. The Effects of Economic Agglomeration on Land Resource Use Efficiency

The idea of the crowding effect argues that economic agglomeration will worsen typ of environmental pollution such as water pollution [37,38] and carbon pollution [39]. It considered that the agglomeration process inevitably generates production and domes waste, leading to increased pollution. From the perspective of the "pollution heav hypothesis", different regions place different emphases on environmental manageme resulting in different levels of pollution from economic agglomeration. To reduce t cost of pollution control, pollution-intensive firms prefer to move to areas with weak environmental controls, thus creating agglomerations [40–42]. From a "race to the botto perspective, local governments may selectively relax environmental regulations for t sake of fiscal revenue. This has led to environmental degradation in the surrounding ar and a vicious cycle process that has led to region-wide environmental degradation [43–4

On the other hand, some scholars argue that economic agglomeration contributes green development. Economic agglomeration will link the upstream and downstream the industrial chain, increasing the efficiency of resource use and indirectly reducing ener consumption and pollution. At the same time, a dense population will reduce the margir consumption of resources [47,48].

Through the literature review, we have identified three main routes through whi economic agglomeration has an impact on LRUE. (1) Production cost saving effect: E nomic agglomeration leads to the gradual improvement of certain industries in terms information transfer, infrastructure sharing, and service systems, thus significantly reducing transaction costs (in terms of management, financing, and marketing) [49,50]. (2) Scale economy effect: The scale economy effect is an important element of the new economic geography and spatial economics [15]. It emphasizes that the division of labor and regional differentiation contribute to the formation of economic agglomerations, which in turn generate multiplier effects. In this process, land resources will gradually reach their optimum value in terms of economies of scale. (3) Knowledge spillover effects: On the one hand, the clustering of similar enterprises helps them to exchange and share knowledge and information, thus creating spillover effects [51]; on the other hand, the clustering of different industries in the same region helps to form a pattern of industrial diversification, thus stimulating knowledge and technology spillover effects [52]. In addition, excessive agglomeration may also lead to resource scarcity, which increases production costs and reduces LRUE [53,54].

2.3. Review and Hypotheses

In summary, although a large number of studies focused on economic agglomeration and its mechanism of influence, little attention has been paid to the issue of whether China's economic agglomeration improves LRUE. There is no unified conclusion about how economic agglomeration influences LRUE. Specifically for land resources (urban land in built-up areas), the carrier of urban development, research on how economic agglomeration affects LRUE is still in the exploratory stage. Urban economic agglomeration directly affects the intensity and direction of resource input (Figure 1): when more of the agglomeration area is devoted to economic development while the ecological environment is ignored, the production of industrial and domestic wastes is facilitated, thus restraining the ecological efficiency of land. However, when the economy develops to a certain stage, the technological innovation ability will be stronger in the region and the industrial structure will be optimized; under the guidance of policies and regulations, the land utilization rate and ecological environment will be gradually improved. In addition, previous studies mostly used a single model on the relationship between economic agglomeration and LRUE, ignoring that there may be both linear and nonlinear, direct and complicated relationships. Based on the above analysis, this thesis puts forward the following hypotheses:

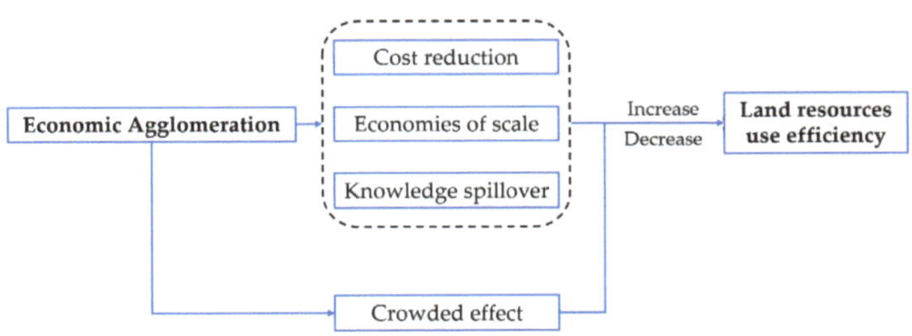

Figure 1. Influencing path of economic agglomeration on land resource use efficiency.

Hypothesis 1. *LRUE has spillover effects.*

Hypothesis 2. *There is a U-shaped relationship between economic agglomeration and LRUE.*

3. Methods and Research Data

3.1. Methods

3.1.1. Moran's I Index

The spatial autocorrelation test is the precondition of spatial econometric models, which determine whether variables are spatially dependent. Spatial autocorrelation tests are divided into overall autocorrelation experiments and partial autocorrelation experiments to analyze whether there is an overall spatial autocorrelation [55]. The range of Moran's

I index is $[-1,1]$. A Moran's I index in the range of (0,1] indicates that each varial has a positive correlation over the whole region; otherwise, the variable has a negat correlation over the whole region. If Moran's I index is equal to 0, the variable is r correlated over the whole region. The Moran's I index is calculated as follows:

$$\text{Moran's I} = \frac{\sum_{i=1}^{n}\sum_{j=1}^{n} W_{ij}(x_i-\overline{x})(x_j-\overline{x})}{s^2\sum_{i=1}^{n}\sum_{j=1}^{n} W_{ij}},$$

x_i represents the observed value of each region; $\overline{x}$ represents the mean of the sample; represents the variance of the sample; W_{ij} represents the binary spatial weight mat (whether the region i borders with region j).

3.1.2. Spatial Econometric Model

According to the first law of geography, the closer the distance, the stronger correlation is. There are interactive effects on the economic behavior of each region: closer the distance is among regions, the stronger the interactions are. Therefore, wh studying the relationship between the regions, the assumption that the economic variab of each region are independent of each other will cause deviation in the results. The spa measurement model fully considers spatial dependence and spatial heterogeneity and c more accurately measure the correlation between economic variables in regional areas. T basic expression form of the spatial panel model is as follows:

$$Y_{it} = \rho\sum_{j=1}^{n} W_{ij}Y_{jt} + \beta X_{it} + \theta\sum_{j=1}^{n} W_{ij}X_{jt} + \gamma_t + \mu_i + \varepsilon_{it},$$

$$\varepsilon_{it} = \lambda\sum_{j=1}^{n} W_{ij}\varepsilon_{it} + \xi_{it},$$

In Equation (2), Y_{it} *and* X_{it} represent the explained variables and explanatory variab in the region period, respectively; Y_{jt} *and* X_{jt} represent the explained variables and expla tory variables in the region period, respectively; W_{ij} is the spatial weight matrix; W_{ij} and $W_{ij}X_{jt}$ are the spatial lag terms of the lag effect of the adjacent region; ρ is the spat autoregression coefficient; β is the regression coefficient of the explanatory variable; θ is t spatial lag term coefficient; γ_t is the temporal effect; μ_i is individual effect; ε_{it} represents t random disturbance term; λ represents the correlation coefficient of the spatial lag fact and ξ_{it} represents a random disturbance term that follows an independent distribution.

When $\rho \neq 0,\ \theta = 0,\ and\ \lambda = 0$, the model is a spatial lag model (SAR), that is, t model contains the spatial lag term of both the explanatory variable and the explain variable, and the error term does not contain the spatial autocorrelation, so the spatial l model (SAR) is selected to test the spatial spillover effect. Considering that the LRUE h interactive effects among regions and that the ecological efficiency of the current period affected by the previous period, the dynamic spatial lag model is adopted to investiga the relationship between economic agglomeration and the LRUE. The dynamic spatial l model is set as follows:

$$Y_{it} = \rho\sum_{j=1}^{n} w_{ij}Y_{it} + \sum_{j=1}^{n}\beta X_{it} + \gamma_i + \mu_t + \varepsilon_{it},$$

The spatial weight matrix represents the network structure matrix of the relationsh among various regions. This paper uses the geographic and economic nested weight matr and it uses the adjacent weight matrix as a comparison.

- The adjacency weight matrix is $w_{ij} = \begin{cases} 1,\ i \neq j \\ 0,\ i = j \end{cases};$
- The geographical distance weight matrix is $w_{ij} = \begin{cases} \frac{1}{d_{ij}^2},\ i \neq j \\ 0,\ i = j \end{cases};$
- The economic distance weight matrix is $w_{ij} = \begin{cases} \frac{1}{|GDP_i - GDP_j|},\ i \neq j \\ 0,\ i = j \end{cases}.$

3.2. Variable Design

This paper uses a spatial econometric model to conduct an empirical study. Table 1 shows the explained variable, explanatory variables, and control variables. See below for details.

Table 1. Description of all variables in the econometric model.

Type	Full Name	Labels	Description
Explained variable *	Labor input	LRUE	Sum of the employment rate of the workplace and the self-employed individuals.
	Capital input		Fixed capital stock calculated by perpetual inventory method.
	Energy input		The annual electricity consumption.
	Natural element input		Urban built-up area.
	Expected output		Real GDP deflator based on 2003.
	Undesired output		Environmental pollution (indicators of industrial waste); Social distribution (ratio of urban to rural disposable income); Social development opportunities (registered urban unemployment rate).
Explanatory variable	Economic agglomeration	Agg	Natural logarithm of non-agricultural output per unit area.
	Square of economic agglomeration	Agg^2	The square term of Agg.
Control variable	Production rate of labor	Lp	Natural logarithm of average labor output.
	Degree of openness	Op	The total volume of import and export/GDP.
	Local industrial structure	Stru	The ratio of the secondary industry output value against the GDP.
	Environmental regulation	Rglt	Comprehensive utilization rate of industrial waste.
	Scientific and technological innovation level	Te	Civic expenditures on urban science and technology/GDP.

* This variable is a composite indicator and is therefore represented by only one label.

3.2.1. Explained Variable

The explained variable is the LRUE. The measure of the LRUE needs to be structured through appropriate methods. Constructing a set of production possibilities that contain the expected output and the undesired output is the first step. Assuming there is a total number of N decision-making units, in each period when $t = 1,2,\cdots,T$, each $DMU_n(n = 1,2,\cdots,N)$ uses many values of x as M, $x = (x_1, x_2, \cdots, x_M) \in R^M_+$, and produces various types of expected output as P, $y^g = (y^g_1, y^g_2, \cdots, y^g_P \in R^P_+)$, and various types of unexpected output as Q, $y^b = \left(y^b_1, y^b_2, \cdots, y^b_Q\right) \in R^Q_+$, the input and output values of DMU_n at period t can be written as $\left(x^t_n, y^{gt}_n, y^{bt}_n\right)$. To meet the conditions of the strong disposability of the cluster, bounded set, input, and expected output, the possible production set under the production technology condition in period t can be expressed as follows:

$$P'(x') = \left\{ \begin{matrix} \left(y^{gt}, y^{bt} : \sum_{n=1}^{N} \lambda^t_n y^{gt}_{np} \geq y^{gt}_p, \forall p; \sum_{n=1}^{N} \lambda^t_n y^{bt}_{nq} = y^{bt}_q, \forall q;\right) \\ \sum_{n=1}^{N} \lambda^t_n x^t_{nm} \leq x^t_m, \forall m; \lambda^t_n \geq 0, \forall n \end{matrix} \right\}, \quad (5)$$

Here, λ^t_n is the density varable, representing the observed value weight in each cross section. Equation (5) is the constant-scale reward condition; when $\sum_{n=1}^{N} \lambda^t_n = 1, \lambda^t_n \geq 0$ is added, and then Equation (5) is the variable-scale reward condition. Because the DEA method measures the relative efficiency, the different inspection periods cannot be directly compared due to the variation of production frontiers. The overall comparison method utilizes all of the DMUs within the production possibility set during the whole sample investigation period to structure the overall production possibility set $P^G(x) = P^1(x^1) \cup P^2(x^2) \cdots P^T(x^T)$. Based on the stability of the unified production frontier and the input–output factors and development objectives of the sample in ea period, the overall efficiency values of DMU are comparable across different sections periods [56].

The slack-based measure (SBM) model is a non-radial distance function; it me the requirements of including pollutants as one of the non-expected outputs during t measurement of green development efficiency, and it also solves the problem of variabl slackness, but as an index assessing relative efficiency, when multiple LRUE values appe to be in the forefront (i.e., many LRUE values are 1), the SBM model will fail in t comparison and sequencing of the efficiency of these cities, which will also affect t accuracy of the subsequent empirical results. The Super-SBM model can solve this proble very well. For this reason, the Super-SBM model is used to structure the LRUE index years from 2004 to 2020 as follows:

$$min\rho = \frac{\frac{1}{M}\sum_{m=1}^{M}\frac{\overline{x_m^t}}{x_{n_0m}^t}}{\frac{1}{P+Q}\left(\sum_{p=1}^{P}\frac{\overline{y_p^{gt}}}{y_{n_0p}^{gt}}+\sum_{q=1}^{Q}\frac{\overline{y_q^{bt}}}{y_{n_0q}^{bt}}\right)},$$

$$\overline{x_m^t} \geq \sum_{n=1,\neq n_0}^{N} x_{nm}^t\lambda_n^t, \overline{x_m^t} = x_{n_0m}^t + s_m^-, \forall m,$$

$$\overline{y_p^{gt}} \leq \sum_{n=1,\neq n_0}^{N} y_{np}^{gt}\lambda_n^t, \overline{y_p^{gt}} = y_{n_0p}^{gt} - s_p^+, \forall p$$

$$\overline{y_q^{bt}} \geq \sum_{n=1,\neq n_0}^{N} y_{nq}^{bt}\lambda_n, \overline{y_q^{bt}} = y_{n_0p}^{bt} + s_p^+, \forall p$$

$$\text{s.t. } \overline{y_q^{bt}} \geq \sum_{n=1,\neq n_0}^{N} y_{nq}^{bt}\lambda_n, \overline{y_q^{bt}} = y_{n_0p}^{bt} + s_p^+, \forall p$$

$$\text{s.t. } \lambda_n^t \geq 0, \forall n(\neq n_0).$$

where s_m^- and s_p^+ are the slack variables indicating input redundancy and insufficie output, respectively.

1. Labor input: the number of employees at the end of the year in each of the select cities is the labor input index, that is, the sum of the employment rate of the workpla and the self-employed individuals; the unit is ten thousand.
2. Capital input: "perpetual inventory method" is used to calculate the fixed capi stock, as the capital input index, and the unit is CNY 10,000. The index is formulat as follows:

$$K_{it} = I_{it} + K_{it-1}(1-\delta), \quad (1$$

K_{it} represents the fixed capital stock amount in the t year of the prefecture-level ci numbered as i; I_{it} represents the total fixed capital amount in the year of the prefectu level city; δ represents the depreciation rate, given as 9.6%. Taking 2003 as the base peri the initial fixed capital stock is defined as $K_{i0} = K_{i1}/(\delta + g_i)$, where g_i represents t average growth rate of investment. Similarly, I_{it} is calculated according to the fixed ass investment price index based on the year 2003, and the fixed asset price index is direc adopted using the fixed asset price index of each province of China.

3. Energy input: the use of energy is related to the national economic construction well as social and ecological environment. Due to the fact that there are no speci energy consumption data available at the prefecture level, this thesis takes the annu electricity consumption as the proxy variable as the energy input index with the un of 10,000 kilowatt-hours.
4. Natural element input: applied land in urban built-up areas.
5. Expected output: the actual GDP of each urban area is used as the expected outp index, and the provincial GDP deflator in 2003 is used for deflation, with the unit CNY 10,000.

6. Undesired output: (1) Negative impacts of economic development on the ecological environment selected are industrial wastewater emission, industrial sulfur dioxide emission, and industrial smoke and dust emission for the "environmental" aspect, and the units are ten thousand tons, tons, and tons, respectively; the annual average density of PM2.5 is selected for the "ecological" aspect, in micrograms/cubic meter. (2) Economic development is going to have a negative impact on social distribution. This thesis selects the income and expenditure gap among urban and rural residents as "social distribution results" (income gap = urban residents' disposable income/disposable income of rural residents; the expenditure gap between urban and rural residents = the expenditure of rural residents) and selects the registered urban unemployment rate for the "social development opportunity" aspect.

3.2.2. Explanatory Variables

Economic agglomeration (Agg) is a process and phenomenon of regional concentration of various economic activities. Based on relevant literature, the study of economic agglomeration is mostly based on analyzing GDP production per unit area.

3.2.3. Control Variables

Considering the influencing forces of LRUE, we take labor productivity (Lp), degree of openness (Op), local industrial structure (Stru), environmental regulations (Rglt), and scientific and technological innovation level (Te) into consideration as control variables.

3.2.4. Data Sources

This thesis takes the panel data of prefecture-level cities in China from 2004 to 2020 as the research object. As of 2022, there are a total of 293 prefecture-level cities[1] in China, among which some of them were canceled and re-grouped under others during the observation period (e.g., Chaohu City, Laiwu City), some of them ranked up or were re-established during the observation period (e.g., Bijie City, Tongren City, Sansha City), and some places in the west region of China do not have full statistical data and the quality of accessible data is low (e.g., Lhasa City, Nyingchi City, Naqu City, Xigaze City, Zhongwei City, Haidong City). Due to the conditions mentioned above, 281 prefecture-level cities were finally selected as samples for the land use eco-efficiency measurement and the subsequent empirical analysis. Data are mainly from the Statistical Yearbook of Chinese Cities from 2005 to 2021, the Statistical Yearbook of each city, the Statistical Bulletin of National Economic and Social Development, the wind database, China's economic and social big data research platform (https://data.cnki.Net/, accessed on 20 December 2022), the Atmospheric Composition Analysis Group, the official website of the University of Washington (https://sites.wustl.edu/acag/datasets/surface-pm2-5/, accessed on 20 December 2022), NTL data, statistical data, and administrative boundaries. Missing data values were replaced by linear interpolation or mean supplementation. Table 2 reports the descriptive statistics of variables.

Table 2. Descriptive statistics of variables.

Variable	Observations	Mean	STDEV	Min	Max
LRUE	4777	0.3526	0.1809	0.0915	1.8263
Agg	4777	6.3964	1.4408	1.5439	11.3351
Lp	4777	3.0461	0.5633	1.0897	5.2671
Op	4777	0.1929	0.3142	0.0016	1.8364
Stru	4777	46.7327	11.0458	17.1000	73.8000
Rglt	4777	79.9185	21.9035	4.5400	100.0000
Te	4777	0.0109	0.0330	0.0000	0.2431
Ups and Downs	4777	0.6734	0.7557	0.0013	3.8138
DN	4777	0.8690	1.8967	0.0027	22.1829

4. Results

4.1. Precondition: Spatial Autocorrelation Test

The precondition for the use of spatial econometric models in this study is the existe of spatial dependence of land resource use efficiency (LRUE), as determined by the spa autocorrelation test. GeoDa software was used to calculate the global Moran index of la use ecological efficiency and economic agglomeration in China following the geograp distance weight matrix (Table 3). The results show that the Moran's I index values of la use ecological efficiency value and economic agglomeration are all greater than 0 and significant at a 1% significance level within the sample period, indicating that land ecological efficiency and economic agglomeration have significant space autocorrelat and that land use ecological efficiency and economic agglomeration are not independ among cities but affected by surrounding areas. Therefore, a spatial econometric model be used in this study.

Table 3. Results of global Moran's I test.

Year	Land Resource Use Efficiency		Economic Agglomeration	
	Moran's I	Z Statistics	Moran's I	Z Statistics
2004	0.060 ***	3.904	0.253 ***	14.984
2005	0.094 ***	6.004	0.254 ***	14.991
2006	0.065 ***	4.024	0.255 ***	15.073
2007	0.076 ***	4.634	0.257 ***	15.202
2008	0.065 ***	4.138	0.257 ***	15.148
2009	0.061 ***	3.851	0.254 ***	14.975
2010	0.114 ***	7.305	0.251 ***	14.783
2011	0 112 ***	7.092	0.244 ***	14.371
2012	0.119 ***	7.411	0.241 ***	14.206
2013	0.108 ***	6.953	0.245 ***	14.410
2014	0.093 ***	5.996	0.248 ***	14.593
2015	0.083 ***	5 249	0.250 ***	14.670
2016	0.067 ***	4.261	0.251 ***	14.744
2017	0.123 ***	7.689	0.254 ***	14.911
2018	0.106 ***	6.642	0.253 ***	14.892
2019	0.103 ***	6.331	0.251 ***	14.781
2020	0 086 ***	5.108	0.255 ***	14.971

*** indicates statistical significance at the 1% level.

4.2. Regression Results of the Spatial Econometric Model

We used MATLAB R2021a software to conduct an empirical analysis based on th spatial weight matrices (geographic distance weight matrix, adjacency weight matrix, a economic distance weight matrix). These results are reported in Table 4. The coeffici under both the static spatial lag model and the dynamic spatial lag model for the spatial term W*LRUE of the explained variables is significantly positive, indicating that the LR has a positive spatial correlation. The spatial lag term coefficient of Agg^2 is significan positive, initially indicating that economic agglomeration may have a positive spat spillover effect on the LRUE. However, merely relying on the regression coefficient of spatial lag model to analyze the local impact of economic agglomeration on the LRUE a its spatial spillover level may lead to deviation [57,58]. The direct and indirect effects independent variables on dependent variables should be broken down further.

Table 4. Regression results of the spatial econometric model.

	Explained Variable: LRUE		
	Adjacent Weight Matrix	**Geographic Distance**	**Economic Distance**
Agg	−0.4629 *** (−18.82)	−0.2252 *** (−10.42)	−0.4997 *** (−20.45)
Agg2	0.0370 *** (29.03)	0.0179 *** (15.33)	0.0392 *** (30.90)
Lp	0.1414 *** (17.03)	0.0880 *** (12.31)	0.1421 *** (16.99)
Op	0.0110 (0.77)	0.0036 (0.30)	0.0068 (0.47)
Stru	−0.0003 (−0.87)	0.0002 (0.66)	−0.0002 (−0.65)
Rglt	−0.0006 *** (−5.27)	−0.0003 *** (−2.65)	−0.0007 *** (−5.28)
Te	0.0000 *** (2.65)	0.0000 *** (3.12)	0.0000 *** (2.92)
W*LRUE	0.1555 *** (7.43)	0.0523 *** (3.31)	0.0657 * (1.92)
R-sq	0.7075	0.7900	0.7034

* and *** indicate statistical significance at the 10% and 1% levels, respectively. Z statistics in parentheses.

Table 5 reports the results of the effect decomposition for Table 4. In both the direct and the indirect effects, the Agg and Agg^2 terms pass the 1% significance test under all three spatial weight matrices, indicating that there is a significant spatial spillover effect of economic agglomeration on LRUE and that the effect of economic agglomeration on LRUE has a U-shaped trend. The above results suggest that Hypothesis 1 and Hypothesis 2 of this paper hold true.

Table 5. Results of the effect decomposition of the spatial econometric model.

Types	Variable	Explained Variable: LRUE		
		Adjacent Weight	**Geographic Distance**	**Economic Distance**
Direct effect	Agg	−0.4636 *** (−19.07)	−0.2253 *** (−10.23)	−0.4992 *** (−20.50)
	Agg^2	0.0371 *** (29.75)	0.0180 *** (15.01)	0.0391 *** (31.18)
	Control	YES	YES	YES
Indirect effect	Agg	−0.0833 *** (−6.16)	−0.0122 *** (−3.14)	−0.0370 *** (−1.84)
	Agg^2	0.0067 *** (6.49)	0.0010 *** (3.21)	0.0029 *** (1.84)
	Control	YES	YES	YES
Gross effect	Agg	−0.5469 *** (−18.40)	−0.2374 *** (−10.45)	−0.5361 ** (−17.36)
	Agg^2	0.0438 *** (28.75)	0.0189 *** (15.58)	0.0420 *** (23.16)
	Control	YES	YES	YES

** and *** indicate statistical significance at the 5% and 1% levels, respectively. Z statistics in parentheses.

4.3. Robustness: Judgment of Spatial Econometric Model Types

Spatial econometric models exist in many forms. The previous section uses a spatial fixed-effects spatial lag model based on common sense. This section goes through a series of validations to re-establish that this model is the most appropriate one for this paper.

Using MATLAB R2021a software, through the Lagrange multiplier (LM), we can test spatial lag model, spatial error model, and LM model to find out which is the best for study of the thesis. The test results are shown in Table 6. The results of the LM test a robust LM test are both significant, fully confirming that the spatial measurement mo is the suitable model for the analysis of the impact of economic agglomeration on la use eco-efficiency. Given that the spatial lag model can reflect both the spatial effect the explanatory variable and the spatial effect of the explained variable as a more gene model, the spatial lag model is selected to expand the analysis. Through the Hausman t whether the model has a fixed effect or a random effect can be determined. Results sh that the p-value is less than 0.01, and a fixed effect should be used. Lastly, the maximu likelihood ratio (LR) test is used to decide if the spatial factor, the temporal factor, or bo should be fixed. The results show that both spatial and temporal factors are significa but the spatial fixed effect coefficient was significantly larger than the temporal fixed eff coefficient. Therefore, the model used in this study is the most appropriate.

Table 6. A series of tests of the spatial econometric model.

Test Object	Statistics	p-Value
LM rest no spatial	34.1785	0.000
Robust LM test no spatial lag	27.2021	0.000
LM test no spatial error	16.0945	0.000
Robust LM test no spatial error	9.1181	0.003
Hausman test	−99.6930	0.000
LR-test joint significance spatial fixed effects	4280.9887	0.000
LR-test joint significance time-period fixed effects	119.7880	0.000

4.4. Robustness: Instrumental Variable (IV) Method

Firstly, the empirical results of the three spatial weight matrices (the geographic distar weight matrix, adjacent weight matrix, and economic distance weight matrix) are genera consistent, indicating that the findings are somewhat robust. Further, in order to allevi the endogeneity problem of the model, the IV method is used for robustness testing.

The first instrumental variable is the topographic relief (Ups and Downs). Topograp is closely related to population agglomeration. Areas with low topographic fluctuations more likely to form population clusters than areas with high topographic fluctuations, whi satisfies the first condition of IV—correlation with the explanatory variables. Topograp height is related to the elevation and area of the region, but not to the LRUE [59]. Usi ArcGIS software, the digital elevation model (SRTM 90 m) data were resampled and reform into a grid of 1 km*1 km, using a grid size of 10 km*10 km as the manipulating unit; t dataset of the topographic grid by kilometers was gradually extracted by using the formu

The second instrumental variable was the number of night lights (DN) in Chine cities, which was satellite-derived night light data [60,61].

Table 7 shows the results of the above. As shown in Table 7, the values generated the F test in Stage I are all greater than 10, which complies with the rule of thumb, indicati that the instrumental variables selected in this paper are not weak instrumental variab The terrain fluctuation rate is significantly correlated with the regression coefficient economic agglomeration in a negative way; that is, when the terrain fluctuation rate is hi the degree of economic agglomeration is low, which is also consistent with reality. In t plateau region, the natural conditions are poor, the soil is not suitable for crop planting, t terrain is rugged, and the traffic costs are high, and these factors lead to a small populati and labor force. The night lighting data showed a significant positive correlation wi the quadratic term of economic agglomeration, which showed frequent economic activ and a high degree of agglomeration in the economically developed area. The regressi results of Stage II confirm that the results of the benchmark regression are robust, on t of the consideration of endogeneity; the impact of economic agglomeration on land u eco-efficiency continues to be U-shaped.

Table 7. Regression results of the spatial econometric model.

	Stage I		Stage II
	Agg	Agg^2	LRUE
Ups and Downs	−0.4139 *** (−20.78)	−4.612 *** (−21.41)	
DN	0.2352 *** (18.85)	4.0846 *** (21.77)	
Agg			−0.2715 *** (−8.95)
Agg^2			03.022 *** (10.53)
Control variables	Yes	Yes	Yes
Constant items	−1.3372 *** (−5.66)	−51.0338 *** (−16.26)	1.2915 *** (13.07)
F	448.31	586.56	
Adjusted R-sq	0.7603	0.7986	0.2586
Cragg–Donald Wald F statistic			318.57

*** indicates statistical significance at the 1% level.

5. Discussion and Policy Recommendations

5.1. Temporal Evolution Trend of Land Resource Use Efficiency

With the help of MaxDEA 8Ultra Software, we examined the overall LRUE evolution trend for 281 prefecture-level cities in China from 2004 to 2020 (Figure 2). It can be found that both nationally and locally, LRUE shows an increasing trend. The upward trend is most obvious in the eastern region, while the upward trend is weakest in the western region. The outcome of this result is different from most of the previous studies as in those an upward trend or a development trend of first rising and then falling is observed [10,31]. Therefore, it is necessary to include energy inputs and undesired outputs in the LRUE evaluation system. This is also related to the importance China has placed on energy conservation and emission reduction in the last 20 years. The importance attached by local governments to energy conservation, emission reduction, and sustainable development is the main reason for the steady rise in LRUE in all regions of China.

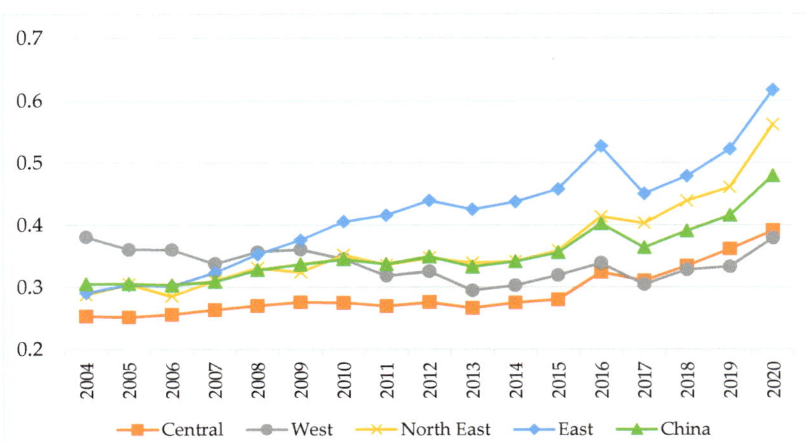

Figure 2. Trends of land resource use efficiency by region in China from 2004 to 2020.

The LRUE of prefecture-level cities is found to go from large to small in easte northeast, central, and western China. This distinctive distribution is related to the top raphy, the economic level, and the policy orientation. The western region, which has ma plateaus, mountains, and deserts, is taken as an example. Due to topographical constrai the cities in this region lack the labor force and consumer base to supply industrial devel ment, thus reflecting a lower LRUE value. At the same time, it can be found that with exception of the eastern region, the rise in LRUE mainly occurred after 2014, which may related to the many energy-saving and emission-reduction policies enacted in China duri this period, such as carbon emission trading scheme and forestry carbon sink polic These policies prompted local governments to guide enterprises in their emission reducti efforts, which in turn reduced the marginal consumption of their output. In terms of LR value, the level remains low in all regions except for the eastern region. This suggests t there is still ample room for upward mobility in China's use of land resources.

5.2. Spatial Evolution Trend of Land Resource Use Efficiency

Using ArcGIS 10.4 software, we mapped the LRUE distribution in China for 20 2006, 2016, and 2020 (Figure 3). Based on the distribution of LRUE values, we classifi them into five categories: low efficiency (≤0.6), medium-low efficiency (0.61–0.1–0 medium efficiency (0.71–0.8), medium-high efficiency (0.81–1.0), and high efficiency (>1 Previous studies focus more on evaluating land resource use efficiency without too mu consideration of the relationship between economic agglomeration and LRUE. High u efficiency (LUE) in these agglomerations demonstrates that "the more intensive the c is, the higher is its use efficiency" [62]. However, our study presents different results af considering the ecological factors.

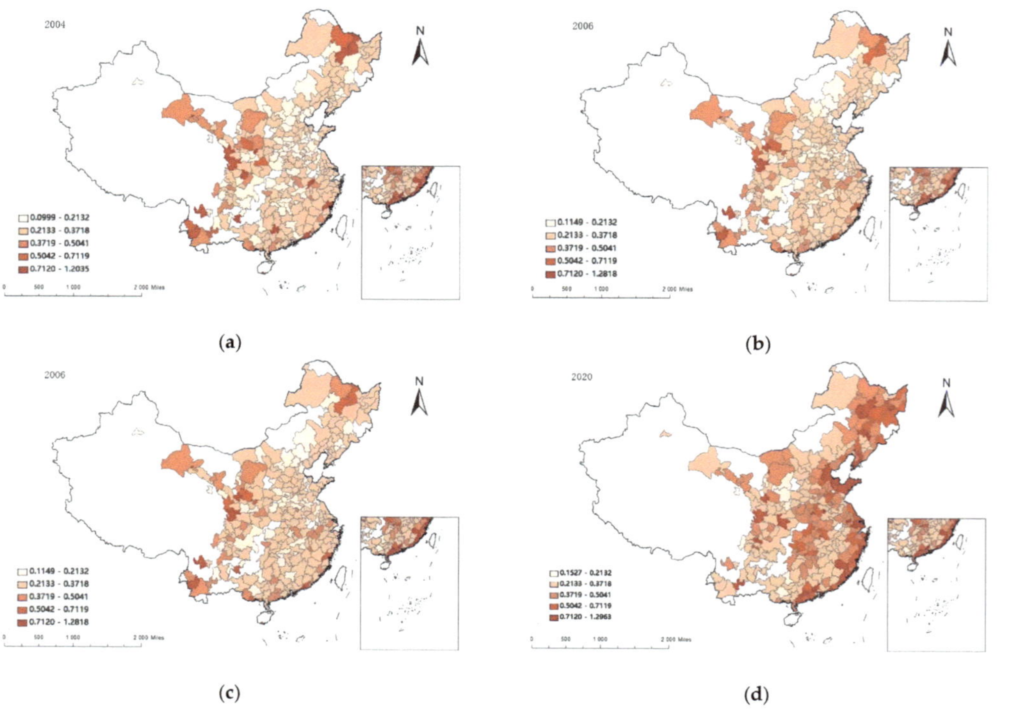

(a) (b)

(c) (d)

Figure 3. Distribution of land resource use efficiency in 4 years: (**a**) land resource use efficiency 2004; (**b**) land resource use efficiency in 2006; (**c**) land resource use efficiency in 2016; (**d**) land resou use efficiency in 2020.

During the observation period, the distribution area of high efficiency changed: In 2004, the high-efficiency areas are mainly distributed in the northern region and a few are scattered around southwest regions. This was not due to the realization of an increased LRUE among these regions, but because of the underdeveloped economy, unexploited ecological environment, and relatively fair social distribution, so comprehensive efficiency is high because of a combination of low input and low undesired output. After 2006, the disadvantages caused by extensive development in these regions gradually appeared. Some cities have shifted from the high-efficiency areas, while a dramatic increase in the efficiency value among other cities is not observed. After 2016, the spatial distribution characteristics of "high in the east, low in the west" appear to be obvious, and the high efficiency values are gradually concentrated in the eastern and northeastern regions and provincial capitals of China. These regions have a decent economic foundation and pay more attention to the improvement of human capital accumulation and technical level, which leads to a continuous increase in the LRUE. The spatial spillover effect is one of the factors that affect the spatial and temporal evolution trend of LRUE. However, there are other reasons that LRUE is probably influenced by the policy implementation efficiency, economic foundation, technological conditions, and environmental awareness in the different regions. Moreover, high land prices force the cities in eastern regions to use land resources more effectively, and the urban built-up area there expanded rapidly.

5.3. Spatial Spillover Phenomenon

This subsection focuses on the spatial spillover phenomenon (direct and indirect effects) of LRUE, as shown in Table 5. The direct effect reflects the impact of economic agglomeration in the region on the LRUE, while the indirect effect reflects the impact of economic agglomeration in other regions (depending on the spatial weight matrix) on the LRUE of the region.

In terms of direct effects, economic agglomeration has a U-shaped relationship with LRUE; i.e., in the early stages of economic agglomeration, the crowding effect is greater than the agglomeration effect, making factors such as high pollution and inequitable distribution worsen LRUE. The economic agglomeration, after a long period of "agglomeration–diffusion–agglomeration", optimizes the allocation of resources, improves regional knowledge and technology, and reduces distributional inequalities. This leads to a reduction in the undesired output of the LRUE, which in turn increases the LRUE.

In terms of indirect effects, there is still a U-shaped relationship between economic agglomeration and LRUE. In the early stages of economic agglomeration, the rapid transfer of economic activities to other regions reduces the production factors in the region, while the pollutants generated in other regions (especially in neighboring regions) will spread to the region, resulting in a reduction in its LRUE. However, when the economic agglomeration of other regions reaches a mature stage, it achieves the optimal allocation of resources and generates knowledge spillover effects. This helps the region to improve its LRUE level through exchange, cooperation, and learning by imitation.

5.4. Policy Recommendations

Based on the above findings, this paper makes the following recommendations:

(1) Reasonably guide economic agglomeration and improve its scope and quality in order to contribute to positive externalities. The U-shaped relationship between economic agglomeration and LRUE shows the following: when the development level of economic agglomeration is low, externalities cannot be revealed; when economic agglomeration develops too fast, it generates a crowding effect, leading to a waste of resources and social injustice, making LRUE decrease; when economic agglomeration develops maturely, resources are fully utilized and LRUE improves. Therefore, the negative effects of economic agglomeration are temporary and can be avoided. Local governments should apply this law of development in a reasonable manner.

(2) Mitigate the problem of over-agglomeration. Some regions in China have the probl of over-agglomeration, which is detrimental to the development of both the regi and neighboring regions. Therefore, local governments should use policy tools guide the level of agglomeration in a reasonable manner.
(3) Develop an economic agglomeration strategy with local characteristics. Curren China's economic agglomeration and LRUE show great differences in spatial c tribution, which implies that localities should tailor their economic agglomerati policies to guide high-quality economic agglomeration behavior according to th own strengths and weaknesses as well as the stage of agglomeration.

6. Conclusions

Based on the panel data of 281 prefecture-level cities in China from 2004 to 2020, t study explores the impact of economic agglomeration on LRUE by using the spatial l model. After performing a series of tests, we obtained robust conclusions: (1) there a positive spatial spillover effect on LRUE, and an increase in LRUE in the region he to improve LRUE in neighboring regions; (2) there is a U-shaped relationship betwe economic agglomeration and LRUE.

There are still some shortcomings in this research due to the limits of time, ener and length. First, the heterogeneity of economic agglomeration can be further discuss As far as different regional economic developments are concerned, it is possible to disc China's eastern, central, and western regions separately. Secondly, in this study, we us prefecture-level city data; the sample size could be larger if county-level city data a used. Thirdly, the transmission mechanism of economic agglomeration and LRUE can further studied.

Author Contributions: Conceptualization, N.Y. and Y.M.; methodology, N.Y.; formal analysis, N and Y.T.; data curation, N.Y.; writing—original draft preparation, N.Y.; writing—review and e ing, Y.T. and Y.M.; supervision, Y.T. All authors have read and agreed to the published version the manuscript.

Funding: This research was funded by the National Natural Science Foundation of China, gra number 71072066, the National Social Science Foundation of China, grant number 22BJY065, t Soft Science in Gansu Province, grant number 22JR4ZA060 and the Sichuan University, grant nu ber 2019hhf-14.

Data Availability Statement: See text for access to all data.

Conflicts of Interest: The authors declare no conflict of interest.

Notes

1 A prefecture-level city refers to a city that ranks below province and above county as an administrative division of China, wh usually governs a couple of subordinated areas including some county-level cities.

References

1. Shu, C.; Xie, H.; Jiang, J.; Chen, Q. Is Urban Land Development Driven by Economic Development or Fiscal Revenue Stimuli China? *Land Use Policy* **2018**, *77*, 107–115. [CrossRef]
2. Lee, C.-C.; Wang, C.-S. Does natural resources matter for sustainable energy development in China: The role of technologi progress. *Resour. Policy* **2022**, *79*, 103077. [CrossRef]
3. Song, W.; Cao, S.; Du, M.; Lu, L. Distinctive roles of land-use efficiency in sustainable development goals: An investigation trade-offs and synergies in China. *J. Clean. Prod.* **2023**, *382*, 134889. [CrossRef]
4. Sueyoshi, T.; Yuan, Y. China's regional sustainability and diversified resource allocation: DEA environmental assessment economic development and air pollution. *Energy Econ.* **2015**, *49*, 239–256. [CrossRef]
5. Shi, T.; Zhang, W.; Zhou, Q.; Wang, K. Industrial structure, urban governance and haze pollution: Spatiotemporal evidence fr China. *Sci. Total. Environ.* **2020**, *742*, 139228. [CrossRef]
6. Li, J.; Xu, C.; Chen, M.; Sun, W. Balanced development: Nature environment and economic and social power in China. *J. Cle Prod.* **2018**, *210*, 181–189. [CrossRef]
7. Li, Y.; Tang, Y.; Wang, K.; Zhao, Q. Environmental Regulation and China's Regional Innovation Output—Empirical Resea Based on Spatial Durbin Model. *Sustainability* **2019**, *11*, 5602. [CrossRef]

Ge, T.; Li, C.; Li, J.; Hao, X. Does neighboring green development benefit or suffer from local economic growth targets? Evidence from China. *Econ. Model.* **2023**, *120*, 106149. [CrossRef]
Wu, C.; Wei, Y.D.; Huang, X.; Chen, B. Economic transition, spatial development and urban land use efficiency in the Yangtze River Delta, China. *Habitat Int.* **2017**, *63*, 67–78. [CrossRef]
Li, Q.; Wei, J.; Gao, W. Spatial differentiation and influencing factors of land eco-efficiency based on low carbon perspective: A case of 287 prefecture-level cities in China. *Environ. Challenges* **2023**, *10*, 100681. [CrossRef]
Hong, W.; Wang, W.; Guo, R. Policies for optimizing land-use layouts in highly urbanized areas: An analysis framework based on construction land clearance. *Habitat Int.* **2022**, *130*, 102697. [CrossRef]
KIJI, S. A study of the industrial area—Agglomeration of industries in an area. *Geogr. Rev. Jpn.* **1952**, *25*, 478–485. [CrossRef]
Starrett, D. Market allocations of location choice in a model with free mobility. *J. Econ. Theory* **1978**, *17*, 21–37. [CrossRef]
Schmutzler, A. The new economic geography. *J. Econ. Surv.* **1999**, *13*, 355–379. [CrossRef]
Krugman, P. Increasing Returns and Economic Geography. *J. Political Econ.* **1991**, *99*, 483–499. [CrossRef]
Head, K.; Mayer, T. Regional wage and employment responses to market potential in the EU. *Reg. Sci. Urban Econ.* **2006**, *36*, 573–594. [CrossRef]
Feldman, M.P.; Audretsch, D.B. Innovation in cities: Science-based diversity, specialization and localized compe-tition. *Eur. Econ. Rev.* **1999**, *43*, 409–429. [CrossRef]
Rosenthal, S.S.; Strange, W.C. The Determinants of Agglomeration. *J. Urban Econ.* **2001**, *50*, 191–229. [CrossRef]
Konishi, H. Formation of Hub Cities: Transportation Cost Advantage and Population Agglomeration. *J. Urban Econ.* **2000**, *48*, 1–28. [CrossRef]
Klaesson, J. Monopolistic competition, increasing returns, agglomeration, and transport costs. *Ann. Reg. Sci.* **2001**, *35*, 375–394. [CrossRef]
Alfaro, L.; Chen, M.X. The global agglomeration of multinational firms. *J. Int. Econ.* **2014**, *94*, 263–276. [CrossRef]
Jaffe, A.B.; Trajtenberg, M.; Henderson, R. Geographic Localization of Knowledge Spillovers as Evidenced by Patent Citations. *Q. J. Econ.* **1993**, *108*, 577–598. [CrossRef]
Audretsch, D.B.; Feldman, M.P. R&d spillovers and the geography of innovation and production. *Am. Econ. Rev.* **1996**, *86*, 630–640. [CrossRef]
Keeble, D.; Nachum, L. Why do business service firms cluster? Small consultancies, clustering and decentralization in London and southern England. *Trans. Inst. Br. Geogr.* **2002**, *27*, 67–90. [CrossRef]
Knack, S.; Barro, R. Determinants of Economic Growth. *South. Econ. J.* **1998**, *65*, 185. [CrossRef]
Feng, Y.; Lee, C.-C.; Peng, D. Does regional integration improve economic resilience? Evidence from urban agglomerations in China. *Sustain. Cities Soc.* **2023**, *88*, 104273. [CrossRef]
Xu, H.; Wang, K.; Li, G.; Zhang, Y. How Officials' Competitive Pressure Affects Sustainable Development Ca-pacity From a Spatial Perspective: Empirical Evidence From China. *Front. Psychol.* **2021**, *12*, 607232. [CrossRef]
Kytzia, S.; Walz, A.; Wegmann, M. How can tourism use land more efficiently? A model-based approach to land-use efficiency for tourist destinations. *Tour. Manag.* **2011**, *32*, 629–640. [CrossRef]
Zhu, X.; Li, Y.; Zhang, P.; Wei, Y.; Zheng, X.; Xie, L. Temporal–spatial characteristics of urban land use efficiency of China's 35mega cities based on DEA: Decomposing technology and scale efficiency. *Land Use Policy* **2019**, *88*, 104083. [CrossRef]
Zhang, X.; Jie, X.; Ning, S.; Wang, K.; Li, X. Coupling and coordinated development of urban land use economic efficiency and green manufacturing systems in the Chengdu-Chongqing Economic Circle. *Sustain. Cities Soc.* **2022**, *85*, 104012. [CrossRef]
Liu, Y.; Sun, H.; Shi, L.; Wang, H.; Xiu, Z.; Qiu, X.; Chang, H.; Xie, Y.; Wang, Y.; Wang, C. Spatial-Temporal Changes and Driving Factors of Land-Use Eco-Efficiency Incorporating Ecosystem Services in China. *Sustainability* **2021**, *13*, 728. [CrossRef]
Tian, Y.; Zhou, D.; Jiang, G. A new quality management system of admittance indicators to improve industrial land use efficiency in the Beijing−Tianjin−Hebei region. *Land Use Policy* **2021**, *107*, 105456. [CrossRef]
Li, J.; Zhu, Y.; Liu, J.; Zhang, C.; Sang, X.; Zhong, Q. Parcel–level evaluation of urban land use efficiency based on multisource spatiotemporal data: A case study of Ningbo City, China. *Trans. GIS* **2021**, *25*, 2766–2790. [CrossRef]
Zhao, Q.; Bao, H.X.; Zhang, Z. Off-farm employment and agricultural land use efficiency in China. *Land Use Policy* **2020**, *101*, 105097. [CrossRef]
Burritt, R.; Schaltegger, S. Eco–efficiency in corporate budgeting. *Environ. Manag. Health* **2001**, *12*, 158–174. [CrossRef]
Wang, X.; Zhang, Y.; Yu, D.; Qi, J.; Li, S. Investigating the spatiotemporal pattern of urban vibrancy and its determinants: Spatial big data analyses in Beijing, China. *Land Use Policy* **2022**, *119*, 106162. [CrossRef]
Hosoe, M.; Naito, T. Trans-boundary pollution transmission and regional agglomeration effects*. *Pap. Reg. Sci.* **2006**, *85*, 99–120. [CrossRef]
Solarek, K.; Kubasińska, M. Local Spatial Plans in Supporting Sustainable Water Resources Management: Case Study from Warsaw Agglomeration—Kampinos National Park Vicinity. *Sustainability* **2022**, *14*, 5766. [CrossRef]
Ahmad, M.; Li, H.; Anser, M.K.; Rehman, A.; Fareed, Z.; Yan, Q.; Jabeen, G. Are the intensity of energy use, land agglomeration, CO_2 emissions, and economic progress dynamically interlinked across development levels? *Energy Environ.* **2020**, *32*, 690–721. [CrossRef]
Walter, I. The pollution content of American trade. *Econ. Inq.* **2010**, *9*, 61–70. [CrossRef]
Copeland, B.R.; Taylor, M.S. North-South Trade and the Environment. *Q. J. Econ.* **1994**, *109*, 755–787. [CrossRef]

42. Millimet, D.; Roy, J. *Four New Empirical Tests of the Pollution Haven Hypothesis When Environmental Regulation is Endogenous*; Tul University: New Orleans, LA, USA, 2013.
43. Wilson, J.D. Capital mobility and environmental standards: Is there a theoretical basis for a race to the bottom? *Fair Trade Harn Prerequisites Free. Trade* **1996**, *1*, 393–427.
44. Cole, M.A.; Elliott, R. FDI and the Capital Intensity of "Dirty" Sectors: A Missing Piece of the Pollution Haven Puzzle. *Rev. I Econ.* **2005**, *9*, 530–548. [CrossRef]
45. Konisky, D.M. Regulatory Competition and Environmental Enforcement: Is There a Race to the Bottom? *Am. J. Politi- Sci.* **2** *51*, 853–872. [CrossRef]
46. Lipscomb, M.; Mobarak, A.M. Decentralization and Pollution Spillovers: Evidence from the Re-drawing of County Border Brazil. *Rev. Econ. Stud.* **2016**, *84*, 464–502. [CrossRef]
47. Kenworthy, J. The eco-city: Ten key transport and planning dimensions for sustainable city development. *Environ. Urban.* **2** *18*, 67–85. [CrossRef]
48. Thissen, M.; Van Oort, F. European place—Based development policy and sustainable economic agglomeration. *Tijdschr. V Econ. En Soc. Geogr.* **2010**, *101*, 473–480. [CrossRef]
49. Graham, D.J.; Kim, H.Y. An empirical analytical framework for agglomeration economies. *Ann. Reg. Sci.* **2007**, *42*, 267– [CrossRef]
50. Combes, P.P.; Duranton, G.; Overman, H.G. Agglomeration and the adjustment of the spatial economys. *Pap. Reg. Sci.* **2005**, 311–349. [CrossRef]
51. Acs, Z.J.; Audretsch, D.B.; Lehmann, E.E. The knowledge spillover theory of entrepreneurship. *Small Bus. Econ.* **2013**, *41*, 757– [CrossRef]
52. Koster, S. Productive Places. The Influence of Technological Change and Relatedness on Agglomeration Externalities—By FRA NEFFKE. *Tijdschr. voor Econ. en Soc. Geogr.* **2010**, *101*, 365–366. [CrossRef]
53. Ke, S. Agglomeration, productivity, and spatial spillovers across Chinese cities. *Ann. Reg. Sci.* **2009**, *45*, 157–179. [CrossRef]
54. Brülhart, M.; Mathys, N.A. Sectoral agglomeration economies in a panel of European regions. *Reg. Sci. Urban Econ.* **2008**, 348–362. [CrossRef]
55. Moran, P.A.P. The Interpretation of Statistical Maps. *J. R. Stat. Soc. Ser. B Methodol.* **1948**, *10*, 243–251. [CrossRef]
56. Pastor, J.T.; Lovell, C.A.K. A global Malmquist productivity index. *Econ. Lett.* **2005**, *88*, 266–271. [CrossRef]
57. Lesage, J.; Pace, R.K. *Introduction to Spatial Econometrics*; Routledge: London, UK, 2008; pp. 19–44. [CrossRef]
58. Elhorst, J. Matlab Software for Spatial Panels. *Int. Reg. Sci. Rev.* **2014**, *37*, 389–405. [CrossRef]
59. Feng, Z.; Tang, Y.; Yang, Y. The relief degree of land surface in China and its correlation with population distribution. *Acta Ge Sin. -Chin. Ed.* **2007**, *62*, 1073. [CrossRef]
60. Raupach, M.; Rayner, P.; Paget, M. Regional variations in spatial structure of nightlights, population density and fossil-fuel C emissions. *Energy Policy* **2010**, *38*, 4756–4764. [CrossRef]
61. Li, W.; Sun, B.; Zhao, J.; Zhang, T. Economic performance of spatial structure in Chinese prefecture regions: Evidence fr night-time satellite imagery. *Habitat Int.* **2018**, *76*, 29–39. [CrossRef]
62. Han, W.; Zhang, Y.; Cai, J.; Ma, E. Does Urban Industrial Agglomeration Lead to the Improvement of Land Use Efficiency China? An Empirical Study from a Spatial Perspective. *Sustainability* **2019**, *11*, 986. [CrossRef]

land

MDPI

ticle

roprietary Varieties' Influence on Economics and ompetitiveness in Land Use within the Hop Industry

uglas MacKinnon [1] and Martin Pavlovič [1,2,*]

[1] Department of Agriculture Economics and Rural Development, Faculty of Agriculture and Life Sciences, University of Maribor, 3211 Hoče, Slovenia
[2] Slovenian Institute of Hop Research and Brewing, 3310 Žalec, Slovenia
* Correspondence: martin.pavlovic@um.si

Abstract: To evaluate changes to hop industry concentration and competitiveness the Herfindahl-Hirschman Index (HHI) was used. The ownership of hop proprietary varieties, their acreage and production were compared with public varieties. Market share for each proprietary hop variety acreage and production was calculated between 2000 and 2020. The quantity of land under centralized control in the U.S. hop industry due to increased proprietary variety acreage between 2000 and 2020 was quantified. Assuming tacit collusion between the participants in the oligopoly, the HHI enabled us to quantify the portion of land under oligopoly control. The HHI analysis of hop acreage and hop production demonstrated that market concentration rose rapidly between the years 2010 (0.0376 and 0.0729) and 2020 (0.4927 and 0.5394). This resulted in decreasing business competitiveness within the market during this period caused primarily by rapid consolidation of ownership during increased proprietary variety acreage and production increases. Calculations revealed that in 2016 a tipping point had been reached concerning market concentration, which resulted in higher sustained season average prices of hops—a key raw material in brewing.

Keywords: hop industry; varieties; market concentration; intellectual property; prices

ation: MacKinnon, D.; Pavlovič, Proprietary Varieties' Influence on nomics and Competitiveness in d Use within the Hop Industry. d 2023, 12, 598. https:// .org/10.3390/land12030598

ademic Editor: Matjaž Glavan

eived: 3 February 2023
ised: 27 February 2023
epted: 27 February 2023
lished: 2 March 2023

1. Introduction

Hops, along with malt and water, are the basic raw materials used for beer production. The basic role of hops is to provide beer with a pleasantly bitter taste and a hoppy aroma [1–3]. Between 2000 and 2020 the proportion of privately owned patented U.S. hop varieties increased. Other countries experienced increases in patented varieties, but these were owned by national associations. In the United States during that time, one variety development company, the Hop Breeding Company, grew to the point where its varieties enjoyed significant market share. The agglomeration of hop farms in Washington, Oregon, and Idaho (i.e., the Pacific Northwest (PNW)) facilitates the exchange of information by reducing monitoring costs thereby increasing market transparency among participants [4]. The increase in proprietary variety acreage and production has a causal effect on hop prices [5]. Tacit collusion results from competitors independently realizing their collective best interests to adjust prices or quantities [6]. The exchange of production-related information including anticipated yields and current prices among farmers may lead to a similar outcome [7].

Since 1913, the United States Department of Agriculture (USDA) has collected and published statistical data regarding the U.S. hop industry. The publication of intellectual property (IP) necessitates the use of symbols for registered trademarks, unregistered trademarks, and copyright, (i.e., "®", "TM" and "©") respectively. The USDA complies with these requirements. Proprietary variety ownership is publicly available through the U.S. Patent and Trademark Office (USPTO). The introduction of proprietary varieties, therefore, enabled the calculation of hop market share by acreage and production for the first time. Market share regarding sales of these varieties to brewers remained unavailable b was not important for calculating influence within the U.S. hop industry.

We used the Herfindahl-Hirschman Index (HHI) to measure changes in hop indus competitiveness by way of measuring market concentration. A similar methodologi approach was used to measure market concentration in the airline industry [8]. Accordi to the 2020 U.S. Federal Register, HHI was used to evaluate the acquisition of the Cr Brew Alliance, Inc. (CBA) by Anheuser-Busch InBev SA/NV ("ABI") and Anheuser-Bus Companies, LLC ("AB Companies"). The results of such analyses can provide insights in industry behavior. Markets with relatively high HHI values, market share inequality, a the presence of major firms are imperfectly competitive. Under such circumstances, mar imperfections are vulnerable to exploitation [9].

The presence of IP introduced constraints into the market where none had previou existed and affected farmer planting decisions on that acreage. More constrained variet were planted at a faster pace than those that were unconstrained [10]. Changes in mark concentration and price-cost margins can be used to determine the direction of comp tiveness [11]. The greater degree of specificity, control, and profit incentivized private h breeding companies to invest further in the development of new intellectual property. Th owners are incentivized by the ability to protect and enforce their rights [12]. Patent la also enabled IP owners to determine production and distribution via licensing agreeme

In 2021, Germany and the U.S. produced 32.91 and 40.87 percent of global hop acrea respectively. These are the two largest hop-producing regions, but each operates unc different business models. In 2020, patented and trademarked varieties in the United Sta represented 70.19% of PNW acreage and 73.44% of PNW production. Between 2009 a 2019, the annual farmgate value of American hops increased by 282% [13]. According the country reports of the International Hop Growers' Convention (IHGC) between 20 and 2019, 70 hop farmers in the PNW received approximately $4.7 billion in farm reven $2.88 billion more than the $1.87 billion the 1087 German hop farmers received during same period [14].

Between 1998 and 2020, USDA data recorded U.S. proprietary variety acreage a production soaring from zero to over 70 percent. Publicly available information regardi proprietary variety ownership enabled us to calculate the U.S. hop market share for the fi time in history. One variety development company, the Hop Breeding Company (HB owned the varieties responsible for over 50 percent of U.S. acreage and production 2020 [13].

The objective of the study was to evaluate changes to the hop industry area a production concentration and competitiveness with respect to the changes in proprieta varieties of hops relative to public varieties. Comi [15] refers to farmers who euphemistica described this process as "decommodification" implying only positive added value for t good of all. Using the Herfindahl-Hirschman Index (HHI), the concentration of acrea producing proprietary varieties between 2000 and 2020 under the control of a cartel-li structure with strict production and sales licensing agreements was quantified. Proprieta variety owners used their IP to create competitive advantages for their companies a those farmers allied with them. Consequently, they denied their competitors and farme they did not favor primary access to their IP. This research analyzed publicly availab industry data to determine the market effects resulting from the increased use of brand proprietary varieties by the craft brewing industry during this time and compared it wi other periods possessing unique characteristics dating back to 1948.

2. Materials and Methods

2.1. Proprietary Hop Variety Supply and Market Share

The United States Department of Agriculture lists each branded proprietary varie together with their respective intellectual property symbols in their publications [16]. T details of patents and trademarks are public information. By tracking the ownership these varieties in patent and trademark records with the U.S. Patent and Trademark Offi

(USPTO) and through the Google Patent Search website, we discerned the influence of individuals and entities over proprietary varieties. Data reported by the USDA included season average prices (SAP), inventory levels, production, and acreage. Wright and Williams [17] suggest that when supply is elastic and demand inelastic (as is the case with hops), the accumulation of stocks is typically damped by a compensating production response. The hop industry suffers from something called the Delayed Surplus Response (DSR). Production is highly elastic when prices and demand increase, but there is a delay of several years when prices and demand decrease. This results in surplus production that negatively affects global prices for hops through recurring boom-and-bust cycles [18]. Data published by the USDA enabled us to calculate the accumulation of aggregate stock levels and the annual market share of acreage by variety. We restricted our research to USDA National Hop Report (NHR) data between 1998 and 2022. That represented the period during which branded proprietary varieties were first reported by the USDA and included the most recently available industry data at the time of our calculations.

The companies that developed proprietary hop varieties own and license the production of multiple proprietary varieties to growers (for production) and sales and distribution of those varieties to merchants in their supply chain, thereby facilitating the management of production and distribution. We calculated the percentage for each proprietary variety produced within the Pacific Northwest (PNW) by the total acreage for the PNW i.e., the total market share. We calculated the market share for each entity owning IP listed by the USDA NASS in the USDA National Hop Report (NHR) by grouping those with common ownership of patented and trademarked products [16].

We expanded the variety-specific acreage market share calculations to group together those varieties that share common ownership to get a better picture of the influence of the five largest variety development companies. One company, the Hop Breeding Company, LLC (HBC) enjoyed increased influence within the industry as its proprietary varieties increased to occupy 51% of the acreage in the PNW. According to the company's website (www.hopbreeding.com, accessed on 18 July 2022), it is a joint venture between John I. Haas, Inc., a hop merchant company, and Yakima Chief Ranches, L.L.C., a company owned by the Smith, Carpenter, and Perrault families. These three families are also shareholders of Yakima Chief Hops, Inc., a hop merchant company. This complicated ownership structure effectively created a duopoly through which the proprietary varieties of the HBC were processed and distributed. Production was handled first through the farming resources to which each of the merchant companies had access. In the case of John I. Haas, Inc., that included the company farm and entities such as Roy Farms Inc., who touts on their website (http://royfarms.com/hops/roy-farms-citra/, accessed on 15 November 2022) their production and direct sales of one popular proprietary HBC variety, Citra®. In the case of Yakima Chief Hops Inc., those resources belonged to their farm owners. According to an article in the beer industry news outlet, Brewbound [18], that number expanded in 2019 from 11 to 15 farms. These varieties created a competitive advantage for the shareholders of the HBC and the other companies with whom it shared common ownership. How the owners of the HBC managed acreage between the two merchant companies to maintain equitable market share remains unknown. Proprietary varieties were distributed worldwide via licensing agreements with select merchants. The influence over such substantial acreage afforded the individuals involved with the HBC a disproportionate amount of influence in the industry. Their patents enabled them to decide via licensing agreements who would produce and sell their varieties. The MacKinnon Report, a hop market report published on Substack.com (https://mackinnonreport.substack.com, accessed on 26 January 2023) detailed that in 2023 the patent owners must reduce proprietary variety acreage by at least 8328 acres (3371 ha.) in response to a massive surplus that began in 2016. Those decisions have the power to create inefficiencies for some farms. How the decisions to reduce acreage will be coordinated by the two companies and which farmers will be told to reduce acreage remains to be seen. Some will be less efficient producers and not be able to compete in the future market.

Data demonstrated that the difference between depletion rates (i.e., the quantity hops shipped from U.S. warehouses) and the total available supply increased by 54 milli pounds between 2016 and 2022 (Figure 1). What we could not ascertain, however, was degree to which inventories managed by the same companies were growing in warehou located outside the purview of the USDA NASS surveys (i.e., every country other th the U.S.). Other countries do not publish similar data, which limited the completeness this study.

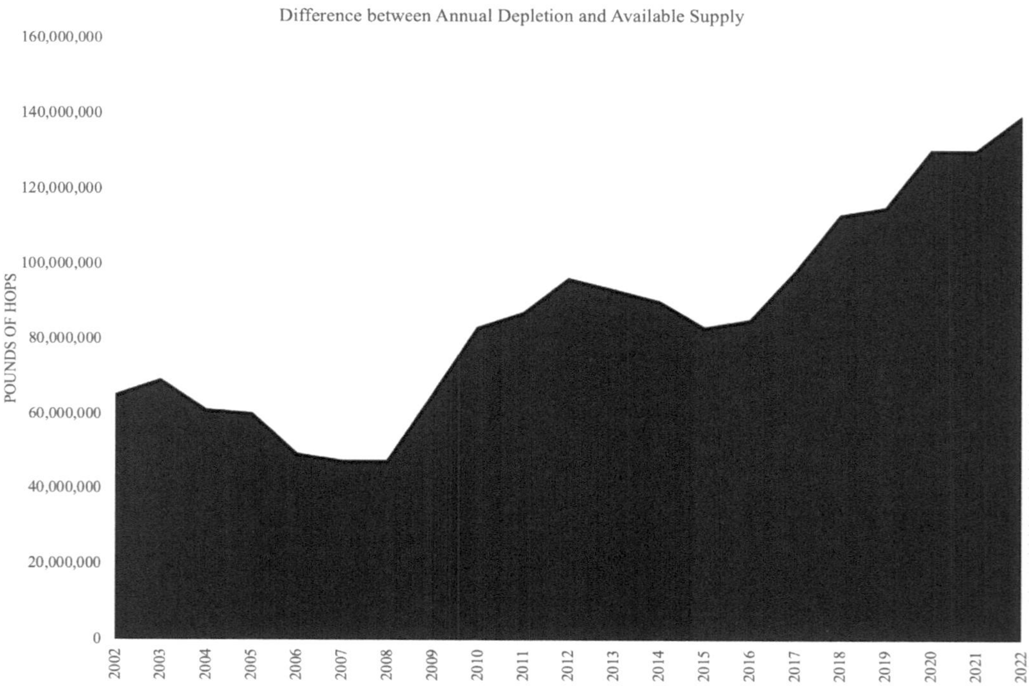

Figure 1. Annual Depletion (Sept n–Sept n-1) Relative to Total Available Supply (Crop (n-1) a Inventory (n-1)) (own study based on [16]).

2.2. *Calculating HHI*

We used the Herfindahl-Hirschman Index (HHI) to evaluate changes to hop indus concentration and competitiveness with respect to the changes in proprietary variet of hops relative to public varieties. A significant portion of PNW acreage appears two aggregate categories called "other" and "experimental". USDA data show that in 20 7.12% of total acreage was categorized as "other" or "experimental". The categories a used to report acreage and production for varieties that do not meet the three-independe grower threshold set by the USDA. Based on historical data, we believe at least half of t acreage was proprietary.

The HHI is a method used also by the United States Department of Justice (USDOJ) measure market concentration during mergers or acquisitions, to evaluate one compe tor's position relative to another, and to uncover potentially anti-competitive practic The HHI values of zero to 0.1500 mean a low market concentration. Values of 0.1500 0.2500 are considered a moderate concentration. Values of 0.2500 and above count as hi concentration. The HHI value will be low when market shares among participants a equal. The value will be high when one firm has a disproportionate share of the market [1 The value of the HHI decreases as the number of firms in the market increases. Mark concentration is inversely proportional to competitiveness [20]. The HHI is responsive to the asymmetry of market shares and is used to evaluate changes in the competitiveness within a single industry over time or to compare one industry to another [19]. In our research, we adopted this method for the first time in the hop industry for the measurement of its market concentration.

The HHI Formula

$$HHI = S1^2 + S2^2 + S3^2 + \ldots Sn^2 \tag{1}$$

where:

n refers to the number of varieties in the market;
S refers to the percent market share for a variety.

3. Results and Discussion

3.1. Acreage and Production Linked to Proprietary Hop Varieties

Calculating the Herfindahl-Hirschman Index (HHI) of the U.S. hop industry based on the market share of hop sales to brewers was a hopeless endeavor as information regarding market share based on sales of hops by merchants to breweries was proprietary information and never shared. We discovered an alternative method for measuring market share. The USDA NASS restrictions related to the reporting of proprietary U.S. acreage and production (i.e., that three or more independent producers must list acreage or production for the corresponding statistics to be reported in aggregate form) to meet the needs of this research.

Acreage, and the infrastructure necessary to harvest that acreage, was the scarcest and most valuable commodity in the hop industry in 2020, not the hops themselves. Acreage was the asset for which there was the greatest competition. The primary method for harvesting hops was via fixed picking machine facilities. Mobile combines exist that harvest cones from the vines in the field. Combines, when they were used, operated in conjunction with the more traditional fixed-picking facilities that could process at least 600 acres (242 hectares) of hops in a season. Combines returned cones harvested in the field to the picking facility to separate leaves, stems, and foreign material from the cones themselves through the picking facility's recleaning equipment. Due to the time-sensitive nature of the harvest, high ambient air temperatures, which could reach over 100 degrees Fahrenheit (37.78 °C) in Washington and Idaho states during harvest, hop growers sought to grow hops on land that was in close proximity (not more than 10–15 min driving time) to their fixed picking facilities to reduce the incidents of hops dehydrating prior to going through the picking machine, which increases cone shatter and reduces quality.

Five companies comprised approximately 70 percent of proprietary U.S. hop acreage and production in the Pacific Northwest in 2022 (Table 1). These variety development companies can license hop merchant companies to sell their varieties. They can license hop farms to produce their varieties. In some cases, the variety of development company ownership and the licensed merchants and farms shared common ownership. Licenses extended beyond companies in which they shared ownership. Previously independent farms were transformed into contract growers. The decision-makers for the five largest variety development companies, therefore, enjoyed a disproportionate influence in the industry and upon the market. The acreage on which a company's proprietary varieties were produced represented the market share of influence of the owners of each variety development company. The market share of influence represented a new and significant measurement possible within the industry all made possible by the growing demand for and reporting of proprietary varieties of hops.

Table 1. Five largest U.S. hop variety development companies and the market share of U.S. acrea and production of their proprietary varieties in 2022.

	Variety Development Company	Market Share of U.S. Total Acreage in 2022	Market Share of U.S Total Production in 2022
1	Hop Breeding Company (HBC)	49.05%	49.12%
2	HopSteiner	7.71%	10.78%
3	Association for the Development of Hop Agronomy (ADHA)	3.27%	3.35%
4	Virgil Gamache Farms (VGF)	3.20%	2.88%
5	CLS Farms	2.06%	2.02%
	TOTAL	65.30%	68.15%

Calculating the market share for each ownership group based on their ownersh of proprietary hop varieties enabled the calculation of the market share of influence ov the scarcest resource in the hop industry, acreage. Branded proprietary varieties a products that enjoy monopoly control by their very nature as patented and trademark products. Seventy percent of the acreage, therefore, was governed by the decision-mak of five entities. Public varieties, in contrast, are available for any grower to produce.

We calculated the market share for each proprietary variety production and acrea relative to total U.S. acreage for the years 2000 through 2020. During this time, market co centration moved from low to high according to the standards set by the U.S. Departme of Justice when evaluating mergers and acquisitions between competitors.

Using the HHI market share data by variety, we calculated the market share for proprietary varieties collectively as the U.S. hop industry resembles what is referred as a complex monopoly in the U.K. [21]. We calculated the increase in market conce tration between 2000 and 2020 of publicly reported U.S. proprietary hop varieties. T increasing HHI values between 2000 and 2020 demonstrated the changes in the degree competitiveness in the industry (Table 2).

Table 2. HHI Values for U.S. Total Proprietary Varieties by Acreage and Production 2000–2020.

Crop Year	HHI Values for Proprietary Varieties by Acreage	HHI Values for Proprietary Varieties by Production
2000	0.0376	0.0729
2001	0.0900	0.1474
2002	0.0961	0.1709
2003	0.0755	0.1416
2004	0.0898	0.1586
2005	0.0904	0.1425
2006	0.0948	0.1791
2007	0.1200	0.2100
2008	0.1533	0.2441
2009	0.1642	0.2593
2010	0.1393	0.1903
2011	0.1496	0.2050
2012	0.1149	0.1618
2013	0.2024	0.2882
2014	0.1822	0.2700

Table 2. *Cont.*

Crop Year	HHI Values for Proprietary Varieties by Acreage	HHI Values for Proprietary Varieties by Production
2015	0.1841	0.2500
2016	0.1832	0.2292
2017	0.2661	0.3170
2018	0.3094	0.3628
2019	0.4058	0.4371
2020	0.4927	0.5394

The U.S. proprietary hop varieties used to calculate market concentration relative to public varieties between 2000 and 2020 listed in alphabetical order: Ahtanum ™, YCR 1, Amarillo ® VGXP01, Apollo ™, Azacca ™ ADHA-483, Bravo ™, Calypso ™, Chelan, Citra ®, HBC 394, Columbus/Tomahawk®/Zeus (also known as: C/T/Z®), Ekuanot ™, HBC 366, El Dorado ®, Eureka ™, IDAHO 7™, Idaho Gem™, Jarrylo ™, ADHA-881, Loral ™, HBC 291, Millennium®, Mosaic ®, HBC 369, Pahto ™, HBC 682, Palisade ®, YCR 4, Pekko ™, ADHA-871, Sabro™, HBC 438, Simcoe ®, YCR 14, Strata™ OR 91331, Summit ™, Super Galena ™, Talus®, Warrior ™, YCR 5, Zappa® (own study based on [13]).

The HHI analysis demonstrates that the market concentration due to the increasing proportion of proprietary varieties rose from a low to a moderate concentration between 2000 and 2010. It remained in the moderate zone until 2016 when it rapidly began to increase through 2020 when a tipping point had been reached (Figures 2 and 3). Official government data documented that in 2017 proprietary varieties represented greater than 50% of U.S. hop acreage [13].

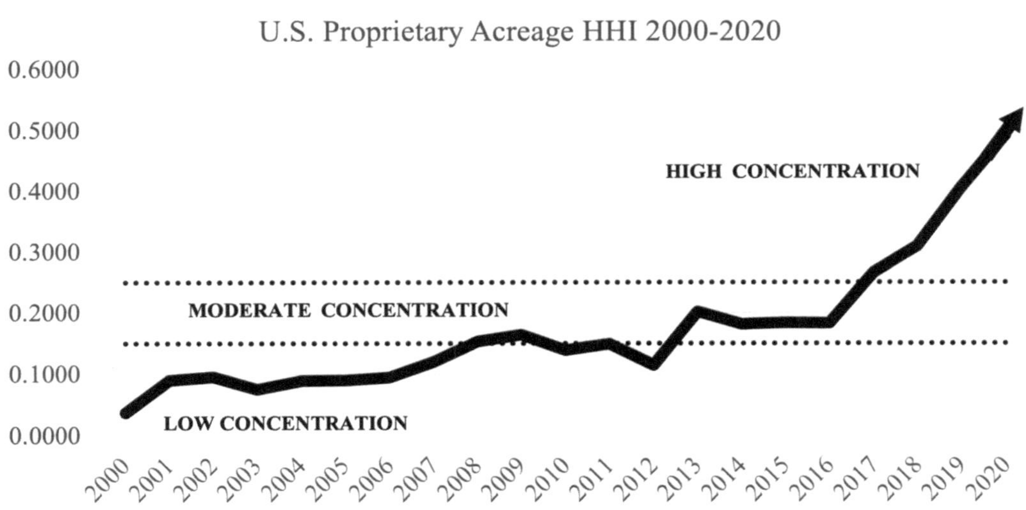

Figure 2. The HHI for total U.S. branded proprietary variety acreage 2000–2020 (own study based on [13]).

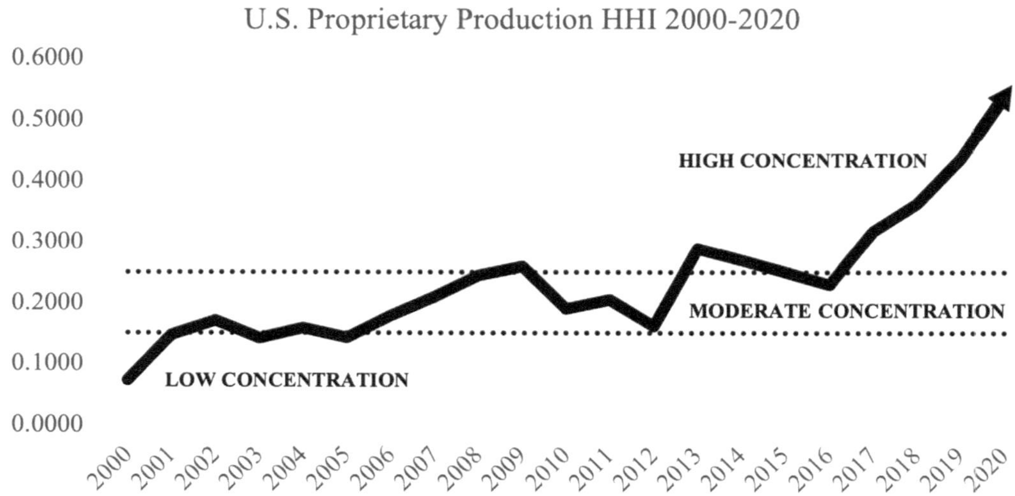

Figure 3. The HHI for total U.S. branded proprietary variety production 2000–2020 (own study bas on [13]).

3.2. Prices of Hops Linked to the Intellectual Property

In the measurement of the effects on price over time, it was necessary to adjust inflation. No appropriate Producer Price Index existed that could be applied to the U.S. h industry. Therefore, we decided to use the Consumer Price Index (CPI) as it reflects chang in the economy and the purchasing power of the U.S. dollar over time. Vermeulen [suggests that U.S. producers adjust their prices as often as retailers. This suggested that t use of the CPI for adjusting prices for inflation would be an appropriate strategy.

Between 2016 and 2022, U.S. season average prices for hops as reported by the USD remained stable at 75-year record high levels as acreage and production of proprieta varieties surpassed 50%. This suggested the existence of a tipping point facilitated proprietary varieties that had been reached with regard to industry influence and its eff on price control. Prices when adjusted for inflation rose rapidly following 2016. The rapic increasing HHI values post-2016 represented rapidly decreasing competitiveness. Reduc competitiveness was both a symptom and a consequence of the predominance of patent products where five entities captured 70% market share. As a result, prices increased two standard deviations above the 75-year average of inflation-adjusted prices of $4. per pound (Figure 4). The parameter of long-term season average prices of hops duri that time was chosen because it demonstrated lower price variability than the previo 36-year period for which data was available [13], a period that included World War I, t Great Depression, U.S. Prohibition, and World War II.

Rhoades (1995) concluded that the results of such analyses can yield useful insigh into industry behavior. Concentration and the degree of competitiveness within an indus can impact price. MacAvoy [11] identified a general hypothesis regarding changes in mark concentration and price-cost margins used to determine the direction of competitivenes

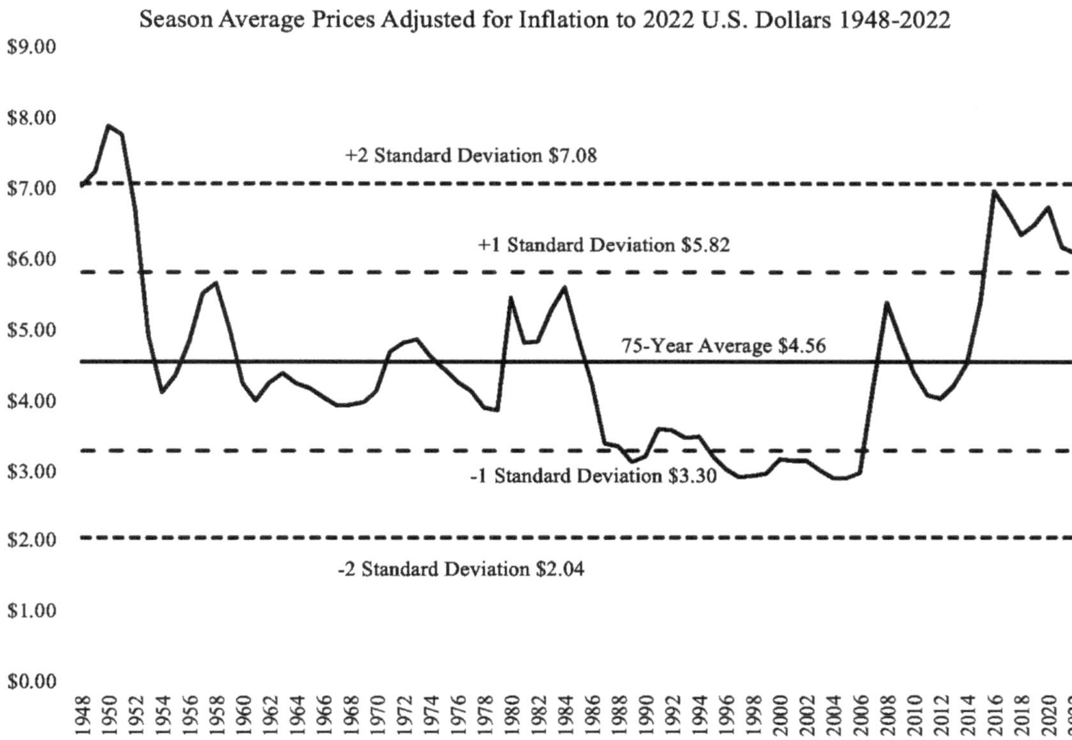

Figure 4. U.S. Season Average Prices 1948–2022 Adjusted for Inflation Using CPI (own study based on [16,23,24]).

Industry concentration reduced price competition as licensing agreements centralized decisions regarding production, sales, and marketing. Standardized quality by controlling harvest timing by IP owners was another result [25,26]. Patents grant the inventor control over the production, sales, and licensing of their invention without government involvement or oversight. Quantity or volume regulation and producer allotments of hops through Federal Marketing Orders have previously led to monopolistic policies [27].

3.3. Discussion

The Herfindahl-Hirschman Index calculations revealed that increases in proprietary variety acreage resulted in increased concentrations of power within the industry. This resulted in reduced competitiveness. Reduced competition in an oligopoly such as that which existed in the hop industry between 2016 and 2022 resulted in inflation-adjusted prices remaining stable at levels between one and two standard deviations higher than the 75-year average of U.S. season average prices, which was $4.56 per pound of hops i.e., $2.06 per kg of hops despite the existence of a growing surplus of hop inventory [13]. This was contrary to the normal tendency for prices to revert to long-term mean values or lower within 2–3 years following a price spike.

Reduced competitiveness within the hop industry during the period 2000–2020, enabled season average prices to remain at elevated levels for a prolonged period as they did between 2016 and 2020. The intrinsic homogeneous traits of branded proprietary varieties of hops such as oil production, which would typically result in symmetrical marginal costs, are overshadowed by extrinsic heterogeneous characteristics such as the perceived value of a brand and the urgency created by artificial scarcity. These characteristics create the perception of additional value for which the brewing industry has been prepared to p handsomely [28]. The premium price and royalties warranted by proprietary varieties c be considered a deadweight loss ultimately born by the beer consumer.

We expanded the variety-specific acreage market share calculations to group the varieties that share common ownership to get a better picture of the influence of t five largest variety development companies. One company, the Hop Breeding Compa had a much greater share than the rest. Common ownership between the entities th create branded proprietary varieties, individual hop farms and hop merchant firms f ther increased market concentration. The individuals who own the entities that crea proprietary varieties have created a competitive advantage for the merchant compani and farms in which they share a financial interest. We concluded that branded prop etary varieties, when their ownership is concentrated in few hands, reduced competiti within the market, encouraged market segmentation and created opportunities for potent anti-competitive behavior.

According to data available between 2009 and 2020 from IHGC economic repo and the Hop Growers of America Statistical Packets, the farmgate value for contract American hops was $2.88 billion greater than German growers. That does not represe the added value that processing, packaging, and reselling add to the price paid by brewe and beer consumers. During this same period, the USDA reported that proprietary varie acreage increased from 40.52 percent to 70.19 percent in the PNW [13].

4. Conclusions

The Herfindahl-Hirschman Index calculations offered a glimpse of changes in prop etary variety market share and the impact these changes have had upon market concent tion and competitiveness within the U.S. hop industry between 2000 and 2022. During th short time, the industry went from one dominated by publicly available varieties to o managed product controlled by a duopoly.

The most relevant of the consequences of increased market concentration of reduc competitiveness was the greater ability of proprietary variety owners to manage the p ception of scarcity of supply for their proprietary products on the market. Artificial scarc created fear among brewers, which led to them signing long-term contracts at high pric (terms dictated by the farmer/merchant entities). Through their efforts, they could redu or eliminate surplus inventory thereby enabling sustained premium prices indefinitely. return to a system based on free market competition rather than one controlled by oligarc would return market-based pricing and eliminate prices set at artificially high levels.

Additional data now suggest a surplus of contracted proprietary varieties develop and grew between 2016 and 2022. Artificially high prices were sustained since free mark forces were not allowed to act. In the face of oversupply, these higher prices artificia increased the cost of production not only for American brewers but for brewers around t globe. The effect of proprietary varieties on the DSR remains to be seen as it is underw in 2023. It appears at the time of this writing the additional opacity created by priva management of approximately 70 percent of the U.S. crop delayed the initial signali period for the DSR to begin.

The effects of such supply management efforts affect not only proprietary U.S. v rieties but public varieties in the U.S., too. The relationship between hop varieties (i hop varieties may be substituted with other varieties) extends the effects of proprieta variety management upon farmers in countries where they are not produced. Additior research regarding the complementary relationship between the U.S. and German prod tion regions is recommended to understand price movements, the disparity of pricing, a perceived value. The reduction in competitiveness within the U.S. industry this resear provides is an important step in furthering the understanding of hop market dynami and the interrelatedness of world markets. Further research might include an examinati of the methods owners of proprietary varieties may use to cooperate with other relat entities to alter supply to determine where the border exists for anti-trust violations. Further investigation into acreage management strategies is warranted.

Author Contributions: Conceptualization, methodology and improving the manuscript, D.M. and M.P.; results interpretation, writing and editing, D.M.; supervision, project administration, M.P. All authors have read and agreed to the published version of the manuscript.

Funding: This research was funded by the Slovenian Research Agency (ARRS) and the University of Maribor (FKBV).

Data Availability Statement: Not applicable.

Conflicts of Interest: The authors declare no conflict of interest.

ferences

Štěrba, K.; Pavel Čejka, P.; Čulík, J.; Jurková, M.; Krofta, K.; Pavlovič, M.; Mikyška, A.; Olšovská, J. Determination of Linalool in Different Hop Varieties Using a New Method Based on Fluidized-Bed Extraction with Gas Chromatographic–Mass Spectrometric Detection. *J. Am. Soc. Brew. Chem.* **2015**, *73*, 151–158. [CrossRef]
Pavlovič, M. Production character of the EU hop industry. *Bulg. J. Agric. Sci.* **2012**, *18*, 233–239.
Pavlovič, M.; Čerenak, A.; Pavlovič, V.; Rozman, Č.; Pažek, K.; Bohanec, M. Development of DEX-HOP multi-attribute decision model for preliminary hop hybrids assessment. *Comput. Electron. Agric.* **2011**, *75*, 181–189. [CrossRef]
Gan, L.; Hernandez, M.A. Making friends with your neighbors? Agglomeration and tacit collusion in the lodging industry. *Rev. Econ. Stat.* **2013**, *95*, 1002–1017. [CrossRef]
MacKinnon, D.; Pavlovič, M. The delayed surplus response for hops related to market dynamics. *Agric. Econ.* **2022**, *68*, 293–298. [CrossRef]
Devlin, A. A proposed solution to the problem of parallel pricing in oligopolistic markets. *Stanf. Law Rev.* **2007**, *59*, 1111–1151.
Rees, R. Tacit Collusion. *Compet. Policy. Oxf. Rev. Econ. Policy* **1993**, *9*, 27–40. [CrossRef]
Johnston, A.; Ozment, J. Concentration in the airline industry: Evidence of economies of scale? *J. Transp. Manag.* **2011**, *22*, 59–74. [CrossRef]
Rhoades, S. Market Share Inequality, the HHI, and Other Measures of the Firm-Composition of a Market. *Rev. Ind. Organ.* **1995**, *10*, 657–674. [CrossRef]
Dixit, A.K.; Stiglitz, J.E. Monopolistic Competition and Optimum Product Diversity. *Am. Econ. Rev.* **1977**, *67*, 297–308.
MacAvoy, P. Testing for Competitiveness of Markets for Long Distance Telephone Services: Competition Finally? *Rev. Ind. Organ.* **1998**, *13*, 295–319. [CrossRef]
Bugos, G.; Kevles, D. Plants as Intellectual Property. *Osiris* **1992**, *7*, 74–104. [CrossRef]
USDA NASS. National Hop Reports (NHR) for Years 2000–2020. [Dataset]. United States Department of Agriculture National Agricultural Statistics Service (USDA NASS). 2020. Available online: https://www.nass.usda.gov/Statistics_by_State/Washington/Publications/Hops/index.php (accessed on 18 July 2022).
MacKinnon, D.; Pavlovič, M. Global hop market analysis within the International Hop Growers' Convention. *Hop Bull.* **2019**, *26*, 99–108.
Comi, M. Other agricultures of scale: Social and environmental insights from Yakima Valley hop growers. *J. Rural Stud.* **2020**, *80*, 543–552. [CrossRef]
USDA NASS. National Hop Reports (NHR) for Years 2000–2022. [Dataset]. United States Department of Agriculture National Agricultural Statistics Service (USDA NASS). 2022. Available online: https://www.nass.usda.gov/Statistics_by_State/Washington/Publications/Hops/index.php (accessed on 22 December 2022).
Wright, B.; Williams, J. The Economic Role of Commodity Storage. *Econ. J.* **1982**, *92*, 596–614. [CrossRef]
Anonymous. Yakima Chief Hops Welcomes Four New Grower-Owners. Brewbound Beer Ind. News 2019. Available online: https://www.brewbound.com/news/yakima-chief-hops-welcomes-four-new-grower-owners/ (accessed on 18 July 2022).
Calkins, S. The New Merger Guidelines and the Herfindahl-Hirschman Index. *Calif. Law Rev.* **1983**, *71*, 402–429. [CrossRef]
Depken, C. Free-Agency and the Competitiveness of Major League Baseball. *Rev. Ind. Organ.* **1999**, *14*, 205–217. [CrossRef]
Brent, R. The meaning of 'Complex Monopoly'. *Mod. Law Rev.* **1993**, *56*, 812–831. [CrossRef]
Vermeulen, F. The Narrative Advantage. *Lond. Bus. Sch. Rev.* **2012**, *23*, 55–57. [CrossRef]
USDA NASS. WA Hops 1915–2013. 2013. Available online: https://www.nass.usda.gov/Statistics_by_State/Washington/Publications/Historic_Data/index.php#hops (accessed on 18 July 2022).
USDA NASS. Hop Stocks Held by Growers, Dealers, and Brewers, United States (1000 Pounds). 2014. Available online: https://www.nass.usda.gov/Statistics_by_State/Washington/Publications/Historic_Data/index.php#hops (accessed on 18 July 2022).
Pavlovič, M.; Pavlovič, V.; Rozman, Č.; Udovč, A.; Stajnko, D.; Wang, D.; Gavrić, M.; Srečec, S. Market value assessment of hops by modeling of weather attributes. *Plant Soil Environ.* **2013**, *59*, 267–272. [CrossRef]

26. Mikyška, A.; Olšovská, J.; Slabý, M.; Štérba, K.; Čerenak, A.; Košir, I.J.; Pavlovič, M.; Kolenc, Z.; Krofta, K. Analytical and sens profiles of Slovenian and Czech hop genotypes in single hopped beers. *J. Inst. Brew.* **2018**, *124*, 209–221. [CrossRef]
27. Folwell, R.J. *The U.S. Hop Marketing Order: The Price of Success is Misunderstanding;* Department of Agricultural Econom Washington State University: Pullman, WA, USA, 1982.
28. Olšovská, J.; Štérba, K.; Pavlovič, M.; Čejka, P. Determination of energy value of beer. *J. Am. Soc. Brew. Chem.* **2015**, *73*, 165– [CrossRef]

land

MDPI

ticle

he Driving Role of Food and Cultivated Land Resource in alancing the Complex Urban System of Socio-Economy and nvironment: A Case Study of Shanghai City in China

ngli Ruan

School of Economics and Management, China University of Mining and Technology, Xuzhou 221116, China; 6420@cumt.edu.cn

Abstract: Food is increasingly seen as a vehicle to address complex sustainability challenges, where the quantitative driving role in balancing the complex urban system of socio-economy and environment is still a gap. To fill this gap, taking Shanghai city as an example, this paper utilizes system dynamics to innovatively set three policy scenarios that aim at adjusting food security and cultivated land resources. The results confirm their positive role in socioeconomic and environmental improvement and coordinated development. In the high-rate grain yield growth scenario, the labor force ratio of the primary industry increases back to the size of 2012 (4.1%), the proportion of the primary industrial investment grows at twice the rate of the current trend, the grain yield per unit area increases back to the capacity of 1997 (798.154 t/km^2), and simultaneously, the occupation of cultivated land resources by the secondary industry and the negative impact of environmental pollution on productivity are mitigated. In that case, the coordination level between the socio-economy and the environment can keep increasing. The results indicate that future urban planning should increase the input of labor force and assets in the primary industry, improve food productivity per unit area through technical means or person training, alleviate the occupation of cultivated land resources by the secondary industry, and mitigate the negative impact of environmental pollution on cultivated land productivity.

Keywords: role of food and cultivated land resource; sustainable development; coordination development; socio-economy and environment; Shanghai city

ation: Ruan, F. The Driving Role Food and Cultivated Land ource in Balancing the Complex ban System of Socio-Economy and ironment: A Case Study of nghai City in China. *Land* **2023**, 905. https://doi.org/10.3390/ d12040905

ademic Editor: Matjaž Glavan

eived: 19 March 2023
vised: 14 April 2023
cepted: 17 April 2023
blished: 18 April 2023

1. Introduction

Humans have entered the urban century [1]. The scale of the urban population worldwide reached 4.22 billion in 2018 and will keep growing to 6.68 billion in 2050 [2]. Urban development simultaneously creates more than 75% of the global GDP and contributes to 75% of energy-based carbon emissions [3]. Therefore, the urban population is the protagonist in the challenge for achieving sustainable development, as it carries most human activities connected with environmental crises and resource depletion. The New Urban Agenda at the Habitat III conference emphasized the role of cities as engines for sustainable development [4]. The knowledge about a sustainable future from an urban perspective should be central, and building a global urban science is essential [5]. Therefore, research on urban-scale development is of great significance for sustainable goals and science.

Further, realizing sustainable goals is not only the independent progress of each goal but, more importantly, to combat undesirable interactions, conflicts, and impacts, which can exactly be reflected in system analysis [6]. The urban system is a complex system and seeks systematic thinking [7]. System approaches can reduce negative surprises and promote integrated planning, management, and governance [8]. Therefore, it has been a research priority for some scholars to fully connect the urban socioeconomic and environmental systems and understand their dynamic interplay and feedback [9].

Among elements in different subsystems of complex urban system models, the source mainly covers water resources [10,11], energy resources [12], nonrenewable a renewable resources [13], land use [14,15], and land development [16]. However, rap urban development and sprawl are also threatening agricultural sustainability [17]. though making cities sustainable is a global determination, little attention has been pa by the government to agriculture due to the belief that the highest benefit use of land not on agriculture; this may result from a lack of the literature on the precise role of urb agriculture in building sustainable cities [18]. Therefore, considering the cultivated la elements in the complex urban system is of great theoretical significance.

A few studies on urban development take cultivated land resources and grain outp into account [19]. Tapia et al. evaluated urban sustainability in Arhus based on an indica framework that includes urban agriculture and food security [20]. Nigussie et al. analyz the capacity and suitability of urban farmland in Ethiopia to contribute to sustainab development [21]. Tong et al. explored the manner to optimize urban food production Tucson, a semi-arid region [22]. Xia et al. quantified the dynamics of green infrastructure agriculture in Zhengzhou from the perspective of ecological security and food security [2 Clerino et al. examined the practices for the sustainability assessment of professional urb agriculture [24]. Gozdziewicz-Biechonska et al. analyzed theoretical frameworks a practices for protecting agricultural land use [25]. These studies consider the element ro of cultivated land and grain in sustainability. However, there is still a gap in their drivi role in the coordinated development of complex urban systems.

Indeed, food is increasingly being regarded as a vehicle to simultaneously addre economic, social, and environmental dilemmas [26]. Urban agriculture is recognized a potential contributor to more sustainable development at the city level as it promotes fo security and food sovereignty [27]. It is also lauded for its potential positive impacts on three pillars of sustainability [18]. However, the effect of agriculture on the urban scale c be conflicting as good or bad [28]. The positive impacts include enhancing dietary divers and improving ecosystem services and social quality [18,29,30], while the negative impa are also stressed by some critics [27,31]. These arguments make it urgent to identify the ro of food and cultivated land resources in the coordinated development of complex urb systems. In addition, although food resources for a metropolis can basically be suppli through trade with neighboring regions, the risk of public health events and natu disasters should be considered, and urban farming should be emphasized to increase fo safety [32].

To sum up, this paper aims to fill the gap of the quantitative driving role of fo and cultivated land resource in balancing the complex urban system of the socio-econor and the environment. Taking Shanghai city as an example, the system dynamics moc is applied to simulate complex urban system development. Different scenarios are set adjust the conditions of food and cultivated land resource and forecast the coordinati level. According to the simulation results, suggestions for more sustainable and coor nated development are proposed from the perspectives of food security and cultivat land protection.

2. Methods and Study Area

2.1. System Dynamics Model

This paper applies system dynamics, which is an effective method for developing understanding of complex systems [33]. This method can incorporate various elemen of different subsystems in an integrated model, and the feedback loop can quickly sei the complex interactions [13]. In addition, scenario simulation is a good tool for poli or scheme analysis [10,34]. Due to these advantages, system dynamics has been appli in broad fields, including the urban system analysis, such as urban water–energy–fo system [32], urban economy–resource–environment system [19], green urban system [3 urban social economy–water environment system [11], urban economy–society–resour environment system [36], etc. Accordingly, this paper takes advantage of system dynam to analyze the food role of urban agriculture in the urban system for coordination development between the socio-economy and the environment.

2.1.1. Model Conceptualization

This paper establishes an urban system, as shown in Figure 1. Firstly, this paper focuses on the food role in the coordination development of the urban system, reflected in 77 loops. For example:

Grain security→Total population→Labor force→Output of the primary industry→Grain security (a positive loop);
Grain security→Total population→Labor force→Output of tertiary/secondary industry→natural land→Grain security (a negative loop);
Grain security→Total population→Air/Water pollution→Pollution index→Output of the primary industry→Grain security (a negative loop);
Grain security→Total population→Labor force→Output of the tertiary/secondary industry→GDP→Output of the primary industry (a positive loop);
Grain security→Total population→Air/Water pollution→Pollution index→Output of the tertiary/secondary industry→GDP→Output of the primary industry (a negative loop).

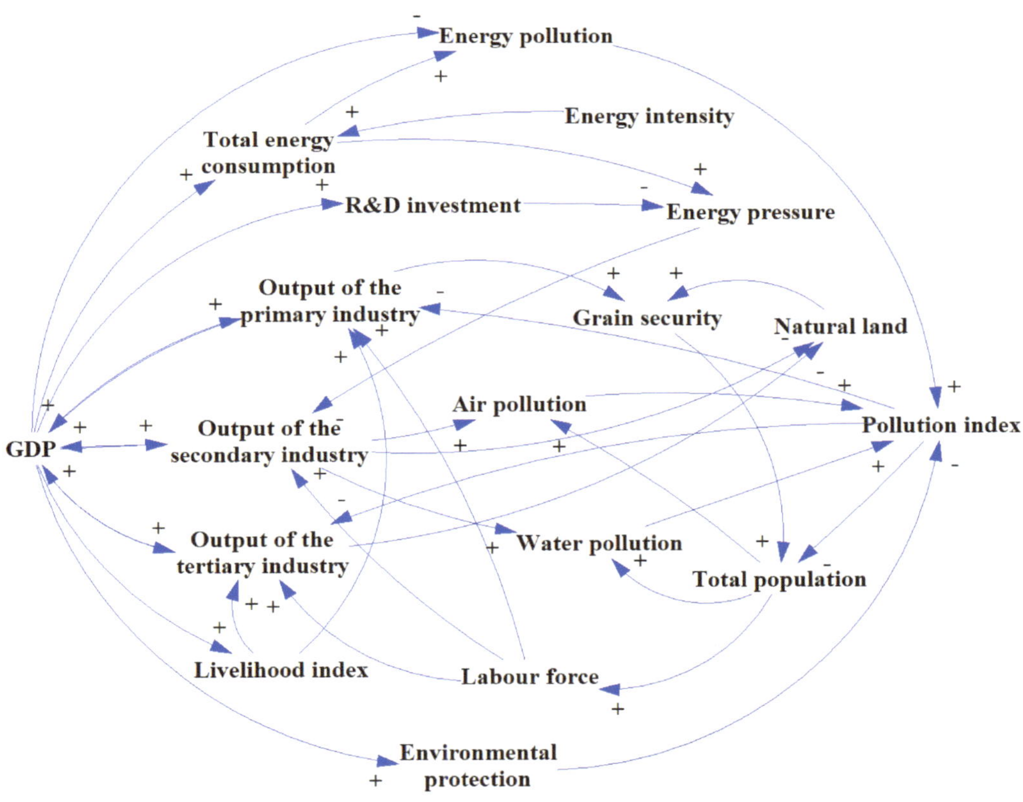

Figure 1. Causal loop diagram of the urban system.

Secondly, this paper hopes to provide a reference for promoting coordination development, considering the subsystems of the socio-economy and the environment. The socio-economy mainly covers the population, livelihood, three industries, and technology investment. The environmental subsystem investigates air and water pollution, energy consumption, and investment in protection.

2.1.2. Model Formulation

Based on the causal loop diagram, this paper further defines the urban system w various elements and interacting formulations (Appendix A). Then, the stock and fl chart of the system model is formulated, as shown in Figure 2. The model builds an urb system and aims to examine how the evolution of grain security (grain yield per capi will influence the system coordination level. The determination methods of parameters the interactions are shown in Appendix B.

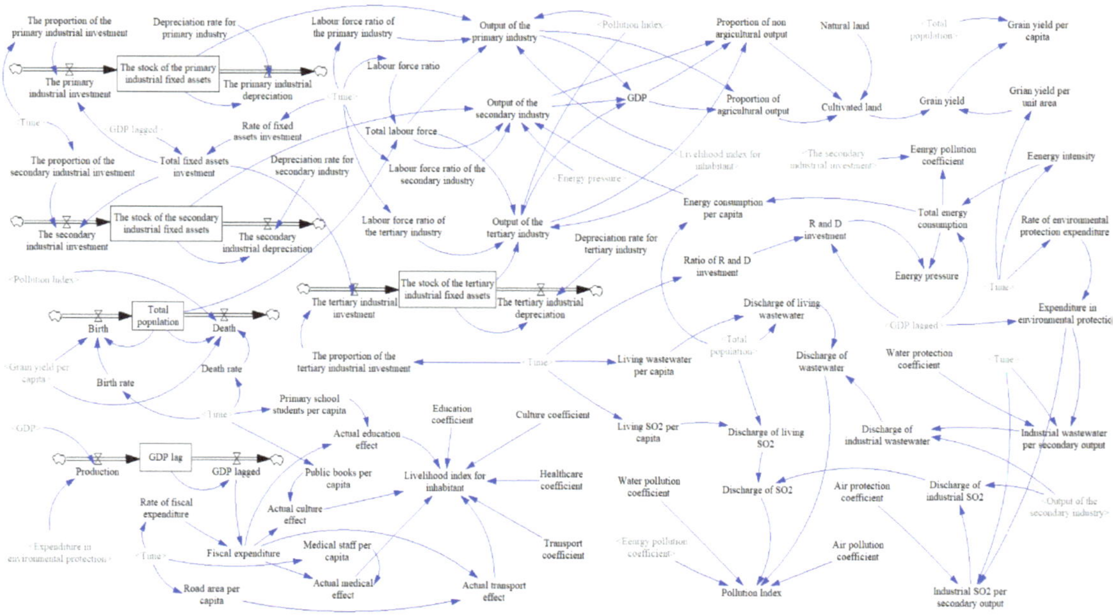

Figure 2. Stock and flow chart of the urban system.

To confirm the model's accuracy, this paper compares the simulated results with t actual data through the following formulas [19].

$$MARE = \frac{1}{N \times T} \sum_{j=1}^{N} \sum_{t=1}^{T} \left| \left(\widehat{Y_{jt}} - Y_{jt} \right) / Y_{jt} \right|$$

MARE represents Mean Absolute Relative Error, and a smaller value inflects a high matching degree and a better simulation validation. Y_{it} is the actual value of indicator j time t, and $\widehat{Y_{jt}}$ is the simulated one.

2.1.3. Scenario Settings

After establishing the urban system, the food role of urban agriculture is examin To this end, three scenarios have been designed, as shown in Table 1. The first is the curre scenario, with all variables evolving as the current trend. The second is the scenario high protection in urban agriculture, with related parameters being adjusted to promo the growth of grain yield at a higher rate than the current value. The third scenario is lo protection in urban agriculture, where the grain yield grows at a lower rate or decrease

Table 1. Parameter setting of three scenarios to examine food role.

Parameters	Scenario 1: Current Trend	Scenario 2: Higher-Rate Grain Yield Growth	Scenario 3: Lower-Rate Grain Yield Growth
LFRP	0.012	0.041	0.006
LFRS	0.2782	0.2492	0.2842
PPII	0.008	0.026	0.002
PSII	0.1308	0.1128	0.1368
GYUA	662	800	602
REPE	0.032	0.037	0.026

Notes: LFRP: labor force ratio of the primary industry; LFRS: labor force ratio of the secondary industry; PPII: proportion of the primary industrial investment; PSII: proportion of the secondary industrial investment; GYUA: grain yield per unit area; REPE: rate of environmental protection expenditure.

2.2. Coupling Coordination Degree Model

To evaluate the coordination level between the urban socio-economy and the environment, this paper applies the coupling coordination degree model, a widely use method for complex systems with various interactions [37–39].

2.2.1. Indicator System and Data Source

Firstly, the indicator system of the urban system is established in consideration of data accessibility and previous studies [35,36,40,41], as shown in Table 2. This paper mainly collects data from Shanghai Statistical Yearbooks, and the simulation span is 1995–2030. To ensure the results are reasonable, this paper applies indicators of per capita value and converts the ones related to currency into values with 2000 as the base year.

Table 2. Indicator system of the urban system.

Subsystem	Indicators	Weight	Direction
Socio-economy	X_{11}: GDP per capita (¥10 thousand)	7.77%	+
	X_{12}: The proportion of agriculture in GDP (%)	15.93%	+
	X_{13}: Fixed assets investment per capita (¥10 thousand)	9.84%	+
	X_{14}: Fiscal expenditure per capita (¥10 thousand)	10.43%	+
	X_{15}: R & D expenditure per capita (¥10 thousand)	23.55%	+
	X_{16}: Birth (10 thousand persons)	32.47%	+
Environment	X_{21}: Cultivated land per capita (km^2/10 thousand persons)	11.95%	+
	X_{22}: Grain yield per capita (t/10 thousand persons)	21.32%	+
	X_{23}: Energy consumption per capita (Ton of standard coal)	11.02%	−
	X_{24}: Discharge of wastewater per capita (t)	13.34%	−
	X_{25}: Discharge of SO_2 per capita (t)	23.62%	−
	X_{26}: Expenditure in environmental protection per capita (¥10 thousand)	18.74%	+

2.2.2. Performance Evaluation of Subsystems

Then, the performance of the socio-economy and environment is evaluated in the following three steps.

Step 1: The indicator values are standardized with the two formulas to eliminate the possible impacts of dimensions.

$$\text{Positive indicators}: \ X'_{ij} = \left(X_{ij} - min\{X_j\}\right) / \left(max\{X_j\} - min\{X_j\}\right) \tag{2}$$

$$\text{Negative indicators}: \ X'_{ij} = (max\{X_j\} - X_{ij}) / (max\{X_j\} - min\{X_j\})$$

where X_{ij} and X'_{ij} represent data before and after standardization, and $max\{X_j\}$ a[illegible] $min\{X_j\}$ represent the maximum and minimum data of indicator j.

Step 2: The weights of indicators are obtained with the entropy method.

$$\text{Proportion of indicator } j \text{ in } i\text{th year}: \ P_{ij} = X'_{ij} / \sum_{i=1}^{m} X'_{ij}$$

$$\text{Entropy redundancy of indicator } j\text{:} \ d_j = 1 - \left(-\frac{1}{ln\ m} \sum_{i=1}^{m} P_{ij} \cdot ln\ P_{ij}\right)$$

$$\text{Weight of indicator } j\text{:} \ w_j = d_j / \sum_{j=1}^{n} d_j$$

Step 3: The performance levels are obtained.

$$S_i = \sum_{j=1}^{n} w_j \cdot X'_{ij}$$

2.2.3. Coordination Evaluation of the Urban System

Finally, the coordination level of the urban system is evaluated with the coupli[illegible] coordination model.

$$\text{The overall level of the urban system}: \ T = \beta_1 \cdot f(SE) + \beta_2 \cdot g(EN)$$

$$\text{The coupling degree}: \ C = 2 \cdot \left[\frac{f(SE) \cdot g(EN)}{(f(SE) + g(EN))^2} \right]^{\frac{1}{2}}$$

$$\text{The coordination level}: \ D = \sqrt{C \cdot T} \quad (1$$

where $f(SE)$ and $g(EN)$ are the performance level of the socio-economy and the envir[illegible] ment, and β_1 and β_2 represent their contributions, respectively. The values of β_1 and β_2 a[illegible] both 0.5 [37].

Based on previous studies [19,37], the coordination degree can be divided into fi[illegible] levels, i.e., seriously unbalanced ($0 \leq D < 0.25$), slightly unbalanced ($0.25 \leq D < 0.5$), bar[illegible] balanced ($0.5 \leq D < 0.75$), and superior balanced ($0.75 \leq D \leq 1$).

2.3. Study Area

Shanghai, a cosmopolitan megacity, is located at 120°52′~122°12′ E and 30°40′~31°53′ It has absorbed about 25 million people with a total area of 6340.5 km^2. It plays a cent[illegible] role in multi-aspects of national development, such as the economy, technology, finance, [illegible] Therefore, it is significant for China and the world to study sustainable and coordinat[illegible] development in Shanghai city. Since 2010, the Shanghai Municipal Government has ma[illegible] many efforts in farmland supplement and reclamation of inefficient construction la[illegible] However, there are still problems, such as insufficient reserve resources of agricultural la[illegible] and contradiction of land use. Thus, agricultural land protection and food security a[illegible] still severe.

3. Results

3.1. Model Accuracy

This paper compares the simulated results with the actual data to confirm the mode[illegible] accuracy. The average relative errors of these representative indicators are all within 5[illegible] as shown in Figure 3; these results mean that the simulation values have good agreeme[illegible] with the actual values [34]. Therefore, the model's behavior is reliable for simulating t[illegible] development of Shanghai.

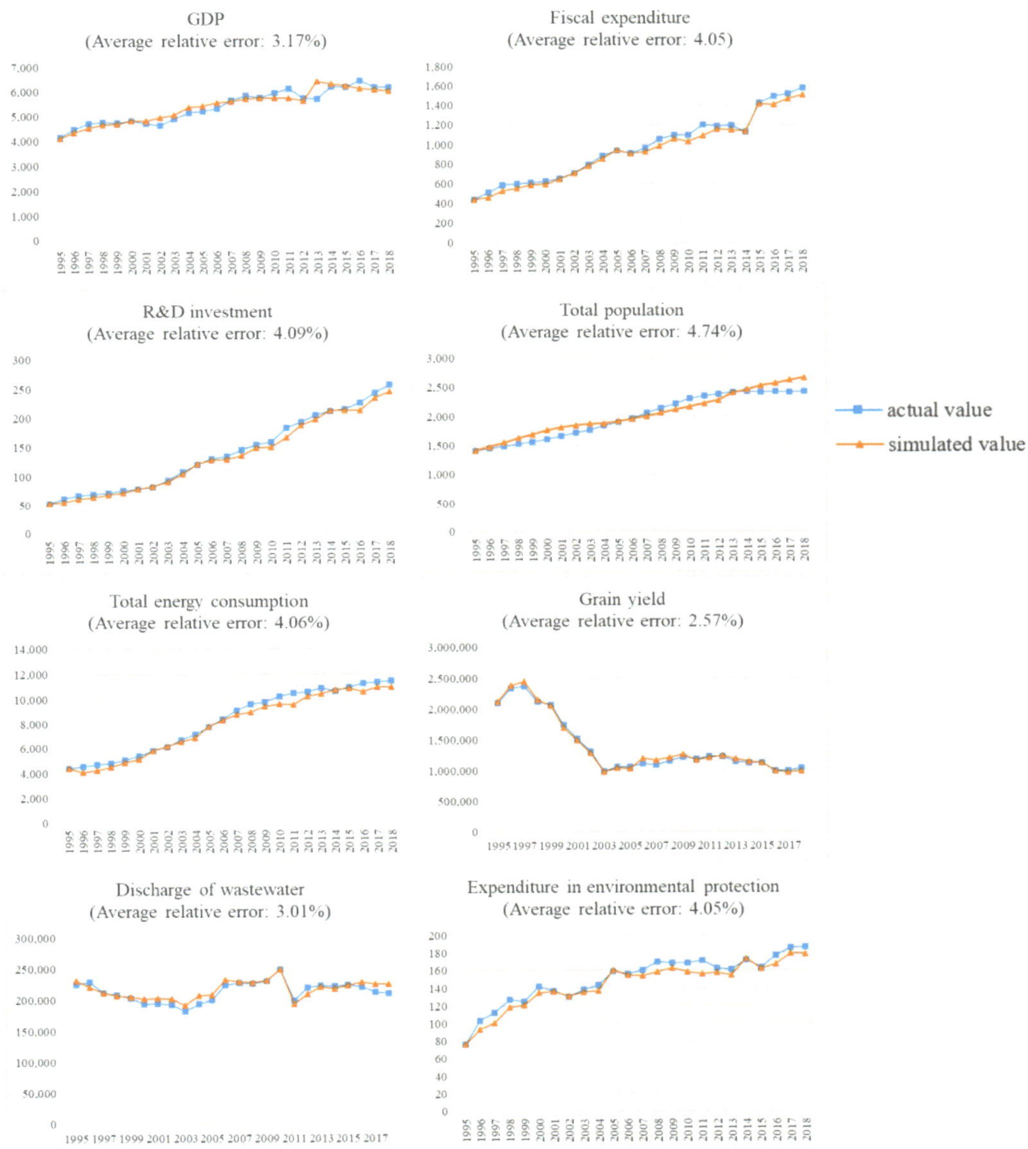

Figure 3. Average relative error of some key indicators in the model.

3.2. Model Results of Three Scenarios

3.2.1. Development of Food in Urban Agriculture

Under different scenarios, cultivated land will perform differently, as shown in Figure 4. If Shanghai follows the current development trend, agricultural land resources will be increasingly severe, and the lower-rate grain yield growth scenario will see a more serious situation. In contrast, the higher-rate grain yield growth scenario can somewhat protect cultivated land and promote resource increase at a gentle rate.

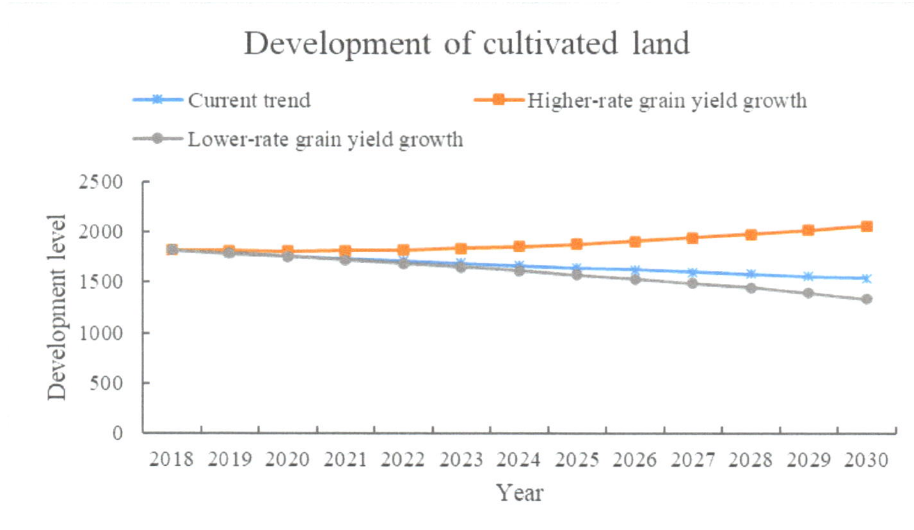

Figure 4. Development of cultivated land under three scenarios.

Similarly, the development of grain yield per capita sees the same development tren as shown in Figure 5. Therefore, the higher-rate grain yield growth scenario that is set this paper can protect cultivated land resources and ensure food security.

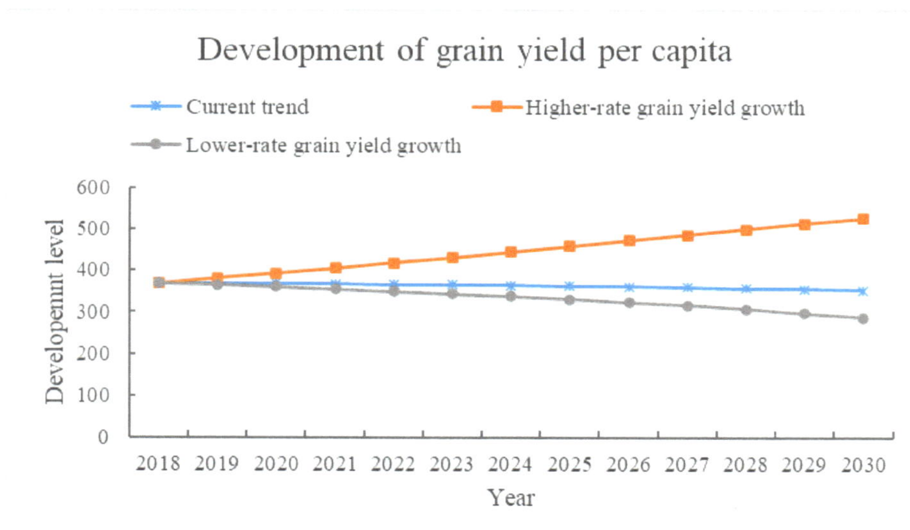

Figure 5. Development of grain yield per capita under three scenarios.

3.2.2. Socioeconomic and Environmental Performance

Under different scenarios, the performance of the socio-economy and the environme in Shanghai differs. As shown in Figure 6, the socio-economy shows an upward trend unc all three scenarios during 2018–2030. The lower-rate grain yield growth scenario sho a minimum rise from 0.32 to 0.55. In contrast, the performance under higher-rate gra yield increases significantly from 0.32 to 0.67. The results imply that ensuring food security benefits socioeconomic development, which can be explained through the indicators' weights, as shown in Table 2. The top three important indicators are birth (32.47%), R & D expenditure per capita (23.55%), and the proportion of agriculture in GDP (15.93%), contributing to over 70% of the socioeconomic growth in total. High grain yield brings high growth in agriculture and promotes population growth, resulting in a high socio-economy. Therefore, the socioeconomic performances ranks are shown below: scenario under higher-rate grain yield growth > current scenario > scenario under lower-rate grain yield growth.

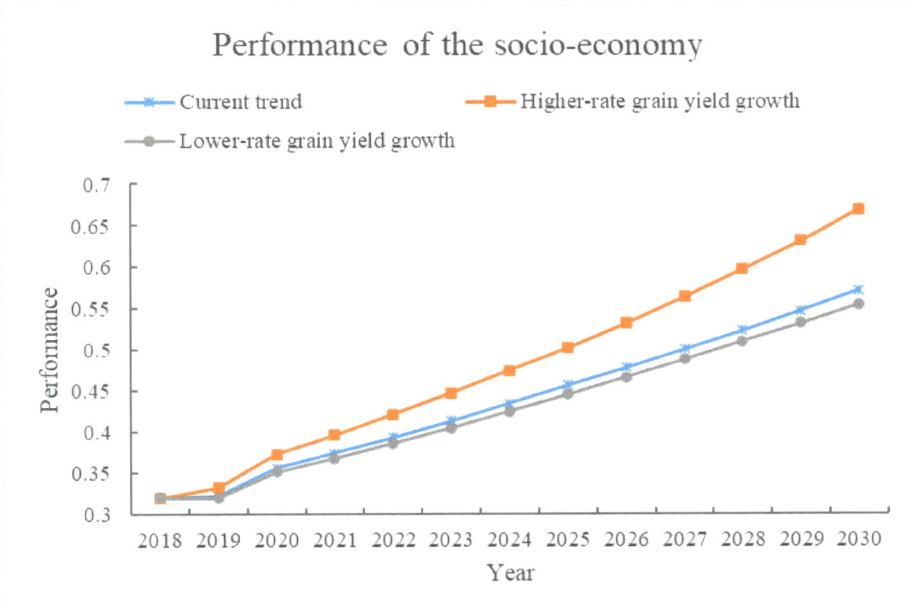

Figure 6. Performance of the socio-economy under three scenarios.

For the environment, the performance shows an opposite trend during 2018–2030, as shown in Figure 7. It will decrease sharply from 0.66 to 0.32 on the current development trend, and the situation under lower-rate grain yield growth will be even more serious. By comparison, the environmental performance can almost maintain the present level under higher-rate grain yield growth. The results imply that ensuring food security can also benefit environmental development, which can be explained through the indicators' weights in Table 2. Grain yield per capita and cultivated land per capita contribute to 23.32% and 11.95% of environmental performance, respectively; this is one-third in total. Therefore, the environmental performance ranks are shown below: scenario under higher-rate grain yield growth > current scenario > scenario under lower-rate grain yield growth.

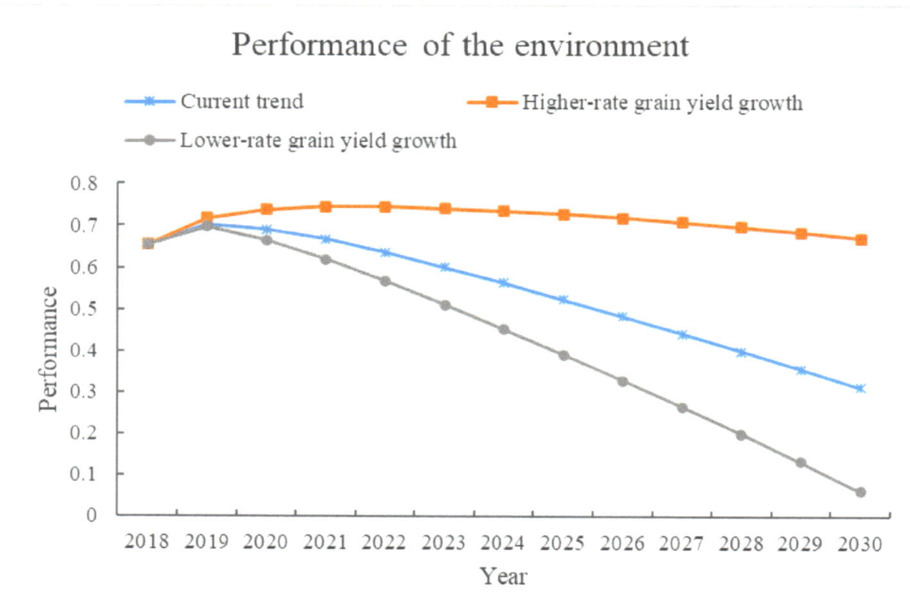

Figure 7. Performance of the environment under three scenarios.

3.2.3. Coupling Coordination Level

The coupling degrees under three scenarios are shown in Figure 8, reflecting t interacting intensity between the socio-economy and the environment. The trends und the current and higher-rate yield growth scenarios are similar, increasing from 0.2 to almo 1. In contrast, the coupling degree under lower-rate grain yield scenario growth sharp declines after 2027 to the original level.

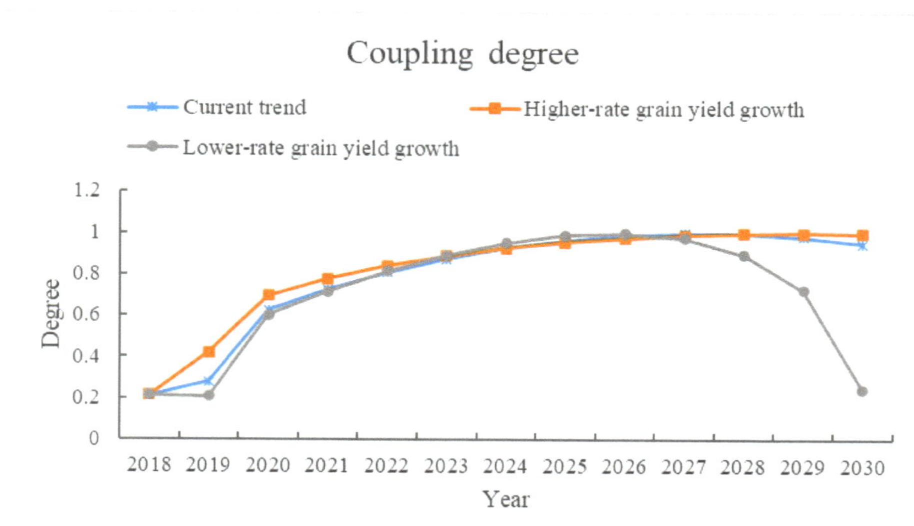

Figure 8. Coupling degree under three scenarios.

The coordination degrees under three scenarios are shown in Figure 9. In general, the coordination levels under current and higher-rate grain yield growth scenarios show an upward trend, while that of the under lower-rate grain yield growth scenario begins to decrease after 2025. Specifically, the coordination level of the current scenario increases gently from 0.3 (slightly unbalanced) to 0.7 (barely balanced). Coordination of higher-rate grain yield growth scenario is the highest and increases significantly to 0.967 (superior balanced). In contrast, that of the lower-rate grain yield growth is the lowest and first increases slowly to slightly unbalanced by 2025, and then, falls back to slightly unbalanced by 2030. These results indicate that neglecting the food role of urban agriculture will lead to uncoordinated development.

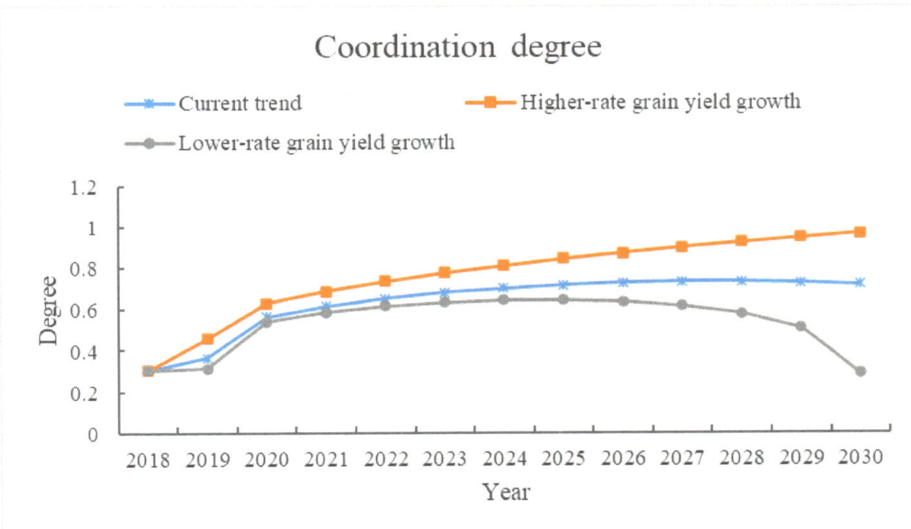

Figure 9. Coordination degree under three scenarios.

4. Discussion

Firstly, we can confirm the food role of urban agriculture in promoting the socio-economy by comparing the prediction results of different scenarios (Figure 6). The high-rate grain yield growth scenario promotes investment in the labor force and assets of the primary industry as well as improves the efficiency of food production. In addition, it alleviates the secondary industry's occupation of cultivated land resources and the negative impact of environmental damage on cultivated productivity. All these adjustments can promote the proportion of agriculture in GDP and have a positive effect on population development, which are essential aspects of the socioeconomic subsystem. These discoveries are in line with previous studies. Strengthening the development of food resources has been proven to be an optimized scenario to improve food security and GDP per capita [34]. Urban agriculture significantly benefits economic stability and physical health [42].

Secondly, we can also confirm the food role of urban agriculture in promoting the environment by comparing the prediction results of different scenarios (Figure 7). The high-rate grain yield growth scenario reduces the investment in the labor force and assets of the secondary industry as well as improves environmental protection, which can reduce pollution emissions and energy consumption. In addition, it protects the cultivated land resources and enhances food security, which has already been recognized as feasible in rich regions [27]. All these adjustments can promote the environmental subsystem. These findings align with other studies that show that more sustainable urban food systems allow energy conservation and emission reduction [42]. The results also provide quantitative evidence for previous views that urban agriculture is essential to deal with environme deterioration [43]. Agriculture presents the potential for meeting sustainable goals reducing adverse environmental impacts [44].

Finally, the food role of urban agriculture in promoting coordinated developm is confirmed by comparing the prediction results of different scenarios (Figures 8 and The high-rate grain yield growth scenario improves the performance of the socioecono subsystem, mitigates the decline in environmental performance, and ensures a high c pling degree, which promotes coordinated development. These discoveries further expa the previous research results, further identifying that improving food security can b efit the development of the socioeconomic and the environmental subsystems and th coordination level.

5. Policy Implications

Based on the simulation results, this paper attaches importance to agriculture a proposes the following suggestions for Shanghai city and other metropolises worldw with tense cultivated land resources and food security capacity to improve sustainabi Food-focused urban planning has already been recommended in some countries [45].

Firstly, increase the input of the labor force and assets in the primary indus to promote its development and improve food security. In the high-rate grain yi growth scenario, the labor force ratio of the primary industry is set to increase back the size of 2012, rather than fall to less than half of the current level; and the proporti of the primary industrial investment will grow at twice the rate of the current grow trend. Comprehensive means, including these measures, can significantly impro the performance of socioeconomic and environmental subsystems and promote th coordinated development.

Secondly, improve food production efficiency through technical means or man training to promote the development of the primary industry and improve food secur In the high-rate grain yield growth scenario, the grain yield per unit area is set to incre back to the capacity of 1997, much faster than the current trend, which has been proven simulation as an effective way to improve food security.

Thirdly, mitigate the occupation of cultivated land resources by the secondary ind try. In the high-rate grain yield growth scenario, the labor force ratio of the second industry is set to decrease at a faster rate than the actual trend; the same is true for t proportion of the secondary industrial investment. Comprehensive means, includi these measures, can also significantly improve the performance of socioeconomic a environmental subsystems and promote their coordinated development.

Last but not least, strengthen environmental protection and mitigate the negati impact of environmental damage on cultivated land production. In the high-rate gr yield growth scenario, the rate of environmental protection expenditure is set to incre faster than the current trend, which has been proven by simulation as an effective way improve food security.

6. Conclusions

This paper applies system dynamics to innovatively analyze the quantitative role food and cultivated land resources in complex urban systems of the socio-economy a the environment. Three scenarios, the current trend scenario, low-rate grain yield grow scenario, and high-rate grain yield growth scenario, are set to adjust food and cultivat land resources to analyze their roles. The coupling coordinated degree model is applied evaluate the coordination level.

The results confirm the positive role of food and cultivated land resources in promoti the performance of socioeconomic and environmental subsystems in Shanghai city a coordinated development. The results indicate that future urban planning should incre the input of labor force and assets in the primary industry, improve food productiv per unit area through technical means or person training, alleviate the occupation cultivated land resources by the secondary industry, and mitigate the negative impact of environmental pollution on cultivated land productivity.

It must be admitted that this study also has some limitations. The first limitation comes from data availability. For example, urban agriculture can also produce water pollution and play a role in promoting metabolism. However, due to the unavailability of data, these factors are not considered in the urban system. In addition to improving data quality, future research can continuously improve the representativeness of complex urban systems, such as incorporating the interactions between other elements as well as between different cities into the comprehensive system, making it more reflective of the actual development pattern.

Funding: This research was funded by the Fundamental Research Funds for the Central Universities (XJ2022003001).

Data Availability Statement: Not applicable.

Conflicts of Interest: The authors declare no conflict of interest.

Appendix A Model Formulations

[1] Air pollution coefficient = 0.81
[2] Air protection coefficient = 0.2553
[3] Birth = Birth rate × Total population × Grain yield per capita^0.3384
[4] Birth rate ([(1995, 0)–(2030, 0.01)], (1995, 0.0055), (1996, 0.0056), (1997, 0.0055), (1998, 0.0052), (1999, 0.0054), (2000, 0.0053), (2001, 0.0043), (2002, 0.0047), (2003, 0.0043), (2004, 0.006), (2005, 0.0061), (2006, 0.006), (2007, 0.0073), (2008, 0.007), (2009, 0.0066), (2010, 0.0071), (2011, 0.0072), (2012, 0.0096), (2013, 0.0076), (2014, 0.0086), (2015, 0.0074), (2016, 0.009), (2017, 0.0081), (2018, 0.0067), (2030, 0.0085))
[5] Cultivated land = Natural land × Proportion of agricultural output^0.2342/Proportion of nonagricultural outputt^8.975
[6] Culture coefficient = 0.3105
[7] Death = Death rate × Total population × (1 + Pollution index × 6.3084)/Grain yield per capita^0.3384
[8] Death rate ([(1995, 0)–(2030, 0.01)], (1995, 0.0075), (1996, 0.007), (1997, 0.0068), (1998, 0.007), (1999, 0.0065), (2000, 0.0072), (2001, 0.0071), (2002, 0.0073), (2003, 0.0075), (2004, 0.0072), (2005, 0.0075), (2006, 0.0072), (2007, 0.0074), (2008, 0.0077), (2009, 0.0076), (2010, 0.0077), (2011, 0.0079), (2012, 0.0054), (2013, 0.0082), (2014, 0.0083), (2015, 0.0086), (2016, 0.0085), (2017, 0.0087), (2018, 0.0086), (2030, 0.0095))
[9] Depreciation rate for primary industry = 0.045
[10] Depreciation rate for secondary industry = 0.0698
[11] Depreciation rate for tertiary industry = 0.045
[12] Discharge of industrial SO_2 = Output of the secondary industry × Industrial SO_2 per secondary output/1000
[13] Discharge of industrial wastewater = Output of the secondary industry × Industrial wastewater per secondary output
[14] Discharge of living SO_2 = Total population × Living SO_2 per capita/1000
[15] Discharge of living wastewater = Total population × Living wastewater per capita
[16] Discharge OF SO_2 = Discharge of industrial SO_2 + Discharge of living SO_2
[17] Discharge of wastewater = Discharge of industrial wastewater + Discharge of living wastewater
[18] Education coefficient = 0.1915
[19] Energy intensity ([(1995, 0.8)–(2030, 2.2)], (1995, 1.0581), (1996, 1.0172), (1997, 0.9991), (1998, 1.0156), (1999, 1.0713), (2000, 1.125), (2001, 1.2388), (2002, 1.3188), (2003, 1.359), (2004, 1.3919), (2005, 1.4852), (2006, 1.5712), (2007, 1.6117), (2008, 1.6427), (2009, 1.6948), (2010, 1.7256), (2011, 1.7102), (2012, 1.8409), (2013, 1.9079), (2014, 1.7114), (2015, 1.7664), (2016, 1.7471), (2017, 1.8431), (2018, 1.855), (2019, 1.7215), (2030, 2.1033))

[20] Energy pollution coefficient = Total energy consumption × 0.63/LN (The seconda industrial investment)
[21] Energy consumption per capita = Total energy consumption/Total population
[22] Energy pressure = Total energy consumption × 0.01/LN (R and D investment)
[23] Expenditure in environmental protection = GDP lagged × Rate of environmen protection expenditure
[24] FINAL TIME = 2030
[25] Fiscal expenditure = GDP lagged × Rate of fiscal expenditure
[26] GDP = Output of the primary industry + Output of the secondary industry + Outp of the tertiary industry
[27] GDP lag = INTEG (Production-GDP lagged,4151.4)
[28] GDP lagged = GDP lag
[29] Grain yield = Cultivated land × Grain yield per unit area
[30] Grain yield per capita = Grain yield/Total population
[31] Grain yield per unit area ([(1995, 370)–(2030, 800)], (1995, 725.517), (1996, 781.17 (1997, 798.154), (1998, 723.485), (1999, 715.366), (2000, 608.604), (2001, 539.629), (20 482.47), (2003, 383.793), (2004, 432.641), (2005, 443.995), (2006, 535.096), (2007, 530.09 (2008, 564.244), (2009, 601.483), (2010, 589.055), (2011, 610.972), (2012, 615.025), (20 607.181), (2014, 597.981), (2015, 590.516), (2016, 519.979), (2017, 520.772), (2018, 539.32 (2030, 762.012))
[32] Healthcare coefficient = 0.129
[33] Industrial SO_2 per secondary output = 53.29 × EXP (−((Time − 1995)/19.46)^5)/(penditure in environmental protection^Air protection coefficient)
[34] Industrial wastewater per secondary output = 1411 × EXP(−0.02637 × (Time 1995))/(Expenditure in environmental protection^Water protection coefficient)
[35] INITIAL TIME = 1995
[36] Labor force ratio([(1995, 0.45)–(2030, 0.6)], (1995, 0.5617), (1996, 0.5838), (1997, 0.563 (1998, 0.5416), (1999, 0.5113), (2000, 0.5082), (2001, 0.4763), (2002, 0.4886), (2003, 0.494 (2004, 0.5521), (2005, 0.5333), (2006, 0.5396), (2007, 0.5334), (2008, 0.5328), (2009, 0.523 (2010, 0.5213), (2011, 0.5126), (2012, 0.4976), (2013, 0.5923), (2014, 0.5709), (2015, 0.552 (2016, 0.5353), (2017, 0.5217), (2018, 0.5095), (2030, 0.5316))
[37] Labor force ratio of the primary industry ([(1995, 0.01)–(2030, 0.14)], (1995, 0.098 (1996, 0.1204), (1997, 0.1271), (1998, 0.1244), (1999, 0.1141), (2000, 0.1077), (2001, 0.1 (2002, 0.1015), (2003, 0.0863), (2004, 0.0688), (2005, 0.063), (2006, 0.055), (2007, 0.052 (2008, 0.0469), (2009, 0.0456), (2010, 0.034), (2011, 0.0338), (2012, 0.041), (2013, 0.03 (2014, 0.0328), (2015, 0.0338), (2016, 0.0333), (2017, 0.0309), (2018, 0.0297), (2030, 0.01
[38] Labor force ratio of the secondary industry ([(1995, 0.25)–(2030, 0.56)], (1995, 0.544 (1996, 0.5226), (1997, 0.491), (1998, 0.4603), (1999, 0.4646), (2000, 0.4431), (2001, 0.398 (2002, 0.3967), (2003, 0.4076), (2004, 0.4535), (2005, 0.4243), (2006, 0.4168), (2007, 0.412 (2008, 0.4027), (2009, 0.3974), (2010, 0.4068), (2011, 0.403), (2012, 0.3944), (2013, 0.350 (2014, 0.3492), (2015, 0.3377), (2016, 0.3285), (2017, 0.3136), (2018, 0.3074), (2030, 0.278
[39] Labor force ratio of the tertiary industry ([(1995, 0.35)–(2030, 0.75)], (1995, 0.356 (1996, 0.357), (1997, 0.3819), (1998, 0.4153), (1999, 0.4213), (2000, 0.4492), (2001, 0.491 (2002, 0.5017), (2003, 0.5061), (2004, 0.4777), (2005, 0.5128), (2006, 0.5282), (2007, 0.535 (2008, 0.5504), (2009, 0.557), (2010, 0.5592), (2011, 0.5632), (2012, 0.5646), (2013, 0.612 (2014, 0.618), (2015, 0.6285), (2016, 0.6382), (2017, 0.6554), (2018, 0.663), (2030, 0.709
[40] Livelihood index for inhabitant = LN (Fiscal expenditure × (Culture coefficient Public books per capita + Education coefficient*Primary school students per capit Healthcare coefficient × Medical staff per capita + Transport coefficient × Road ar per capita))
[41] Living SO_2 per capita ([(1994, 0)–(2030, 12)], (1994, 6.402), (1995, 10.792), (1996, 5.30 (1997, 4.856), (1998, 6.417), (1999, 5.888), (2000, 8.585), (2001, 10.348), (2002, 7.10 (2003, 7.627), (2004, 6.736), (2005, 7.28), (2006, 6.808), (2007, 6.463), (2008, 6.917), (20

6.317), (2010, 4.121), (2011, 1.272), (2012, 1.46), (2013, 1.778), (2014, 1.351), (2015, 2.733), (2016, 0.283), (2017, 0.241), (2018, 0.033), (2019, 0.041), (2030, 0.131))

[42] Living wastewater per capita ([(1994, 60)–(2030, 95)], (1994, 61.2117), (1995, 76.662), (1996, 78.8422), (1997, 74.6138), (1998, 77.3412), (1999, 75.0479), (2000, 75.345), (2001, 76.124), (2002, 74.257), (2003, 68.5793), (2004, 74.66), (2005, 78.6243), (2006, 89.3075), (2007, 86.7248), (2008, 84.9603), (2009, 85.6561), (2010, 91.8367), (2011, 65.6157), (2012, 72.0773), (2013, 73.5404), (2014, 73.0833), (2015, 73.3747), (2016, 76.1157), (2017, 74.6071), (2018, 74.5462), (2019, 74.1606), (2030, 77.1461))

[43] Medical staff per capita([(1995, 0.005)–(2030, 0.011)], (1995, 0.0078), (1996, 0.0075), (1997, 0.0073), (1998, 0.0071), (1999, 0.0069), (2000, 0.0067), (2001, 0.0063), (2002, 0.0059), (2003, 0.0058), (2004, 0.0055), (2005, 0.0055), (2006, 0.0056), (2007, 0.0059), (2008, 0.006), (2009, 0.0059), (2010, 0.0059), (2011, 0.0059), (2012, 0.0062), (2013, 0.0065), (2014, 0.0068), (2015, 0.007), (2016, 0.0074), (2017, 0.0078), (2018, 0.0085), (2019, 0.0084), (2030, 0.0107))

[44] Natural land = 6341

[45] Output of the primary industry = 0.086/Pollution index^0.8284 × Livelihood index for inhabitant^0.1163 × The stock of the primary industrial fixed assets^0.91 × (Labor force ratio of the primary industry × Total labor force)^0.5272

[46] Output of the secondary industry = 0.3843/Energy pressure^0.4549 × Energy consumption per capita^0.6016 × The stock of the secondary industrial fixed assets^0.8047 × (Total labor force*Labor force ratio of the secondary industry) ^0.3084

[47] Output of the tertiary industry = 23.0808/Pollution index^0.8284 × Livelihood index for inhabitant^0.1163 × The stock of the tertiary industrial fixed assets^0.162 × (Total labor force × Labor force ratio of the tertiary industry) ^0.7136

[48] Pollution index = LN (Energy pollution coefficient × (Air pollution coefficient × Discharge of SO2+Water pollution coefficient × Discharge of wastewater)/(Discharge of SO2+Discharge of wastewater))

[49] Primary school students per capita([(1995, 0.02)–(2030, 0.08)], (1995, 0.0776), (1996, 0.0734), (1997, 0.0688), (1998, 0.063), (1999, 0.0556), (2000, 0.049), (2001, 0.0433), (2002, 0.0393), (2003, 0.0367), (2004, 0.0293), (2005, 0.0283), (2006, 0.0272), (2007, 0.0258), (2008, 0.0276), (2009, 0.0304), (2010, 0.0305), (2011, 0.0312), (2012, 0.032), (2013, 0.0328), (2014, 0.0331), (2015, 0.0331), (2016, 0.0326), (2017, 0.0325), (2018, 0.0342), (2019, 0.034), (2030, 0.0415))

[50] Production = GDP-Expenditure in environmental protection

[51] Proportion of agricultural output = Output of the primary industry/GDP

[52] Proportion of nonagricultural output = (Output of the secondary industry + Output of the tertiary industry)/GDP

[53] Public books per capita([(1995, 1)–(2030, 4)], (1995, 1.1216), (1996, 1.1392), (1997, 3.2102), (1998, 3.1493), (1999, 3.0989), (2000, 3.4191), (2001, 3.3974), (2002, 3.3959), (2003, 3.3378), (2004, 3.1891), (2005, 3.2005), (2006, 3.0866), (2007, 3.03), (2008, 2.9865), (2009, 2.9833), (2010, 2.9566), (2011, 2.9369), (2012, 3.0261), (2013, 2.9975), (2014, 3.035), (2015, 3.1337), (2016, 3.1719), (2017, 3.2146), (2018, 3.2567), (2019, 3.3208), (2030, 3.8485))

[54] R and D investment = GDP lagged × Ratio of R and D investment

[55] Rate of environmental protection expenditure ([(1995, 0)–(2030, 0.1)], (1995, 0.01846), (1996, 0.02309), (1997, 0.02376), (1998, 0.02666), (1999, 0.02642), (2000, 0.02949), (2001, 0.02909), (2002, 0.02802), (2003, 0.02815), (2004, 0.02782), (2005, 0.03057), (2006, 0.02933), (2007, 0.02843), (2008, 0.02906), (2009, 0.02925), (2010, 0.02833), (2011, 0.02788), (2012, 0.02827), (2013, 0.02814), (2014, 0.0277), (2015, 0.02636), (2016, 0.02756), (2017, 0.03015), (2018, 0.03027), (2019, 0.02829), (2030, 0.035))

[56] Rate of fiscal expenditure ([(1995, 0.1)–(2030, 0.35)], (1995, 0.1064), (1996, 0.115), (1997, 0.1238), (1998, 0.1255), (1999, 0.1294), (2000, 0.1294), (2001, 0.1382), (2002, 0.1515), (2003, 0.1621), (2004, 0.1723), (2005, 0.1805), (2006, 0.1711), (2007, 0.1709), (2008, 0.1801), (2009, 0.1899), (2010, 0.1844), (2011, 0.1956), (2012, 0.2073), (2013, 0.2096), (2014, 0.1815), (2015, 0.2303), (2016, 0.2315), (2017, 0.2464), (2018, 0.2556), (2030, 0.3096))

[57] Rate of fixed assets investment ([(1995, 0.1)–(2030, 0.7)], (1995, 0.6361), (1996, 0.654 (1997, 0.5707), (1998, 0.5129), (1999, 0.4397), (2000, 0.3885), (2001, 0.3794), (2002, 0.377 (2003, 0.3604), (2004, 0.3807), (2005, 0.3852), (2006, 0.3703), (2007, 0.3462), (2008, 0.332 (2009, 0.335), (2010, 0.2968), (2011, 0.2532), (2012, 0.2604), (2013, 0.2614), (2014, 0.238 (2015, 0.2363), (2016, 0.226), (2017, 0.2366), (2018, 0.2334), (2030, 0.1108))

[58] Ratio of R and D investment ([(1995, 0.01)–(2030, 0.06)], (1995, 0.0129), (1996, 0.013 (1997, 0.0144), (1998, 0.0145), (1999, 0.0151), (2000, 0.0159), (2001, 0.0168), (2002, 0.017 (2003, 0.0189), (2004, 0.021), (2005, 0.0232), (2006, 0.0244), (2007, 0.0239), (2008, 0.024 (2009, 0.0269), (2010, 0.0269), (2011, 0.0299), (2012, 0.0337), (2013, 0.036), (2014, 0.034 (2015, 0.0348), (2016, 0.0351), (2017, 0.0393), (2018, 0.0416), (2030, 0.054))

[59] Road area per capita ([(1995, 4)–(2030, 18)], (1995, 4.01), (1996, 4.46), (1997, 4.91), (19 6.04), (1999, 8.52), (2000, 8.68), (2001, 13.6), (2002, 11.6), (2003, 12.46), (2004, 15.3 (2005, 11.78), (2006, 11.84), (2007, 15.4), (2008, 15.7), (2009, 17.54), (2010, 11.12), (20 11.18), (2012, 11.24), (2013, 11.3), (2014, 11.51), (2015, 11.83), (2016, 12.09), (2017, 12.3 (2018, 12.49), (2019, 12.7), (2030, 14.63))

[60] SAVEPER = TIME STEP

[61] The primary industrial depreciation = The stock of the primary industrial fix assets × Depreciation rate for primary industry

[62] The primary industrial investment = Total fixed assets investment × The proporti of the primary industrial investment

[63] The proportion of the primary industrial investment ([(1995, 0)–(2030, 0.012)], (19 0.0061), (1996, 0.0107), (1997, 0.004), (1998, 0.0033), (1999, 0.0044), (2000, 0.0044), (20 0.0035), (2002, 0.0025), (2003, 0.0018), (2004, 0.0018), (2005, 0.0017), (2006, 0.0037), (20 0.002), (2008, 0.0018), (2009, 0.0022), (2010, 0.0032), (2011, 0.0038), (2012, 0.0023), (20 0.0034), (2014, 0.0021), (2015, 0.0007), (2016, 0.0006), (2017, 0.0003), (2018, 0.0007), (20 0.0003))

[64] The proportion of the secondary industrial investment ([(1995, 0)–(2030, 0.4)], (19 0.3226), (1996, 0.3314), (1997, 0.3354), (1998, 0.3333), (1999, 0.3323), (2000, 0.329 (2001, 0.3427), (2002, 0.332), (2003, 0.3291), (2004, 0.3275), (2005, 0.3055), (2006, 0.30 (2007, 0.3135), (2008, 0.2942), (2009, 0.2707), (2010, 0.2699), (2011, 0.2557), (2012, 0.246 (2013, 0.2199), (2014, 0.1924), (2015, 0.1509), (2016, 0.1455), (2017, 0.1426), (2018, 0.158 (2030, 0.075))

[65] The proportion of the tertiary industrial investment ([(1995, 0.6)–(2030, 1)], (19 0.6713), (1996, 0.6582), (1997, 0.6611), (1998, 0.6636), (1999, 0.6635), (2000, 0.666 (2001, 0.654), (2002, 0.6656), (2003, 0.6692), (2004, 0.6708), (2005, 0.693), (2006, 0.687 (2007, 0.6847), (2008, 0.7041), (2009, 0.7271), (2010, 0.727), (2011, 0.7406), (2012, 0.751 (2013, 0.7768), (2014, 0.8057), (2015, 0.8484), (2016, 0.8539), (2017, 0.8572), (2018, 0.840 (2030, 0.922))

[66] The secondary industrial depreciation = The stock of the secondary industrial fix assets × Depreciation rate for secondary industry

[67] The secondary industrial investment = Total fixed assets investment × The proporti of the secondary industrial investment

[68] The stock of the primary industrial fixed assets = INTEG (The primary industr investment-The primary industrial depreciation,464.362)

[69] The stock of the secondary industrial fixed assets = INTEG (The secondary industr investment-The secondary industrial depreciation,8072)

[70] The stock of the tertiary industrial fixed assets = INTEG (The tertiary industr investment-The tertiary industrial depreciation,5501.27)

[71] The tertiary industrial depreciation = The stock of the tertiary industrial fixed sets*Depreciation rate for tertiary industry

[72] The tertiary industrial investment = Total fixed assets investment × The proportion the tertiary industrial investment

[73] TIME STEP = 1

[74] Total energy consumption = Energy intensity × GDP lagged

[75] Total fixed assets investment = GDP lagged × Rate of fixed assets investment
[76] Total labor force =Total population × Labor force ratio
[77] Total population = INTEG (Birth-Death,1414)
[78] Transport coefficient = 0.2399
[79] Water pollution coefficient = 0.35
[80] Water protection coefficient = 0.7555

Appendix B

Table A1. The Determination Methods of Parameters.

No.	Parameters	Values	Methods
1	Air pollution coefficient	Appendix A	[19]
2	Air protection coefficient	Appendix A	Regression analysis
3	Elastic coefficient of Grain yield per capita for Birth	0.3384	Regression analysis
4	Birth rate	Appendix A	Table function
5	Elastic coefficient of Proportion of agricultural output for Cultivated land	0.2342	Regression analysis
6	Elastic coefficient of Proportion of nonagricultural output for Cultivated land	8.975	Regression analysis
7	Culture coefficient	Appendix A	Regression analysis
8	Elastic coefficient of Pollution index for Death	6.3084	Regression analysis
9	Elastic coefficient of Gain yield per capita for Birth	0.3384	Regression analysis
10	Death rate	Appendix A	Table function
11	Depreciation rate for primary industry	Appendix A	[19]
12	Depreciation rate for secondary industry	Appendix A	Regression analysis
13	Depreciation rate for tertiary industry	Appendix A	[19]
14	Education coefficient	Appendix A	Regression analysis
15	Energy intensity	Appendix A	Table function
16	Elastic coefficient of Total energy consumption for energy pollution coefficient	0.63	Regression analysis
17	Elastic coefficient of Total energy consumption for Energy pressure	0.01	Regression analysis
18	Grain yield per unit area	Appendix A	Table function
19	Healthcare coefficient	Appendix A	Regression analysis
20	Industrial wastewater per secondary output	Appendix A	Regression analysis
21	Labor force ratio	Appendix A	Table function
22	Labor force ratio of the primary industry	Appendix A	Table function
23	Labor force ratio of the secondary industry	Appendix A	Table function
24	Labor force ratio of the tertiary industry	Appendix A	Table function
25	Living SO_2 per capita	Appendix A	Table function
26	Living wastewater per capita	Appendix A	Table function
27	Medical staff per capita	Appendix A	Table function
28	Natural land	Appendix A	Constant
29	Output of the primary industry	Appendix A	Regression analysis
30	Elastic coefficient of Pollution index for Output of the primary industry	0.8284	Regression analysis
31	Elastic coefficient of Livelihood index for inhabitant for Output of the primary industry	0.1163	Regression analysis
32	Elastic coefficient of the stock of the primary industrial fixed assets for Output of the primary industry	0.91	Regression analysis
33	Elastic coefficient of the labor force for Output of the primary industry	0.5272	Regression analysis
34	Output of the secondary industry	Appendix A	Regression analysis

Table A1. *Cont.*

No.	Parameters	Values	Methods
35	Elastic coefficient of Energy pressure for Output of secondary industry	0.4549	Regression analysis
36	Elastic coefficient of Energy consumption per capita for Output of secondary industry	0.6061	Regression analysis
37	Elastic coefficient of the stock of the secondary industrial fixed assets for Output of the secondary industry	0.8047	Regression analysis
38	Elastic coefficient of the labor force for Output of the secondary industry	0.3084	Regression analysis
39	Output of the tertiary industry	Appendix A	Regression analysis
40	Elastic coefficient of Pollution index for Output of the tertiary industry	0.8284	Regression analysis
41	Elastic coefficient of Livelihood index for inhabitant for Output of the secondary industry	0.1163	Regression analysis
42	Elastic coefficient of the stock of the tertiary industrial fixed assets for Output of the tertiary industry	0.162	Regression analysis
43	Elastic coefficient of the labor force for Output of the tertiary industry	0.7136	Regression analysis
44	Primary school students per capita	Appendix A	Table function
45	Public books per capita	Appendix A	Table function
46	Rate of environmental protection expenditure	Appendix A	Table function
47	Rate of fiscal expenditure	Appendix A	Table function
48	Rate of fixed assets investment	Appendix A	Table function
49	Ratio of R and D investment	Appendix A	Table function
50	Road area per capita	Appendix A	Table function
51	The proportion of the primary industrial investment	Appendix A	Table function
52	The proportion of the secondary industrial investment	Appendix A	Table function
53	The proportion of the tertiary industrial investment	Appendix A	Table function
54	Transport coefficient	Appendix A	Regression analysis
55	Water pollution coefficient	Appendix A	(Xing et al., 2019)
56	Water protection coefficient	Appendix A	Regression analysis

References

1. Elmqvist, T.; Andersson, E.; Frantzeskaki, N.; McPhearson, T.; Olsson, P.; Gaffney, O.; Takeuchi, K.; Folke, C. Sustainability a resilience for transformation in the urban century. *Nat. Sustain.* **2019**, *2*, 267–273. [CrossRef]
2. United Nations. *World Urbanization Prospects: The 2018 Revision*; Department of Economic and Social Affairs Affairs, Ed.; Unit Nations: New York, NY, USA, 2019.
3. Ruan, F.-L.; Yan, L. Interactions among electricity consumption, disposable income, wastewater discharge, and economic grow Evidence from megacities in China from 1995 to 2018. *Energy* **2022**, *260*, 124910. [CrossRef]
4. United Nations. The New Urban Agenda. In Proceedings of the Housing and Sustainable Urban Development (Habitat I Quito, Ecuador, 17–20 October 2016.
5. Acuto, M.; Parnell, S.; Seto, K.C. Building a global urban science. *Nat. Sustain.* **2018**, *1*, 2–4. [CrossRef]
6. Liu, J.; Mooney, H.; Hull, V.; Davis, S.J.; Gaskell, J.; Hertel, T.; Lubchenco, J.; Seto, K.C.; Gleick, P.; Kremen, C.; et al. Syster integration for global sustainability. *Science* **2015**, *347*, 1258832. [CrossRef]
7. Kutty, A.A.; Abdella, G.M.; Kucukvar, M.; Onat, N.C.; Bulu, M. A system thinking approach for harmonizing smart a sustainable city initiatives with United Nations sustainable development goals. *Sustain. Dev.* **2020**, *28*, 1347–1365. [CrossRef]
8. Liu, J.; Hull, V.; Godfray, H.C.J.; Tilman, D.; Gleick, P.; Hoff, H.; Pahl-Wostl, C.; Xu, Z.; Chung, M.G.; Sun, J.; et al. Nex approaches to global sustainable development. *Nat. Sustain.* **2018**, *1*, 466–476. [CrossRef]
9. Seto, K.C.; Golden, J.S.; Alberti, M.; Turner, B.L. Sustainability in an urbanizing planet. *Proc. Natl. Acad. Sci. USA* **2017**, *1* 8935–8938. [CrossRef] [PubMed]
10. Mou, Y.; Luo, Y.; Su, Z.; Wang, J.; Liu, T. Evaluating the dynamic sustainability and resilience of a hybrid urban system: Case Chengdu, China. *J. Clean. Prod.* **2021**, *291*, 125719. [CrossRef]
11. Cui, D.; Chen, X.; Xue, Y.; Li, R.; Zeng, W. An integrated approach to investigate the relationship of coupling coordinati between social economy and water environment on urban scale—A case study of Kunming. *J. Environ. Manag.* **2019**, *234*, 189–1 [CrossRef]
12. Feng, Y.Y.; Chen, S.Q.; Zhang, L.X. System dynamics modeling for urban energy consumption and CO_2 emissions: A case stu of Beijing, China. *Ecol. Model.* **2013**, *252*, 44–52. [CrossRef]

Guan, D.; Gao, W.; Su, W.; Li, H.; Hokao, K. Modeling and dynamic assessment of urban economy–resource–environment system with a coupled system dynamics–geographic information system model. *Ecol. Indic.* **2011**, *11*, 1333–1344. [CrossRef]
Lu, X.-H.; Ke, S.-G. Evaluating the effectiveness of sustainable urban land use in China from the perspective of sustainable urbanization. *Habitat Int.* **2018**, *77*, 90–98. [CrossRef]
Everest, T.; Koparan, H.; Sungur, A.; Özcan, H. An important tool against combat climate change: Land suitability assessment for canola (a case study: Çanakkale, NW Turkey). *Environ. Dev. Sustain.* **2022**, *24*, 13137–13172. [CrossRef]
Tan, S.; Liu, Q.; Han, S. Spatial-temporal evolution of coupling relationship between land development intensity and resources environment carrying capacity in China. *J. Environ. Manag.* **2022**, *301*, 113778. [CrossRef] [PubMed]
Liu, Y.; Yang, Y.; Li, Y.; Li, J. Conversion from rural settlements and arable land under rapid urbanization in Beijing during 1985–2010. *J. Rural. Stud.* **2017**, *51*, 141–150. [CrossRef]
Azunre, G.A.; Amponsah, O.; Peprah, C.; Takyi, S.A.; Braimah, I. A review of the role of urban agriculture in the sustainable city discourse. *Cities* **2019**, *93*, 104–119. [CrossRef]
Xing, L.; Xue, M.; Hu, M. Dynamic simulation and assessment of the coupling coordination degree of the economy-resource-environment system: Case of Wuhan City in China. *J. Environ. Manag.* **2019**, *230*, 474–487. [CrossRef] [PubMed]
Tapia, C.; Randall, L.; Wang, S.; Borges, L.A. Monitoring the contribution of urban agriculture to urban sustainability: An indicator-based framework. *Sustain. Cities Soc.* **2021**, *74*, 103130. [CrossRef]
Nigussie, S.; Liu, L.; Yeshitela, K. Towards improving food security in urban and peri-urban areas in Ethiopia through map analysis for planning. *Urban For. Urban Green.* **2021**, *58*, 126967. [CrossRef]
Tong, D.; Crosson, C.; Zhong, Q.; Zhang, Y. Optimize urban food production to address food deserts in regions with restricted water access. *Landsc. Urban Plan.* **2020**, *202*, 103859. [CrossRef]
Xia, H.; Ge, S.; Zhang, X.; Kim, G.; Lei, Y.; Liu, Y. Spatiotemporal Dynamics of Green Infrastructure in an Agricultural Peri-Urban Area: A Case Study of Baisha District in Zhengzhou, China. *Land* **2021**, *10*, 801. [CrossRef]
Clerino, P.; Fargue-Lelièvre, A.; Meynard, J.-M. Stakeholder's practices for the sustainability assessment of professional urban agriculture reveal numerous original criteria and indicators. *Agron. Sustain. Dev.* **2023**, *43*, 3. [CrossRef]
Goździewicz-Biechońska, J.; Brzezińska-Rawa, A. Protecting ecosystem services of urban agriculture against land-use change using market-based instruments. A Polish perspective. *Land Use Policy* **2022**, *120*, 106296. [CrossRef]
Moragues-Faus, A.; Marsden, T.; Adlerová, B.; Hausmanová, T. Building Diverse, Distributive, and Territorialized Agrifood Economies to Deliver Sustainability and Food Security. *Econ. Geogr.* **2020**, *96*, 219–243. [CrossRef]
Badami, M.G.; Ramankutty, N. Urban agriculture and food security: A critique based on an assessment of urban land constraints. *Glob. Food Secur.* **2015**, *4*, 8–15. [CrossRef]
Nitya, R.A.O.; Patil, S.; Singh, C.; Roy, P.; Pryor, C.; Poonacha, P.; Genes, M. Cultivating sustainable and healthy cities: A systematic literature review of the outcomes of urban and peri-urban agriculture. *Sustain. Cities Soc.* **2022**, *85*, 104063. [CrossRef]
Fanfani, D.; Duží, B.; Mancino, M.; Rovai, M. Multiple evaluation of urban and peri-urban agriculture and its relation to spatial planning: The case of Prato territory (Italy). *Sustain. Cities Soc.* **2022**, *79*, 103636. [CrossRef]
Walters, S.A.; Stoelzle Midden, K. Sustainability of Urban Agriculture: Vegetable Production on Green Roofs. *Agriculture* **2018**, *8*, 168. [CrossRef]
Horst, M.; McClintock, N.; Hoey, L. The Intersection of Planning, Urban Agriculture, and Food Justice A Review of the Literature. *J. Am. Plan. Assoc.* **2017**, *83*, 277–295. [CrossRef]
Li, X.; Zhang, L.; Hao, Y.; Zhang, P.; Xiong, X.; Shi, Z. System dynamics modeling of food-energy-water resource security in a megacity of China: Insights from the case of Beijing. *J. Clean. Prod.* **2022**, *355*, 131773. [CrossRef]
Forrester, J.W. *Industrial Dynamics*; MIT Press: Cambridge, MA, USA, 1961.
Wen, C.; Dong, W.; Zhang, Q.; He, N.; Li, T. A system dynamics model to simulate the water-energy-food nexus of resource-based regions: A case study in Daqing City, China. *Sci. Total. Environ.* **2022**, *806*, 150497. [CrossRef] [PubMed]
Liu, H.; Liu, Y.; Wang, H.; Yang, J.; Zhou, X. Research on the coordinated development of greenization and urbanization based on system dynamics and data envelopment analysis—A case study of Tianjin. *J. Clean. Prod.* **2019**, *214*, 195–208. [CrossRef]
Tan, Y.; Jiao, L.; Shuai, C.; Shen, L. A system dynamics model for simulating urban sustainability performance: A China case study. *J. Clean. Prod.* **2018**, *199*, 1107–1115. [CrossRef]
Zhang, Y.; Zhao, F.; Zhang, J.; Wang, Z. Fluctuation in the transformation of economic development and the coupling mechanism with the environmental quality of resource-based cities–A case study of Northeast China. *Resour. Policy* **2021**, *72*, 102128. [CrossRef]
Tomal, M. Evaluation of coupling coordination degree and convergence behaviour of local development: A spatiotemporal analysis of all Polish municipalities over the period 2003–2019. *Sustain. Cities Soc.* **2021**, *71*, 102992. [CrossRef]
Chen, Y.; Zhang, D. Multiscale assessment of the coupling coordination between innovation and economic development in resource-based cities: A case study of Northeast China. *J. Clean. Prod.* **2021**, *318*, 128597. [CrossRef]
Fan, Y.; Fang, C.; Zhang, Q. Coupling coordinated development between social economy and ecological environment in Chinese provincial capital cities-assessment and policy implications. *J. Clean. Prod.* **2019**, *229*, 289–298. [CrossRef]
Ruan, F.; Yan, L. Challenges facing indicators to become a universal language for sustainable urban development. *Sustain. Dev.* **2022**, *30*, 41–57. [CrossRef]

42. Fantini, A. Urban and peri-urban agriculture as a strategy for creating more sustainable and resilient urban food systems a facing socio-environmental emergencies. *Agroecol. Sustain. Food Syst.* **2023**, *47*, 47–71. [CrossRef]
43. Yan, D.; Liu, L.; Liu, X.; Zhang, M. Global Trends in Urban Agriculture Research: A Pathway toward Urban Resilience a Sustainability. *Land* **2022**, *11*, 117. [CrossRef]
44. Ayambire, R.A.; Amponsah, O.; Peprah, C.; Takyi, S.A. A review of practices for sustaining urban and peri-urban agricult Implications for land use planning in rapidly urbanising Ghanaian cities. *Land Use Policy* **2019**, *84*, 260–277. [CrossRef]
45. Cohen, N. Roles of Cities in Creating Healthful Food Systems. *Annu. Rev. Public Health* **2022**, *43*, 419–437. [CrossRef] [PubM

land

MDPI

ticle

valuation of the Contribution of Farmland Attributes) the Total Benefit from Its Contamination emediation: Evidence from Taiwan

i-Ing Wu [1,*], Ching-En Su [2], Je-Liang Liou [3] and Ta-Ken Huang [4]

1 Department of Agricultural Economics, National Taiwan University, Taipei 10617, Taiwan
2 Agricultural Policy Research Center, Agricultural Technology Research Institute, Taipei 106018, Taiwan
3 The Center for Green Economy, Chung-Hua Institution for Economic Research, Taipei 106220, Taiwan
4 Department of Water Resources and Environmental Engineering, Tamkang University, New Taipei City 251301, Taiwan
* Correspondence: piwu@ntu.edu.tw; Tel.: +886-2-3366-2663

Abstract: This study fills the gap in the existing literature by developing a two-stage quantile spatial Durbin model to evaluate the benefit from the cancellation of controlled contaminated farmlands. The results of monetary benefit are to identify the contribution of farmland attributed to the change in the total benefit resulting from the cancellation of contaminated farmlands. The results show that the significant impacts are the attributes resulting from the size of the transacted farmland, the distance between the transacted farmland and the main traffic artery, and the price of the construction site where the transacted farmlands are located. The results indicate that for every 1000 square meter increase in farmland size, the farmland price increases by about 45–105% in the non-agricultural planning zone, the Taoyuan Aerotropolis life circle, and decreases by about 81–131% in the agricultural planning zones. Moreover, for the price quantiles of 50% and above, the total benefit from the announcement of contamination cancellation to the ensuing transaction is reflected by an increase in the transaction price of 1.67–12.98% of the total benefit for non-agricultural Taoyuan Aerotropolis life circle zoning. By contrast, the total benefit from the same action taken for the other three agricultural development life circles is reflected by a reduction in the transaction price of 1.89–134.89%. These results indicate that the cancellation of highly priced contaminated farmlands is not anticipated if they are planned for agricultural purposes.

Keywords: endogeneity; geographic information system; not-in-my-backyard; remediation of contaminated farmland; two-stage quantile regression; spatial Durbin model; spatial land planning

ation: Wu, P.-I.; Su, C.-E.; Liou, .; Huang, T.-K. Evaluation of the ntribution of Farmland Attributes he Total Benefit from Its ntamination Remediation: dence from Taiwan. *Land* **2023**, *12*, '. https://doi.org/10.3390/ d12050967

ademic Editor: Matjaž Glavan

ceived: 3 April 2023
vised: 19 April 2023
cepted: 24 April 2023
blished: 26 April 2023

1. Introduction

Agricultural production relies to a high degree upon natural conditions, with climate being the most important, and soil and water resources being essential factors for the cultivation of agricultural products. The quality of water and soil has a significant impact on the quantity of agricultural production and the related output values. This is usually not easy to detect and thus tends to result in the soil being contaminated by different types of heavy metal pollutants. Similarly, various kinds of pollutants lead to the contamination of groundwater. The groundwater is an important water source for irrigation in agricultural production. The soil and groundwater contamination mainly comes from the improper application of nitrogen [1]. In Egypt, water is a scarce resource and thus it is preserved underground for recyclable use. If the groundwater is contaminated, the impact is not only felt in the specific region, sector, or industry, but throughout the whole country [2]. In addition, the method of cultivation and the types and application frequencies of chemical fertilizers and pesticides also have a key role to play in the heavy metal contamination of farmlands [3]. It can thus be concluded that soil and groundwater contamination impacts not only the amount and value of agricultural products, but also farmland transactio food security, human health, and ecosystem services [4–7].

To eliminate the hazards that might arise from contaminated farmland sites, the U Environmental Protection Agency passed the Comprehensive Environmental Respor Compensation, and Liability Act (CERCLA), also referred to as the Superfund, in 19 About 70% of sites with soil or related contamination are the responsibility of industri and there are no definite responsible targets for the remaining 30% of sites [8]. Cleani these contaminated sites poses a huge financial burden for the U.S. government. T Environmental Protection Administration in Taiwan (Taiwan EPA) started to impleme the Soil and Groundwater Pollution Remediation Act (Remediation Act hereafter) in 20 Along with the progress in pollution investigation, the polluted sites increase year by ye The most recent amendment to the Remediation Act classifies all soil-contaminated sites "controlled sites", "remediated sites", "cancellation of controlled sites", or "cancellation remediated sites" [9].

There are six municipalities in Taiwan, and their major cities account for 71.53% of total population in Taiwan [10]. Among these, Taoyuan city has the smallest populati and has the largest number of soil-contaminated farmland controlled sites. It, however, h the highest value of crops, with its share of the whole country being 5.83% [11]. These cro have a high probability of being grown on contaminated farmlands. Under the propo of *Taiwan's Comprehensive National Spatial Land Planning Act* (*Spatial Planning Act* hereaft legislated in 2016, each city and county has its own planning for industrial, agricultu residential, and natural conservation zoning development [12].

In order to have a proper arrangement of different zoning developments, determini the characteristic benefits of rectifying the contaminated sites is essential. The benefits a commonly explored with the hedonic price method (HPM) by connecting the farmla price and all types of characteristics of farmlands. In terms of research topics in th regard, some studies have focused on the zone classification in Ontario, Canada, an New York state, US [13,14]; other studies explored the factors that have major impacts farmland prices in Argentina, the Netherlands, Aragón, Illinois in the US, and New Sou Wales [15–19]; and yet a few other studies concentrated on the specification of function forms with the analysis of five corn belt states in the US [20]. In terms of where resear is conducted, in large countries such as the United States, studies have been conduct to find the farmland price relationship in a setting with agricultural zones or agricultu productivity of Illinois and Ohio farmlands in the US [18,21], and similar studies have al been conducted in Canada [13]. Impact studies on factors influencing farmland prices we conducted by [19] in Australia. Factors influencing farmland prices were also explored the Netherlands by [16], and in Argentina by [15].

As for Taiwan, a study conducted by [22] sought to find the factors influenci farmland prices using a simple regression without accounting for the spatial differentiati Moreover, this study mainly focused on the factors influencing farmland prices in ru areas without the concern of adding or deleting any new or existing element on or arou farmland. There have been many related studies since 1 August 2012, when the actu details of real estate transactions were compiled and made known to the public by Taiwa Ministry of the Interior [23]. Ref. [24] explored the impact of solar panel installation the prices of farmland. Ref. [25] evaluated the not-in-my-back-yard (NIMBY) and yes-my-back-yard (YIMBY) effects from cropland open spaces while accounting for spat differentiation among farmlands.

It can be concluded that studies using the HPM in different parts of the world explore the issues in relation to farmland remediation and their prices are mainly affect by the characteristics of farmland per se, such as the types of crops grown on the farmlar the production conditions for the farmland, the size of the farmland, and the surroundi characteristics of the farmland. However, most of the studies stated above do not accou for the spatial issues related to the farmlands. If farmlands exhibit spatial dependence, th spatial HPM has to be considered. Spatial types of HPM have become accessible since t obstacles due to software operations have been reduced. The application of spatial HPM to farmland prices is discussed in [26–28].

Other types of studies consider the impact of a change in a specific farmland attribute on farmland prices. Examples are found in [29] regarding Czech farmland buyers. Different types of buyers place different emphases on different farmland characteristics. This, in turn, will have different impacts on farmland prices. Similarly, examples can be found in the study conducted by [30] of Germany's farmland prices due to the impact of government intervention, and that by [31] of farmland prices in the US from farm program payments. A study by [32] explored the prices of Belgian farmland with cadmium contamination, and another study by [33] considered the impact of natural amenities on farmland prices. The results from all these studies indicate that the changes in human-made policies or natural characteristics of farmland have different impacts on farmland prices. Thus, a quantile regression is much more appropriate than ordinary least squares regression when the relationship between farmland prices and all the potential explanatory characteristics is established [34].

It can thus be reasonably assumed that when looking into the factors that influence farmland prices, one should not only account for the spatial differentiation of farmland, but should also consider the divergent impacts of all types of farmland characteristics on farmland prices. As a result, the HPM has to combine both spatial differences and the dissimilar impacts on farmland prices through quantile regression. When spatialization occurs both in relation to the farmland prices and all types of attributes of farmland, the spatial Durbin model (SDM) should be employed [35–37]. It can be clearly seen that accounting for the spatial problem in relation to the dependent variable, namely, the farmland price, and/or the explanatory variables of farmland attributes involves considering the factors that have essential impacts on the farmland price when endogeneity might exist among the explanatory variables with adjustments in the spatial farmland price. Under such circumstances, the adoption of ordinary quantile SDM is not sufficient. When combining a two-stage quantile regression proposed by [38] with SDM by developing a new two-stage quantile, SDM is an appropriate method that can be used to resolve the variation in the spatial impact on the dependent variable of the farmland price and endogeneity among the spatial explanatory factors.

The benefits from the remediation of farmlands in different cities or counties are essential information for the land planning of a specific city or county and for the country as a whole. To the best of our knowledge, there is no study evaluating the monetary benefit via a change in the transacted farmland price due to farmland attributes that have a potential impact on the remediation of contaminated sites. The benefit generated by each farmland attribute is the foundation for prioritizing the cancellation order for all types of contaminated farmlands. Moreover, the results of the evaluated benefits can be used as a compensation guideline when zones are classified for agricultural purposes, and this is deemed to be a beneficial and a fair action for agricultural development [39,40].

As a result, the purpose of this study is to evaluate the monetary benefits of cancelling contaminated farmlands using a two-stage quantile SDM, newly developed in this study. The results identify the contribution of each farmland attribute or group of attributes to the change in the total benefit due to the cancellation of contaminated farmlands. The results can be used by related agencies to command the potential benefits of farmlands located in different cities or counties to determine the priority of cancelling contaminated farmlands.

The remaining sections of this paper are arranged as follows. Section 2 explains the conceptual framework of benefit evaluation for contaminated farmland remediation. Section 3 presents the evaluation methods for the two-stage quantile SDM. Section 4 discusses the data sources and selection of characteristics and their treatment. Section 5 provides a specification of the empirical model and an analysis of the results. The final section is the conclusion of this study.

2. Method and Models

2.1. A Panoramic View of Contaminated Farmlands and Spatial Planning of the Study Area

The major crops grown in Taoyuan city are rice and various kinds of vegetab These contaminated farmlands have a high possibility of having a negative impact on production of all types of agricultural products. This, in turn, has a further negative imp on the revenues from such production. Agricultural development is without a do ruined by contaminated farmlands. According to the records compiled by the Soil a Groundwater Remediation Fund Management Board of the Taiwan EPA [9], it is known t the latest available data from 31 December 2020, when this study started, show that 38.7 of the farmland in Taoyuan city consists of controlled contaminated farmland sites, a 29.75% of the area and each percentage is ranked the highest among all six municipalit in Taiwan. The contaminated farmland hot spots in the six municipalities of Taiwan shown in Figure 1.

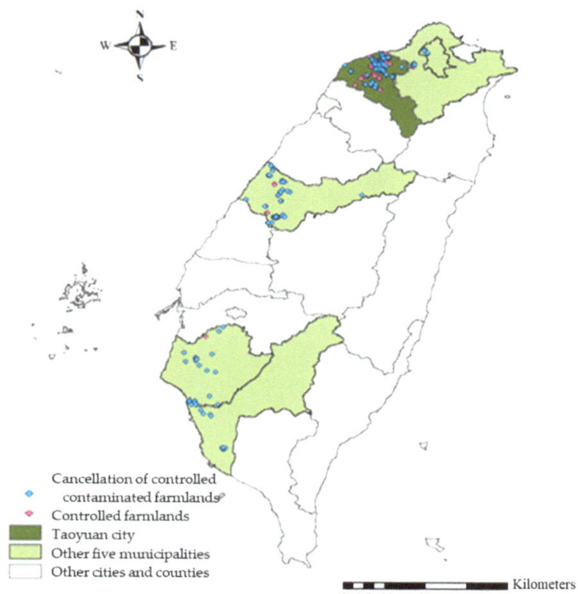

Figure 1. The contaminated farmland hot spots in the six municipalities of Taiwan.

The newest land zoning plan in Taiwan under the *Spatial Planning Act* has been p posed to maintain the agricultural development zone and non-agricultural zones to ensu food security and protect the infrastructure for the production areas. Thus, the plann farmlands should not be fragmented under the *Spatial Planning Act*. It is a comprehensi policy-oriented, and objective guideline for zoning in each city and county. In order achieve the appropriate zoning, data collection for existing administrative areas, data inv tigation, comprehensive analyses of the land spatial planning, and related arrangeme are necessary tasks for each city or county.

Although the contaminated farmland sites and areas for Taoyuan city are not rank the highest in Taiwan, the share of production in terms of total crop quantities in Taoyu city is 12.29% among the six municipalities [11]. The food security and safety of these cro in relation to Taoyuan city and the nearby cities, namely, the capital of Taiwan, Taipei, a the city with the largest population in Taiwan, New Taipei city, have become a major foc of attention. Thus, it is essential to command the remediation benefit of planned farmlan from the existing agricultural development zone and non-agricultural development zor in Taoyuan city. The benefits are not only to be used to prioritize the remediation of all contaminated sites, but also for the compensation of areas zoned as farmland, which is deemed to have a lower price. Past research considered the farmer's personal situation, such as age or cultivated area, for compensation [41]. The results of the evaluation can be used to determine the amount of compensation due to the possible reduction in the farmland price from zoning for agricultural purposes.

2.2. Evaluation Methods

In order to achieve the purpose of this study, an appropriate estimation model is required. The hedonic price model is the link between the farmland prices and all types of characteristics that have different influences on the farmland price. Assume that a piece of farmland h has various characteristics $L = L_h(\mathbf{S}, \mathbf{O}, \mathbf{C})$, where $\mathbf{S}$ is the characteristics matrix of the farmland per se, such as the size of the transacted farmland, and $\mathbf{O}$ is the matrix of the surrounding attributes of the farmland, such as the distance between the transacted farmland and the main traffic artery, the nearby construction price, and other NIMBY and/or YIMBY objects. The issue of most concern in this study is the matrix related to the contamination of the farmland, $\mathbf{C}$. The price of farmland is PL_h. The differentiation of the farmland price is reflected by the characteristics of the farmland. The marginal implicit price of a specific characteristic can then be derived from the linkage of the farmland price and all the characteristics [42]. The hedonic price function is written as Equation (1) below:

$$PL_h = PL_h(S_1, \ldots, S_i, O_1, \ldots, O_j, C_1 \ldots, C_k) \quad (1)$$

In Equation (1), i is the number of variables related to the attributes of the transacted farmland per se, j is the number of surrounding characteristics of the transacted farmland, and k is the number of related contaminated attributes of the transacted farmland.

Since for each piece of farmland located in different places there might exist spatial correlation, without accounting for the spatial relationship, the evaluation results will be biased and inefficient. Furthermore, the fluctuation in farmland prices could be high, and this is especially significant in a country such as Taiwan with a high population density. Thus, the impact of a particular farmland attribute on the farmland price is expected to be different. Quantile regression involving the estimation of different percentiles can reflect such a phenomenon. The HPM should combine both the spatial and quantile attributes in the hedonic price function estimation. Taiwan's EPA classifies the status of contaminated farmlands as "controlled sites", "remediated sites", "cancellation of controlled sites", and "cancellation of remediated sites". The change in a specific characteristic is expected to have a different impact on farmland prices due to their different statuses.

The hedonic spatial quantile model accounts for the spatial dependence of explanatory variables and independent farmland prices. The change in a specific characteristic, however, also has different impacts on the farmland prices. The impact on the farmland price of a change in a specific characteristic for a particular piece of farmland announced as a "controlled site" or as a "remediated site" is expected to be different, since for some potential contaminated farmlands, more time is needed to evaluate their impacts on farmland prices. Furthermore, Taoyuan city is close to the Taipei metropolitan life circle. This results a wide gap among farmland prices in Taoyuan city. Quantile regression is a method used to catch the impact of each farmland characteristic on a specific level of farmland price [34]. Most importantly, by using quantile regression, we can observe the impact of extreme farmland prices. Such drastically high and extremely low farmland prices normally give rise to controversies in zone planning for agricultural development.

2.3. Hedonic Quantile SDM

In addition to accounting for the different farmland price impacts from a specific change in farmland attributes, it is frequently observed that both the explanatory variables and dependent variables related to farmland prices give rise to the problem of spatial dependence. The spatial adjustment both for the explanatory variables and dependent variable is achieved by using the SDM [35,36]. The application of SDM is very limited, a most studies in this regard have been conducted in recent years [43–45]. By accounting the farmland price differentiation, the combination of the SDM and quantile model h resulted in the quantile SDM [36,37,45–47].

This model can be expressed as (2) below:

$$PL_h = \rho_\theta \mathbf{W} PL_h + \alpha_{0\theta} + \boldsymbol{\alpha}_{I\theta} \boldsymbol{S} + \boldsymbol{\alpha}_{J\theta} \boldsymbol{O} + \boldsymbol{\alpha}_{K\theta} \boldsymbol{C} + \boldsymbol{\beta}_\theta \mathbf{WZ} + \nu_\theta \qquad \nu \overset{i.i.d}{\sim} N(0, \sigma^2)$$

The left-hand side of Equation (2) is the farmland price, the right-hand side **W** the spatial weight matrix, and θ is the designed percentile and ranges between 0 a 1. Moreover, ρ is the coefficient with the modification of the spatial matrix. $\boldsymbol{S}$ is t matrix of farmland attributes per se; $\boldsymbol{C}$ is the matrix of contamination-related variables transacted farmland; $\boldsymbol{O}$ is the matrix of other explanatory variables, such as the surroundi characteristics of the transacted farmland; $\boldsymbol{\alpha}_{I\theta}$, $\boldsymbol{\alpha}_{J\theta}$, and $\boldsymbol{\alpha}_{K\theta}$ are the coefficient matrices variables $\boldsymbol{S}$, $\boldsymbol{O}$, and $\boldsymbol{C}$, respectively; α_0 is the intercept term; and $\boldsymbol{\nu}_\theta$ is the error term. β_θ is t coefficient matrix for those explanatory variables of $\boldsymbol{S}$, $\boldsymbol{O}$, and/or $\boldsymbol{C}$ with spatial dependen

2.4. Two-Stage Hedonic Quantile SDM

2.4.1. The First-Stage Hedonic Quantile SDM

The hedonic quantile SDM adjusts the spatial dependence of the dependent variable adding the weighted farmland price variable **W**PL. Since this might result in endogene with all other explanatory variables, i.e., $\boldsymbol{S}$, $\boldsymbol{O}$, and $\boldsymbol{C}$, there are two methods proposed deal with the endogeneity problem in spatially related models. One of these was develop by [38], and was extended to become a two-stage hedonic quantile SDM. The other w proposed by [48] and involves instrumental quantile regression. Similarly, their model h been extended to deal with the endogeneity when the spatially related variables are add to adjust for the spatial issues.

The method proposed by [48] to adjust the endogeneity is less likely to work in pract because it is difficult to transform their theoretical concept into an empirical operation. (the contrary, when the endogeneity originates from the spatially adjusted farmland pri variable and the differences in farmland prices are accounted for, the method propos by [38] is much more feasible. In order to apply the quantile SDM, θ is classified as vario farmland price levels [34]. The model proposed by [38] to circumvent the endogene problem involves the operation of a two-stage estimation quantile SDM. The first stage the estimation involves combining the SDM in (2), and the quantile regression turns out be as follows:

$$\min_{\boldsymbol{\alpha},\, \boldsymbol{\beta}_\theta} \left[\sum_{PL_h \geq (\boldsymbol{\alpha}_\theta \mathbf{X} + \boldsymbol{\beta}_\theta \mathbf{Z})} \theta |PL_h - \boldsymbol{\alpha}_\theta \mathbf{X} - \boldsymbol{\beta}_\theta \mathbf{Z}| \sum_{PL_h < (\boldsymbol{\alpha}_\theta \mathbf{X} + \boldsymbol{\beta}_\theta \mathbf{Z})} (1-\theta) |PL_h - \boldsymbol{\alpha}_\theta \mathbf{X} - \boldsymbol{\beta}_\theta \mathbf{Z}| \right]$$

where the θ s are the divisions of the percentiles for the first-stage quantile SDM estin tion. The same notations that appear in previous equations have similar definitions. is designated as the non-spatial characteristics matrix of $\boldsymbol{S}$, $\boldsymbol{O}$, and $\boldsymbol{C}$ stated above, and refers to the corresponding characteristics' explanatory variables with spatial dependen $\boldsymbol{\alpha}_\theta$ and $\boldsymbol{\beta}_\theta$ are the matrices of the corresponding estimated coefficients for the non-spat explanatory attributes and spatial explanatory characteristics stated above. The predict farmland price matrices $\widehat{PL_h}$ are then obtained from Equation (3).

2.4.2. The Second-Stage Hedonic Quantile SDM

The results predicted for $\widehat{PL_h}$ from (3) account for the spatial correlation both f several of the explanatory variables and the dependent variable, as well as the pri differences in the dependent variable for farmland prices. These predicted values of $\widehat{P}$

are one of the explanatory variables in the second-stage quantile regression [38]. The second-stage hedonic quantile SDM is estimated in Equation (4) as follows:

$$\min_{\hat{\alpha}_\theta}\left[\sum_{PL_h \geq \widehat{PL_h}} \theta\left|PL_h - \widehat{PL_h}\right| \sum_{PL_h < \widehat{PL_h}} (1-\theta)\left|PL_h - \widehat{PL_h}\right|\right] \quad (4)$$

From the estimation results, the marginal effect (price) for every unit of contaminated farmland that is remediated can be computed. Thus, the total remediation benefit for a certain size of contaminated farmland accompanied by a set of characteristics is consequently obtained.

Under the estimation of a two-stage quantile SDM, the marginal benefit for a specific characteristic g with spatial dependence is the marginal implicit price referred to as the marginal willingness-to-pay (MWTP) can thus be computed as (5):

$$\text{MWTP}_g = \frac{\alpha_{g\theta} + \beta_{g\theta}}{(1-\rho_\theta)} \quad (5)$$

The total benefit for the corresponding characteristic g is then computed as in (6) below:

$$\text{TB}_g = \int \text{MWTP}_g dg \quad (6)$$

Equation (6) is the benefit contributed by a certain characteristic g to the total benefit of the contaminated farmland cancellation for different zonal divisions under the *Spatial Planning Act*.

3. Data Sources and Selection of Characteristics and Their Treatments

All transacted farmland-related data from 1 August 2012 to 31 December 2020 were collected from the Web Service of Actual Real Transactions of Real Estate [23]. There were 26,695 pieces of farmland with completed transactions. A total of 3 pieces of farmland were then eliminated, with 0 transacted prices and 599 transacted prices including parking lots and buildings, and 3767 pieces of farmland transacted for special transactions between relatives or employees or urgent selling and urgent buying, etc. A further 446 pieces of farmland with the highest and the lowest 1% of prices were also excluded to avoid possible outliers. As a result, there were 21,880 pieces of farmland transacted. Among these transacted farmlands, 1141 pieces of transacted farmlands were cancelled contaminated controlled sites. All these farmland prices were deflated by the consumer price index of 2020 to remain on the same base for further analysis [49].

3.1. *Planning for the Development of a New Life Circle under the Spatial Planning Act*

Taoyuan city is not a major agricultural production base in Taiwan. Since it belongs to the metropolitan life circle of Taipei, the production of agricultural products makes it one of the local suppliers of ingredients to more than 10 million people in the Taipei metropolitan area. As such, the quality of these products is essential for about half of the population in Taiwan. Taoyuan city has verified the conditions of environmental resources, current land use, population, distribution of the ethnic groups, and the industrial structural development under "*The Draft of the Spatial Planning Act of Taoyuan City*" (*Spatial Planning Act of Taoyuan City* hereafter) to replan the city. Accordingly, the city is being planned to be divided into six life circles including the Zhongli metropolitan life circle, ecological leisure life circle, Taoyuan Aerotropolis life circle, Taoyuan metropolitan urban life circle, rural development life circle, and new town life circle [50]. Each life circle currently covers two to four administrative districts of the city. Since these are planned life circles, a shape file and overlay map under ArcGIS were employed for further analysis.

3.2. Selection and Price Treatment for Farmlands with Contamination of Controlled Sites

The detection of farmland pollution began in 2002 before data collection by the M istry of Interior through the Web Service of Actual Real Transactions of Real Estate o August 2012. The function Generate near the table for ArcGIS was used to identify whet the transacted farmland had been controlled, contaminated, and cancelled. The plots transactions for all farmlands, the announced controlled farmlands, and the cancellatior contaminated controlled farmlands are shown in Figure 2.

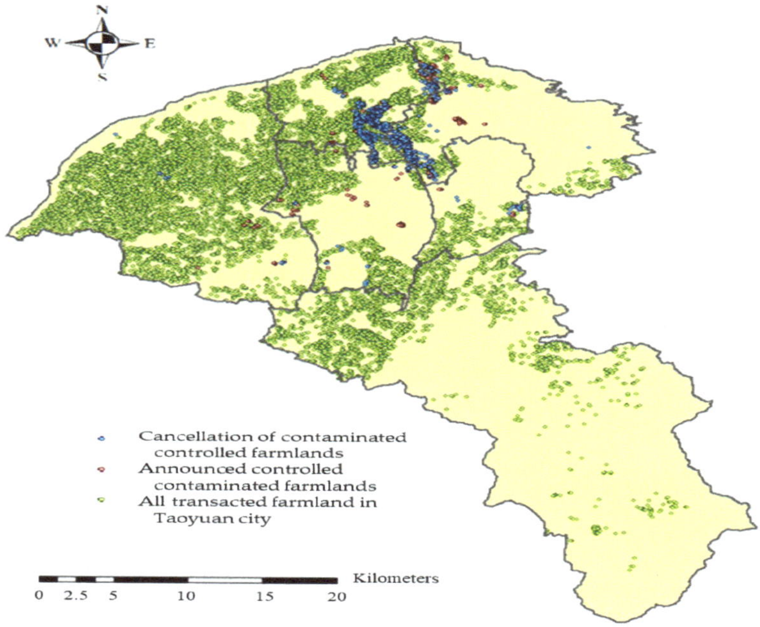

Figure 2. The plots of all cancelled contaminated controlled farmlands, announced controll farmlands, and all transacted farmlands in Taoyuan city from 1 January 2002 to 31 December 202

3.3. Selection of Farmland Characteristics Based on the Price for the Cancellation of Contamina Farmlands

The characteristics that impact farmland prices are less complex than those impacti housing prices. From the past literature, it is known that the size of the farmland, t characteristics of the farmland per se, and the surrounding characteristics of the farmla have impacts on farmland prices [14–18,20,24]. Three types of characteristics were select to reflect their potential impacts on farmland prices. The first type refers to the char teristics directly related to the transacted farmland itself. The second type refers to t contamination of controlled sites surrounding the transacted farmland. The third ty refers to the surrounding characteristics of the transacted farmland. The specific factors each type of characteristic and the collection of the data are discussed in the subsectio listed below.

3.3.1. Characteristics of Transacted Contaminated Controlled Farmlands

The first type of characteristics that influence farmland prices is the farmland its It includes the size of the farmland (AreaT) and the attributes in regard to the transact farmland and the process of cancelling its contamination. It can be observed in two stag How long does it take from the announcement that a piece of farmland is a "controlled si to the "cancellation of the controlled site"? (DmT). Furthermore, how long does it take transact a "cancellation of controlled farmland"? (SmT). The longer it takes for the Dn indicates that the farmland is seriously polluted. A longer period of time is required to confirm and detect if the pollution is eliminated based on a regular schedule. The severity of the degree of contamination of the farmland is represented by the length of time in months.

The reason for using months as the time length representation is a reasonable and differential measurement based on the data from the Web Service of Actual Real Transactions of Real Estate. The overall number of months for 1141 cancellations of controlled farmland to their transactions is 28.63 months, as shown in Table 1. It took 57.51 months for all these farmlands to be announced as "controlled sites". It takes a longer period of time to announce a particular piece of farmland as a "controlled site" than to transact it once the controlled contamination is cancelled. Since the controlled sites are not allowed to trade for the duration between the cancellation of the site and the transaction, this might indirectly imply how zealous a farmland owner is to remedy the contaminated farmland.

Table 1. The definitions, notations, mean values, and standard deviations of all variables in the estimation.

Notation of Variable	Variable Definition	Mean Value *	Standard Deviation
	Dependent Variable		
PL	Average price of 1141 transacted cancellations of controlled farmland until 31 December 2020 (USD)	956,526.86	803,719.82
	Characteristics of transacted cancellations of contaminated controlled farmlands per se		
AreaT	Average size of all cancellations of controlled farmlands (1000 m^2)	1.24	1.00
Area2T	The average of the square size of all transacted farmlands (1000 m^2)	2.55	3.92
TT1	Dummy variable of 1 if transacted farmland is in the Taoyuan Aerotropolis life circle; 0 otherwise	0.82	0.38
TT2	Dummy variable of 1 if transacted farmland is in the Zhongli metropolitan, ecological leisure, Taoyuan metropolitan urban, new town, or rural development life circle; 0 otherwise	0.18	0.38
DmT	Average months of transacted farmlands from being announced a controlled site to the cancellation of control	57.51	29.90
SmT	Average months of transacted farmlands from the cancellation of control to the transaction (months)	28.63	28.83
	Characteristics of contamination of controlled sites surrounding the transacted farmland		
NdmT	The average duration of the nearest announced controlled site prior to the date of a specific piece of transacted farmland (months)	84.82	31.38
NsmT	The average duration of the nearest announced cancelled controlled site prior to the date of a specific piece of transacted farmland (months)	28.22	28.55
TFd11	The distance between a transacted cancellation site of farmland and its nearest controlled farmland (meters)	58.54	34.13
TFd21	The distance between a transacted cancelled site of farmland and the nearest cancellation of another piece of controlled farmland (meters)	67.05	37.18
	Surrounding characteristics of the transacted farmland		
HighT	Distance between transacted farmland and its nearest main traffic artery (1000 m)	1.26	0.69
High2T	Square for the distance between transacted farmland and its nearest main traffic artery (1000 m)	2.05	2.25
TC	The price of the construction site for the life circle where the transacted farmland is located (USD/m^2)	1424.65	76.48

Note: * The exchange rate for USD to NTD in 2012–2020 is 29.609, 29.770, 30.382, 31.927, 32.301, 30.421, 30.189, 30.924, and 29.567. This results in an average exchange rate in 2012–2020 of 30.566.

All the transactions fall within four life circles, and the number of transactions in ea life circle is shown in Table 2. It is observed that among these transactions, about 60% the total contaminated farmlands were transacted within 2.5 years of the cancellation control. It is also found that among the 1141 pieces of transacted farmland, 82% of t contaminated controlled sites cancelled are located in the Taoyuan Aerotropolis life cir Such a dramatic difference in the number of pieces of transacted farmland implies that t expectation that farmland will be switched to other uses is high in the Taoyuan Aerotropo life circle. The Taoyuan Aerotropolis life circle is planned for industrial and resident use, and is a hub for all types of communication and transportation. Thus, this life circle less likely to involve the development of agriculture. In order to differentiate the farmla price in different life circles, a dummy variable is used to represent the transactions in t Taoyuan Aerotropolis life circle, TT1 = 1 from the transactions in all the other five l circles, TT2 = 1.

Table 2. The months taken between the cancellation of controlled farmlands and the transaction different life circles in 2012–2020 [a].

Months between the Cancellation of a Controlled Site and Its Transaction	Life Circle [b]				Total Transacted Farmlands
	Zhongli Metropolitan Life Circle	Taoyuan Aerotropolis Life Circle	Taoyuan Metropolitan Urban Life Circle	Rural Development Life Circle	
<10 months	7	422	7	3	439
10~19 months	36	147	8	1	192
20~29 months	4	15	27	2	48
30~39 months	5	24	17	0	46
40~49 months	1	60	12	0	73
50~59 months	3	189	13	0	205
60~69 months	1	60	2	0	63
70~79 months	0	22	8	1	31
>80 months	1	1	42	0	44
Total farmlands	58	940	136	7	1141
Average months from cancelled controlled site until transaction	18.34	24.90	59.21	20.86	28.63

Notes: [a]: Since the *Web Service of Actual Real Transactions of Real Estate* from [23] started collecting data 1 August 2012, data for the year 2012 cover only 1 August to 31 December. All other years include a full year transactions. [b]: There is no transaction for the ecological leisure life circle and new town life circle among the life circles in Taoyuan for the period covered in this study.

3.3.2. Characteristics of the Contamination of Controlled Sites Surrounding the Transact Farmland

The characteristics of the contamination of controlled sites surrounding the transact farmland might also have a potential impact on the price of the designated transact cancellation of contaminated farmland. This includes the date on which the nearest co taminated controlled site is announced prior to the date of the transaction involving t contaminated cancelled farmland. For reasons similar to those stated earlier, the time diff ence is measured in months and is shown as NdmT. From this attribute, it can be observ how the price of transacted cancelled contaminated controlled farmland changes when nearest contaminated controlled site is announced much earlier than the transaction. T outcome indicates whether a piece of transacted farmland is damaged by its noxious si i.e., a NIMBY facility, or benefited by an innocuous nearby site, i.e., a YIMBY facility.

Similarly, one can also observe whether the price of transacted cancelled contaminat controlled farmland might change when its nearest contaminated controlled farmland cancelled much earlier than when it is transacted. The difference in months is designated NsmT. Moreover, the distance between these transactions and other announced controll sites and the cancellation of other contaminated controlled farmlands is accounted f by the dummy variables TFD11 and TFD21. These represent the different statuses of the surrounding farmlands with different potential impacts on the price of the transacted farmlands. The details of all variables stated above are listed in Table 1.

3.3.3. Surrounding Characteristics of Transacted Cancelled Contaminated Controlled Farmlands

The farmland price is expected to be higher if there is a main traffic artery surrounding the farmland for the convenient transportation of agricultural products. The data on the entrance and exit in relation to surrounding interchanges and traffic data were accessed from the web shape files compiled by the Ministry of the Interior, Taiwan [51]. The Generate near table in ArcGIS test was conducted to obtain the distance between each piece of farmland and the entrance of the exit for its closest interchange. It was found that there are seven main traffic interchanges surrounding Taoyuan city. The average distance between the 1141 pieces of transacted cancelled contaminated controlled farmland and their nearest interchange (HighT) is 1260 m. As Taoyuan city is a metropolitan satellite city of Taipei, the capital of Taiwan, many pieces of farmland are mainly located in the urban–rural border areas. The impacts on the farmland price are not only based on the distance between a particular piece of farmland and the main traffic arteries, but also the prices of the surrounding real estate, such as the price of housing or construction sites. The high price of such real estate normally drives a high degree of speculation for farmland conversion [24,27,52].

Farmland located in different life circles planned under the *Spatial Planning Act of Taoyuan City* indicates its potential for being switched to non-farming use. The higher the construction price surrounding the transacted cancelled controlled contaminated farmland (TC), the higher the price of the farmland will be once its control is cancelled. The average price of these construction sites in the period from 1 August 2012 to 31 December 2020 was computed for each life circle to stand for the level of the speculation related to switching the farmland for other purposes.

4. Specification of Empirical Model and Estimation Results

4.1. Specification for the Transacted Cancelled Contaminated Controlled Farmland Price

Before the estimation was conducted, Moran's I value was employed in ArcGIS to test the existence of spatial dependence for the farmland price and each explanatory variable. The test of Moran's I is used to determine if the spatial issue exists in the dependent and/or independent variable, and its value ranges between −1 and 1. Any non-zero value implies the variable concerned will have a spatial problem. The value of Moran's I for the dependent variable of farmland is 0.588. The value of Moran's I for any explanatory variable greater than 0.5 will then need a corresponding spatial adjustment. All the explanatory variables with Moran's I value exceed this threshold except for the squared term for the size of the transacted farmland in the Taoyuan Aerotropolis life circle, Area2T × TT1. Thus, the specification for each explanatory variable has an additional variable to adjust its spatial problem. The estimation was then conducted to find the factors that have different impacts on the prices of all transacted farmlands. The specification of the first-stage quantile SDM is shown in (7) below:

$$
\begin{aligned}
\mathrm{PL_j} = {} & \rho_\theta \mathrm{W} \times \mathrm{PL_j} + \alpha_{\theta 0} + \alpha_{\theta 1}\mathrm{AreaT_j} \times \mathrm{TT1_j} + \beta_{\theta 2}\mathrm{W} \times \mathrm{AreaT_j} \times \mathrm{TT1_j} + \alpha_{\theta 3}\mathrm{Area2T_j} \times \mathrm{TT1_j} \\
& + \alpha_{\theta 4}\mathrm{AreaT_j} \times \mathrm{TT2_j} + \beta_{\theta 5}\mathrm{W} \times \mathrm{AreaT_j} \times \mathrm{TT2_j} + \alpha_{\theta 6}\mathrm{Area2T_j} \times \mathrm{TT2_j} + \beta_{\theta 7}\mathrm{W} \times \mathrm{Area2T_j} \times \mathrm{TT2_j} \\
& + \alpha_{\theta 8}\mathrm{DmT_j} \times \mathrm{TT1_j} + \beta_{\theta 9}\mathrm{W} \times \mathrm{DmT_j} \times \mathrm{TT1_j} + \alpha_{\theta 10}\mathrm{DmT_j} \times \mathrm{TT2_j} + \beta_{\theta 11}\mathrm{W} \times \mathrm{DmT_j} \times \mathrm{TT2_j} \\
& + \alpha_{\theta 12}\mathrm{SmT_j} \times \mathrm{TT1_j} + \beta_{\theta 13}\mathrm{W} \times \mathrm{SmT_j} \times \mathrm{TT1_j} + \alpha_{\theta 14}\mathrm{SmT_j} \times \mathrm{TT2_j} + \beta_{\theta 15}\mathrm{W} \times \mathrm{SmT_j} \times \mathrm{TT2_j} \\
& + \alpha_{\theta 16}\mathrm{NdmT_j} + \beta_{\theta 17}\mathrm{W} \times \mathrm{NdmT_j} + \alpha_{\theta 18}\mathrm{NsmT_j} + \beta_{\theta 19}\mathrm{W} \times \mathrm{NsmT_j} + \alpha_{\theta 20}\mathrm{TFd11_j} + \beta_{\theta 21}\mathrm{W} \times \mathrm{TFd11_j} \\
& + \alpha_{\theta 22}\mathrm{TFd21_j} + \beta_{\theta 23}\mathrm{W} \times \mathrm{TFd21_j} + \alpha_{\theta 24}\mathrm{TC_j} \times \mathrm{TT1_j} + \beta_{\theta 25}\mathrm{W} \times \mathrm{TC_j} \times \mathrm{TT1_j} + \alpha_{\theta 26}\mathrm{TC_j} \times \mathrm{TT2_j} \\
& + \beta_{\theta 27}\mathrm{W} \times \mathrm{TC_j} \times \mathrm{TT2_j} + \alpha_{\theta 28}\mathrm{HighT_j} \times \mathrm{TT1_j} + \beta_{\theta 29}\mathrm{W} \times \mathrm{HighT_j} \times \mathrm{TT1_j} + \alpha_{\theta 30}\mathrm{High2T_j} \times \mathrm{TT1_j} \\
& + \beta_{\theta 31}\mathrm{W} \times \mathrm{High2T_j} \times \mathrm{TT1_j} + \alpha_{\theta 32}\mathrm{HighT_j} \times \mathrm{TT2_j} + \beta_{\theta 33}\mathrm{W} \times \mathrm{HighT_j} \times \mathrm{TT2_j} + \alpha_{\theta 34}\mathrm{High2T_j} \times \mathrm{TT2_j} \\
& + \beta_{\theta 35}\mathrm{W} \times \mathrm{High2T_j} \times \mathrm{TT2_j} + \upsilon_j, \quad j = 1, \ldots, 1141, \quad \theta = 10, 25, 50, 75, 90
\end{aligned}
\tag{7}
$$

All the α_θs and ρ_θ in (7) are coefficients to be estimated. The 10%, 25%, 50%, 75 and 90% price levels were selected to observe the impact differences for all types of ch acteristics. The estimation from (7) has already adjusted the potential spatial and pr differentiation issues. The estimation results of (7) are shown in Table 3. The prices p dicted, $\widehat{\mathrm{PL}}$, from (7) for each quantile can further adjust the possible endogeneity proble for the explanatory variables and spatial dependent price variable. Thus, a second-sta quantile SDM estimation, shown in (8), was required:

$$
\begin{aligned}
\widehat{\mathrm{PL_i}} = {} & \rho'_\theta \mathrm{W} \times \widehat{\mathrm{PL_i}} + \alpha'_{\theta 0} + \alpha'_{\theta 1}\mathrm{AreaT_j} \times \mathrm{TT1_j} + \beta'_{\theta 2}\mathrm{W} \times \mathrm{AreaT_j} \times \mathrm{TT1_j} + \alpha'_{\theta 3}\mathrm{Area2T_j} \times \mathrm{TT1_j} \\
& + \alpha'_{\theta 4}\mathrm{AreaT_j} \times \mathrm{TT2_j} + \beta'_{\theta 5}\mathrm{W} \times \mathrm{AreaT_j} \times \mathrm{TT2_j} + \alpha'_{\theta 6}\mathrm{Area2T_j} \times \mathrm{TT2_j} + \beta'_{\theta 7}\mathrm{W} \times \mathrm{Area2T_j} \times \mathrm{TT2_j} \\
& + \alpha'_{\theta 8}\mathrm{DmT_j} \times \mathrm{TT1_j} + \beta'_{\theta 9}\mathrm{W} \times \mathrm{DmT_j} \times \mathrm{TT1_j} + \alpha'_{\theta 10}\mathrm{DmT_j} \times \mathrm{TT2_j} + \beta'_{\theta 11}\mathrm{W} \times \mathrm{DmT_j} \times \mathrm{TT2_j} \\
& + \alpha'_{\theta 12}\mathrm{SmT_j} \times \mathrm{TT1_j} + \beta'_{\theta 13}\mathrm{W} \times \mathrm{SmT_j} \times \mathrm{TT1_j} + \alpha'_{\theta 14}\mathrm{SmT_j} \times \mathrm{TT2_j} + \beta'_{\theta 15}\mathrm{W} \times \mathrm{SmT_j} \times \mathrm{TT2_j} \\
& + \alpha'_{\theta 16}\mathrm{NdmT_j} + \beta'_{\theta 17}\mathrm{W} \times \mathrm{NdmT_j} + \alpha'_{\theta 18}\mathrm{NsmT_j} + \beta'_{\theta 19}\mathrm{W} \times \mathrm{NsmT_j} + \alpha'_{\theta 20}\mathrm{TFd11_j} \\
& + \beta'_{\theta 21}\mathrm{W} \times \mathrm{TFd11_j} + \alpha'_{\theta 22}\mathrm{TFd21_j} + \beta'_{\theta 23}\mathrm{W} \times \mathrm{TFd21_j} + \alpha'_{\theta 24}\mathrm{TC_j} \times \mathrm{TT1_j} \\
& + \beta'_{\theta 25}\mathrm{W} \times \mathrm{TC_j} \times \mathrm{TT1_j} + \alpha'_{\theta 26}\mathrm{TC_j} \times \mathrm{TT2_j} + \beta'_{\theta 27}\mathrm{W} \times \mathrm{TC_j} \times \mathrm{TT2_j} \\
& + \alpha'_{\theta 28}\mathrm{HighT_j} \times \mathrm{TT1_j} + \beta'_{\theta 29}\mathrm{W} \times \mathrm{HighT_j} \times \mathrm{TT1_j} + \alpha'_{\theta 30}\mathrm{High2T_j} \times \mathrm{TT1_j} \\
& + \beta'_{\theta 31}\mathrm{W} \times \mathrm{High2T_j} \times \mathrm{TT1_j} + \alpha'_{\theta 32}\mathrm{HighT_j} \times \mathrm{TT2_j} + \beta'_{\theta 33}\mathrm{W} \times \mathrm{HighT_j} \times \mathrm{TT2_j} \\
& + \alpha'_{\theta 34}\mathrm{High2T_j} \times \mathrm{TT2}_i + \beta'_{\theta 35}\mathrm{W} \times \mathrm{High2T_j} \times \mathrm{TT2_j} + \mu'_i, \quad \mathrm{j} = 1, \ldots, 1141, \quad \theta = 10, 25, 50, 75, 90
\end{aligned}
$$

All the α'_θs and ρ'_θ in (8) are also coefficients to be estimated. Similar to the first-sta estimation, five quantiles for the second-stage SDM were employed for the final results. T estimation results are listed in Table 4, and all the analyses were taken from the two-sta quantile SDM estimation.

4.2. Analyses of the Estimations

The second-stage estimation results clearly show that there are more significant planatory variables in the estimation than those in the first stage, as the second-sta estimation adjusts for not only the spatial and price differentiation but also the endogene problem. All the explanatory variables have a dimension of spatial one and non-spat one, except for the squared term of the size of the transacted farmland. The specificati was set for some explanatory variables in quadratic form, such as the distance betwe two objects and the size of the farmland. The signs of the different quantiles in the seco stage quantile SDM estimation for these variables do not necessarily have to be the sar What matters is the total impact from the spatial and non-spatial influence of a speci explanatory variable on the transacted farmland price.

As with the transaction of cancelled contaminated controlled farmlands in the Taoyu Aerotropolis life circle, the results show that the sign of the linear term for farmland size positive and its quadratic term is negative for all quantiles. This indicates that the larg the size of the transacted farmland is, the higher the farmland price will be. The transact farmland price decreases after the adjustment of the spatial problem for farmland size. T results for the other three life circles do not have a consistent impact on the farmland pri for all quantiles. This is because there are only 200 pieces of transacted farmland for t other three life circles. The locations of these farmlands are very dispersed. Thus, it is ha to observe impacts on farmland prices that are consistent with the farmland size with without spatial considerations.

In addition to the size of the transacted farmland, the impact of all other variables the farmland price is a combination of a specific explanatory variable with its adjustment the spatial problem. That is, the marginal effect through (5) for each explanatory variab was computed for this purpose.

Variable	θ = 10		θ = 25		θ = 50		θ = 75		θ = 90	
	Estimated Coefficient	Standard Error	Estimated Coefficient	Standard Error	Estimated Coefficient	Standard Error	Estimated Coefficient	Standard Error	Estimated Coefficient	Standard Error
W × PL	0.27969 ***	0.062	0.32699 ***	0.062	0.32802 ***	0.063	0.23591 ***	0.062	0.29139 ***	0.099
AreaT × TT1	2144.53929 ***	76.979	2219.49001 ***	76.340	2643.95958 ***	77.302	3026.44436 ***	76.795	3333.88032 ***	122.651
W × AreaT × TT1	−753.77099 ***	156.566	−824.73965 ***	155.268	−821.79514 ***	157.223	−656.46156 ***	156.192	−721.75108 ***	249.458
Area2T × TT1	−57.08680 ***	19.305	−23.89028	19.145	−73.86038 ***	19.386	−89.83783 ***	19.259	−116.28741 ***	30.760
AreaT × TT2	1122.74642 ***	187.844	1146.16862 ***	186.287	1453.87683 ***	188.632	1921.91428 ***	187.395	2009.06092 ***	299.294
W × AreaT × TT2	−610.68094 *	315.341	−463.06473	312.726	−360.50165	316.663	−255.33356	314.587	−371.36050	502.435
Area2T × TT2	4.32667	40.569	40.55915	40.232	53.90735	40.739	−35.18512	40.472	112.94204 *	64.638
W × Area2T × TT2	38.88052	73.192	−6.63868	72.585	−34.02774	73.499	−36.27863	73.017	−32.05499	116.617
DmT × TT1	0.93664	2.522	4.19233 *	2.501	5.03103 **	2.532	3.02552	2.516	6.40637	4.018
W × DmT × TT1	−2.12633	5.555	−5.24183	5.509	0.17818	5.578	−3.01206	5.541	8.75418	8.850
DmT × TT2	−3.80623	2.973	−3.02136	2.949	−2.65582	2.986	−2.12289	2.966	−0.67860	4.738
W × DmT × TT2	−1.94780	5.694	−0.67135	5.647	3.16638	5.718	1.64758	5.680	18.60552 **	9.072
SmT × TT1	7.17177	5.945	8.67244	5.896	8.36727	5.970	11.03534 *	5.931	9.53037	9.472
W × SmT × TT1	−0.04704	13.148	−8.83640	13.039	−6.24354	13.203	−6.44615	13.116	15.64050	20.949
SmT × TT2	−1.54079	5.945	−2.48724	5.896	−2.46295	5.970	2.81570	5.931	3.25217	9.472
W × SmT × TT2	7.01458	13.008	3.72989	12.900	2.89319	13.063	0.89679	12.977	20.51948	20.726
NdmT	4.20647 **	2.123	2.91134	2.105	1.72694	2.132	0.85875	2.118	0.44864	3.383
W × NdmT	−1.92237	4.919	0.46857	4.878	−3.76235	4.939	−0.43874	4.907	−17.53501 **	7.837
NsmT	−5.55217	5.344	−2.32690	5.300	−0.74934	5.367	−5.71724	5.331	−4.03591	8.515
W × NsmT	−3.79705	11.777	−0.74078	11.679	2.61899	11.826	1.36889	11.749	−2.22539	18.764
TFd11	2.08126	1.492	2.86060 *	1.479	1.56983	1.498	1.44959	1.488	0.34875	2.377
W × TFd11	2.05125	2.678	−0.76596	2.656	−1.10499	2.690	0.11969	2.672	−0.26821	4.268
TFd21	−1.83134	1.285	−1.24215	1.275	−0.05067	1.291	−0.85056	1.282	0.19335	2.048
W × TFd21	−2.21531	2.446	−1.17893	2.426	−0.36057	2.456	−1.28722	2.440	−0.56339	3.897
TC × TT1	−0.02809	0.027	−0.01109	0.027	−0.01850	0.027	−0.01797	0.027	−0.03854	0.043
W × TC × TT1	0.10212 ***	0.038	0.06967 *	0.038	0.02398	0.038	0.01901	0.038	0.04302	0.060
TC × TT2	0.00086	0.026	0.03418	0.026	0.02859	0.026	0.01848	0.026	0.00613	0.041
W × TC × TT2	0.06677 *	0.036	0.02044	0.036	−0.01399	0.036	−0.01635	0.036	−0.00577	0.058
HighT × TT1	491.05389	541.201	−386.34171	536.713	−693.98820	543.471	−350.11166	539.907	176.81742	862.300
W × HighT × TT1	−595.45319	570.049	265.08674	565.322	593.92725	572.440	169.84652	568.686	−428.78431	908.265
HighT × TT2	−247.74887	941.149	−2084.82326 **	933.344	−2122.27123 **	945.096	−1594.14662 *	938.898	−870.07419	1499.541
W × HighT × TT2	1660.38584	1143.653	2538.03692 **	1134.170	1928.43550 *	1148.450	1432.68647	1140.919	591.58764	1822.194
High2T × TT1	−193.40553	121.897	60.20723	120.886	104.38330	122.408	49.61242	121.606	−4.10463	194.220
W × High2T × TT1	225.07601 *	132.474	−17.43810	131.376	−68.70747	133.030	−2.35215	132.158	68.53098	211.073
High2T × TT2	165.64340	319.076	752.11291 **	316.430	703.67106 **	320.414	507.26719	318.313	280.59356	508.386
W × High2T × TT2	−728.00787 *	398.782	−932.30394 **	395.475	−643.83970	400.454	−473.59254	397.828	−188.31585	635.382
Constant	−3266.65921 ***	998.009	−2629.95589 ***	989.733	−472.56818	1002.194	131.21710	995.622	105.70152	1590.136
R^2	0.6480		0.7087		0.7437		0.7671		0.7769	
$n = 1141$										

Note: Numbers with *, **, and *** indicate that the estimated coefficients are significantly different from zero at 10%, 5%, and 1% significance levels, respectively.

Table 4. The second-stage quantile SDM estimation results for the transacted cancellation of contaminated controlled farmlands.

Variable	θ = 10		θ = 25		θ = 50		θ = 75		θ = 90	
	Estimated Coefficient	Standard Error	Estimated Coefficient	Standard Error	Estimated Coefficient	Standard Error	Estimated Coefficient	Standard Error	Estimated Coefficient	Standard Error
W × $\widehat{PL}$	0.69963 ***	0.034	0.82800 ***	0.029	0.76764 ***	0.031	0.60845 ***	0.038	0.59382 ***	0.039
AreaT × TT1	2132.17230 ***	11.530	2204.44804 ***	9.768	2607.53071 ***	10.580	3018.47241 ***	10.840	3310.46467 ***	14.488
W × AreaT × TT1	−1471.40834 ***	64.346	−1822.32887 ***	61.173	−1870.66198 ***	72.379	−1698.75253 ***	100.406	−1726.26050 ***	112.483
Area2 × TT1	−52.65457 ***	2.902	−21.30718 ***	2.456	−64.03781 ***	2.659	−86.60906 ***	2.732	−110.42939 ***	3.635
AreaT × TT2	1128.73282 ***	28.088	1120.25737 ***	23.769	1425.40262 ***	25.693	1916.85119 ***	26.381	2021.76451 ***	35.062
W × AreaT × TT2	−959.35733 ***	57.819	−995.08767 ***	52.782	−1062.60650 ***	64.050	−1102.61202 ***	87.217	−1067.52242 ***	97.149
Area2T × TT2	1.21243	6.065	44.20168 ***	5.133	59.45443 ***	5.549	−31.09312 ***	5.699	112.26524 ***	7.573
W × Area2T × TT2	39.43325 ***	10.937	−18.63587 **	9.266	−51.15672 ***	10.028	6.94172	10.540	−92.83453 ***	14.149
DmT × TT1	1.77761 ***	0.377	4.78522 ***	0.320	5.26758 ***	0.345	3.45832 ***	0.354	5.86725 ***	0.473
W × DmT × TT1	−1.52087 *	0.830	−5.23210 ***	0.704	−1.70639 **	0.774	−2.45510 ***	0.782	5.15025 ***	1.089
DmT × TT2	−3.15926 ***	0.445	−2.49186 ***	0.376	−2.67401 ***	0.406	−1.58211 ***	0.417	−1.45869 ***	0.564
W × DmT × TT2	−0.17341	0.864	0.64507	0.724	3.09811 ***	0.779	1.48060 *	0.799	13.19676 ***	1.097
SmT × TT1	9.17411 ***	0.888	9.92087 ***	0.753	9.71141 ***	0.813	11.36121 ***	0.835	8.71693 ***	1.111
W × SmT × TT1	−4.01732 **	1.973	−11.15248 ***	1.668	−8.32943 ***	1.807	−7.93912 ***	1.874	11.01435 ***	2.523
SmT × TT2	−0.49368	0.889	−2.02028 ***	0.752	−1.18903	0.812	3.18498 ***	0.834	2.16210 *	1.112
W × SmT × TT2	4.15303 **	1.945	1.37751	1.646	1.07958	1.779	−0.78934	1.835	15.53052 ***	2.461
NdmT	4.08806 ***	0.317	2.42449 ***	0.268	1.51230 ***	0.290	0.26471	0.298	1.57965 ***	0.409
W × NdmT	−3.06441 ***	0.742	−1.29147 **	0.628	−3.37794 ***	0.672	−0.71588	0.691	−13.87550 ***	0.954
NsmT	−6.69853 ***	0.799	−2.82151 ***	0.676	−1.88626 ***	0.731	−5.63121 ***	0.750	−4.28454 ***	0.997
W × NsmT	1.08292	1.773	2.53721 *	1.493	3.41758 **	1.611	3.10276 *	1.668	−2.03879	2.202
TFd11	1.98192 ***	0.223	3.01874 ***	0.189	1.50736 ***	0.204	1.55436 ***	0.209	0.14291	0.278
W × TFd11	0.82869 **	0.420	−2.15791 ***	0.350	−1.17792 ***	0.370	−0.90475 **	0.386	−0.06465	0.503
TFd21	−1.61056 ***	0.193	−1.23379 ***	0.163	0.05565	0.176	−0.87115 ***	0.181	0.42089 *	0.240
W × TFd21	−1.39639 ***	0.388	−0.04232	0.320	−0.73106 **	0.336	−0.21179	0.355	−0.67298	0.460
TC × TT1	−0.02745 ***	0.004	−0.01393 ***	0.003	−0.01924 ***	0.004	−0.01564 ***	0.004	−0.03414 ***	0.005
W*TC × TT1	0.08446 ***	0.006	0.05201 ***	0.005	0.02322 ***	0.005	0.01600 ***	0.005	0.03178 ***	0.007
TC × TT2	0.00488	0.004	0.03380 ***	0.003	0.02992 ***	0.004	0.02047 ***	0.004	0.00978 **	0.005
W*TC × TT2	0.04854 ***	0.005	−0.00143	0.005	−0.02320 ***	0.005	−0.01816 ***	0.005	−0.01160 *	0.007
HighT × TT1	659.04644 ***	80.780	−431.05289 ***	68.403	−675.84204 ***	73.977	−327.30231 ***	75.906	262.68773 ***	100.950
W*HighT*TT1	−735.30497 ***	85.018	405.53827 ***	72.300	624.24798 ***	78.312	245.68931 ***	80.541	−414.47119 ***	106.455
HighT × TT2	−360.01226 **	140.645	−2233.61067 ***	119.095	−2116.94733 ***	128.667	−1475.33421 ***	132.103	−653.07420 ***	175.611
W × HighT × TT2	1478.06867 ***	174.189	2877.20797 ***	144.778	2380.19809 ***	156.233	1563.92686 ***	160.403	613.85302 ***	213.316
High2T × TT1	−212.17173 ***	18.220	78.41233 ***	15.419	109.28209 ***	16.670	51.38790 ***	17.112	−14.21774	22.750
W × High2T × TT1	244.04416 ***	19.769	−67.56086 ***	16.864	−86.71288 ***	18.233	−28.59628	18.736	52.39476 **	24.837
High2T × TT2	251.28484 ***	47.702	816.08639 ***	40.383	707.57398 ***	43.632	487.55109 ***	44.800	251.26357 ***	59.555
W × High2T × TT2	−707.13719 ***	60.786	−1059.47273 ***	50.456	−802.98503 ***	54.402	−538.00267 ***	55.888	−261.10911 ***	74.319
Constant	−2572.53058 ***	166.772	−1665.01516 ***	144.739	−304.98999 **	138.993	16.29999	138.967	263.19113	183.934
R^2	0.9480		0.9537		0.9630		0.9765		0.9750	
n = 1141										

Note: Numbers with *, **, and *** indicate that the estimated coefficients are significantly different from zero at 10%, 5%, and 1% significance levels, respectively.

5. Discussion for the Monetary Benefit of Characteristics

5.1. Attribute with the Largest Contribution to the Total Benefit for Contaminated Farmlands

The marginal effect for each factor is shown in Table 5. The total benefit of contamination remediation of controlled farmlands contributed by each factor was computed and is presented in Table 6. The results show that for every 1000 square meter increase in farmland size, the farmland price in the Taoyuan Aerotropolis life circle increases for every quantile of the farmland price. On the other hand, the transactions of farmland of the same size increasing in all three other life circles results in the farmland price decreasing for all levels of farmland price. The marginal effect of the transacted farmland size results in the highest share of the contribution to the total benefit, either positive or negative, for the transactions involving cancelled contaminated controlled farmlands. The marginal effect and the total benefit for the attribute of transacted farmland size are shown in Figure 3. A comparison is made between two different life circles and different quantiles for farmland prices. The total farmland price increases by about 45–105% for an average 1000 m^2 increase in the farmland size for non-agricultural use in the Taoyuan Aerotropolis life circle and decreases by about 81–131% for the same size of increase in the other three agricultural-development-purpose life circles. The transacted farmland increases by the scale of its average size, while the total benefit declines for those transactions occurring in the other three life circles. This is because these life circles cover most of the cultivated lands in Taoyuan city. Thus, the farmland price via the change in total benefit will not increase instantly after the farmland's contamination is cancelled. The total benefit of the farmland price in the Taoyuan Aerotropolis life circle increases for all price levels once the contaminated controlled site is cancelled. Although the cancellation of the control releases more farmland for cultivation, the rise in the farmland price is inevitable due to a high degree of expectation that the farmland will be converted for another purpose. Thus, the cancellation of contaminated farmlands in the Taoyuan Aerotropolis life circle may not benefit the development of agriculture.

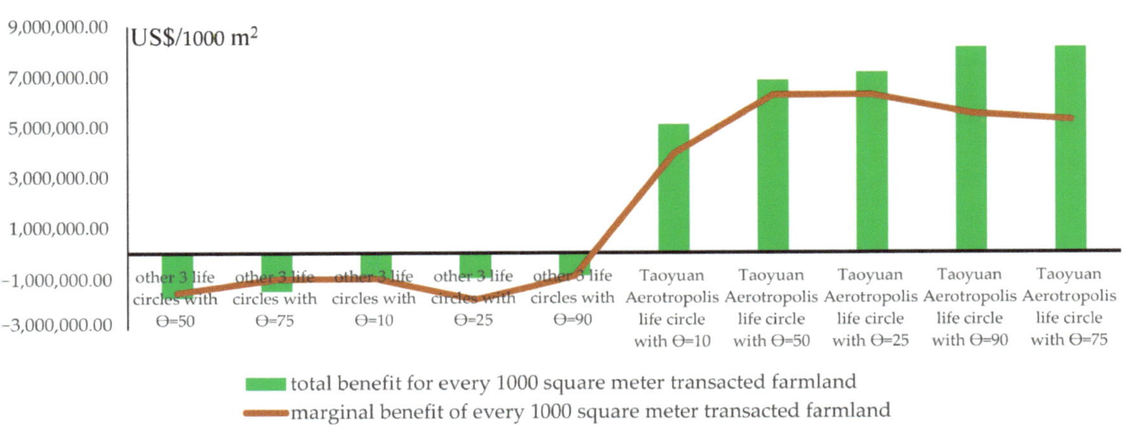

Figure 3. The relationship between the marginal benefit and total benefit of every 1000 square meter change in the transacted farmland size for different life circles with different price quantiles.

5.2. Monetary Benefit of Other Characteristics of Contaminated Farmland per Se and Its Surroundings

Table 6 indicates that the number of months from announcing a controlled contaminated farmland to its cancellation in the Taoyuan Aerotropolis life circle is lower than that for the other three life circles. Similarly, the length of time for a cancelled piece of farmland to complete its transaction in the Taoyuan Aerotropolis life circle is also much shorter than that in the other three life circles once the control has been cancelled. The total benefit for a contaminated controlled farmland from the announcement to its cancellation and from its cancellation to the transaction are shown in Table 6 for each quantile of farmla price, respectively. A comparison of each part of the marginal benefit and the total ben is displayed in Figure 4. It is observed that the total benefit contributed positively these two attributes is higher for farmlands in the Taoyuan Aerotropolis life circle than counterpart price quantile in the other three life circles.

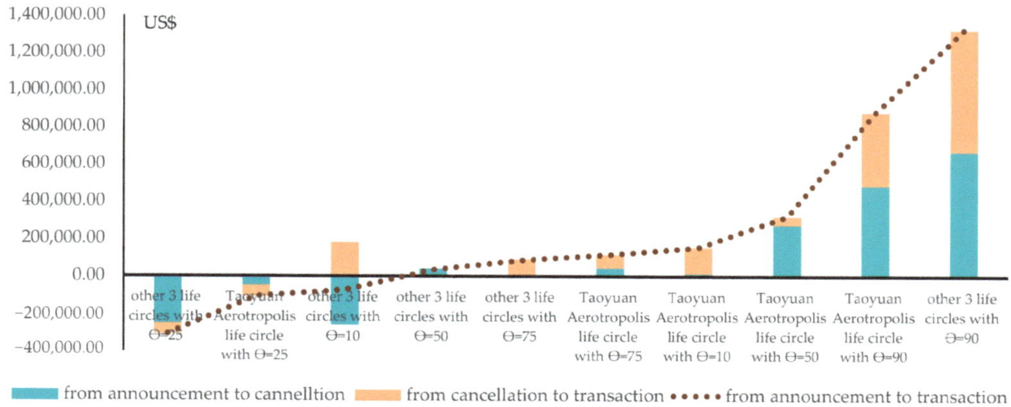

Figure 4. Total benefit for a piece of transacted farmland from announcing its cancellation to transaction.

Table 5. Marginal benefit for the transaction of cancelled contaminated controlled farmland.

Factor on Farmland Price	Quantile				
	$\theta = 10$	$\theta = 25$	$\theta = 50$	$\theta = 75$	$\theta = 90$
Size of transacted farmland in Taoyuan Aerotropolis life circle (USD/1000 m^2)	3,920,751.43	6,274,969.18	6,263,082.59	5,272,919.47	5,516,051.8
Size of transacted farmland in the other 3 life circles (USD/1000 m^2)	−1,017,560.41	−1,840,228.23	−1,563,949.08	−1,032,273.13	−918,841.9
Average months of transacted farmland in Taoyuan Aerotropolis life circle from announcing a controlled site to the cancellation of the site (USD/month)	279.64	−850.01	5014.12	838.25	8874.13
Average months of transacted farmland in other 3 life circles from announcing a controlled site to the cancellation of the site (USD/month)	−3629.92	−3512.78	597.13	−84.82	9454.52
Average months of transacted farmland in Taoyuan Aerotropolis life circle from the cancellation of a site to the transaction (USD/month)	5616.74	−2342.64	1945.82	2859.34	15,892.71
Average months of transacted farmland in the other 3 life circles from the cancellation of a site to the transaction (USD/month)	3985.74	−1222.61	−154.11	2001.68	14,250.66
The average duration of the nearest announced controlled site prior to the date of a specific piece of transacted farmland (USD/month)	1114.95	2155.12	−2626.80	−376.98	−9903.79
The average duration of the nearest cancelled controlled site prior to the date of a specific piece of transacted farmland (USD/month)	−6116.48	−540.77	2156.09	−2112.65	−5093.18
The distance between the transacted farmland and the nearest controlled farmland (USD/meter)	3061.30	1637.38	463.85	542.78	63.03
The distance between the transacted farmland and the nearest cancelled contaminated controlled farmland (USD/meter)	−3275.15	−2427.29	−950.97	−904.86	−203.05
The price of a construction site in Taoyuan Aerotropolis life circle (USD/meter)	62.10	72.43	5.60	0.30	−1.90
The price of a construction site in the other 3 life circles (USD/meter)	58.19	61.57	9.46	1.93	−1.47
The distance between transacted farmland in Taoyuan Aerotropolis life circle and its nearest main traffic artery (USD/1000 m)	2,854,043.26	2,904,022.75	821,320.68	110,342.66	−198,448.0
The distance between transacted farmland in the other 3 life circles and its nearest main traffic artery (USD/1000 m)	−1,452,537.24	−1,366,066.75	−370,619.40	−107,227.67	6995.91

Table 6. Total benefit contributed by each factor for the cancellation of contaminated farmland transaction price.

Impact Factor of Total Benefit [a]	Quantile [b]				
	θ = 10	θ = 25	θ = 50	θ = 75	θ = 90
Benefit of the average size of transacted farmland in the Taoyuan Aerotropolis life circle (1300 m^2)	5,096,977.03 (45.03%)	8,157,459.27 (55.02%)	8,142,007.46 (85.58%)	6,854,796.18 (97.80%)	7,170,866.32 (105.20%)
enefit of the average size of transacted farmland in the other 3 life circles (970 m^2)	−976,856.64 (104.31%)	−1,766,619.77 (131.24%)	−1,501,390.43 (80.67%)	−990,983.45 (85.93%)	−882,087.94 (89.81%)
Benefit of the average months of transacted farmland in aoyuan from announcing a controlled site to the cancellation of the site (54.68 months)	15,291.50 (0.14%)	−46,479.75 (−0.31%)	274,170.65 (2.88%)	45,835.24 (0.65%)	485,238.50 (7.12%)
Benefit of the average months of transacted farmland in the other 3 life circles from announcing a controlled site to the cancellation of the site (70.75 months)	−256,818.03 (27.42%)	−248,527.78 (18.46%)	42,246.29 (−2.27%)	−6000.13 (0.52%)	668,906.63 (−68.10%)
Benefit of the average months of transacted farmland in the aoyuan Aerotropolis life circle from the cancellation of a site to the transaction (24.92 months)	139,969.25 (1.24%)	−58,378.59 (−0.39%)	48,488.52 (0.51%)	71,255.64 (1.02%)	396,047.90 (5.81%)
Benefit of the average months of transacted farmland in the other 3 life circles from the cancellation of a site to the transaction (46.03 month)	183,462.02 (−19.59%)	−56,278.22 (4.18%)	−7,092.85 (0.38%)	92,138.32 (−7.99%)	655,957.60 (−66.79%)
Benefit of the average duration of the nearest announced controlled site prior to the date of a specific transacted armland in the Taoyuan Aerotropolis life circle (78.54 months)	87,567.89 (0.77%)	169,263.23 (1.14%)	−206,307.66 (−2.17%)	−29,608.06 (−0.42%)	−777,844.66 (−11.41%)
Benefit of the average duration of the nearest announced controlled site prior to the date of a specific transacted farmland in the other 3 life circles (114.19 months)	127,314.66 (−13.59%)	246,093.70 (−18.28%)	−299,954.20 (16.12%)	−43,047.83 (3.73%)	−1,130,913.43 (115.14%)
Benefit of the average duration of the nearest cancelled contaminated controlled site prior to the date of a specific transacted farmland in the Taoyuan Aerotropolis life circle (24.55 months)	−150,160.31 (−1.33%)	−13,276.19 (−0.09%)	52,931.36 (0.56%)	−51,864.82 (−0.74%)	−125,037.62 (−1.83%)
Benefit of the average duration of the nearest cancelled contaminated controlled site prior to the date of a specific transacted farmland in the other 3 life circles (45.37 months)	−277,504.42 (29.63%)	−24,533.80 (1.82%)	97,821.11 (−5.26%)	−95,851.60 (8.31%)	−231,077.01 (23.53%)
Benefit of the distance between a transacted farmland in the aoyuan Aerotropolis life circle and its nearest controlled site (56.30 m)	172,351.63 (1.52%)	92,184.13 (0.62%)	26,113.98 (0.27%)	30,560.10 (0.44%)	3549.70 (0.05%)
Benefit of the distance between a transacted farmland in the other 3 life circles and its nearest controlled site (69.00 m)	211,228.16 (−22.56%)	112,978.47 (−8.39%)	32,006.15 (−1.72%)	37,453.38 (−3.25%)	4347.97 (−0.44%)
Benefit of the distance between a transacted farmland in the Taoyuan Aerotropolis life circle and its nearest cancelled contaminated controlled site (65.14 m)	−213,341.62 (−1.88%)	−158,113.59 (−1.07%)	−61,947.92 (−0.65%)	−58,941.31 (−0.84%)	−13,227.12 (−0.19%)
Benefit of the distance between a transacted farmland in the other 3 life circles and its nearest cancelled contaminated controlled site (75.61 m)	−247,634.63 (26.44%)	−183,527.45 (13.63%)	−71,903.42 (3.86%)	−68,415.89 (5.93%)	−15,353.66 (1.56%)
Benefit of the price of a construction site in the Taoyuan Aerotropolis life circle (US$1431.22)	2,716,452.92 (24.00%)	3,168,716.87 (21.37%)	245,167.83 (2.58%)	13,168.23 (0.19%)	−83,154.49 (−1.22%)
Benefit of the price of a construction site in the other 3 life circles (US$1393.94)	2,479,117.32 (−264.72%)	2,623,388.73 (−194.88%)	403,127.66 (−21.66%)	82,241.71 (−7.13%)	−62,448.47 (6.36%)
Benefit of the distance between a transacted farmland in the Taoyuan Aerotropolis life circle and its nearest main traffic artery (1210 m)	3,453,392.66 (30.51%)	3,513,868.35 (23.70%)	993,797.03 (10.45%)	133,514.36 (1.90%)	−240,123.01 (−3.52%)
Benefit of the distance between a transacted farmland in the other 3 life circles and its nearest main traffic artery (1500 m)	−2,178,806.52 (232.65%)	−2,049,100.31 (152.22%)	−555,928.16 (29.87%)	−160,842.11 (13.95%)	10,495.32 (−1.07%)
Total benefit of the Taoyuan Aerotropolis life circle (total %)	11,318,500.95 (100.00%)	14,825,243.73 (100.00%)	9,514,421.25 (100.00%)	7,008,715.56 (100.00%)	6,816,315.52 (100.00%)
Total benefit of the other 3 life circles (total %)	−936,498.08 (100.00%)	−1,346,126.43 (100.00%)	−1,861,067.85 (100.00%)	−1,153,307.60 (100.00%)	−982,172.99 (100.00%)

Notes: [a]: The magnitudes in parentheses under the column of the impact factor are the average of each factor in the related transaction life circle. [b]: The numbers in the parentheses under each quantile are the percentages of each factor provided to the total benefit of different life circles.

For the 50% price quantile and above, the total benefit from the announcement of the contamination cancellation to its transaction results in an increase in the transaction price of 1.67–12.98% of the total benefit for non-agricultural Taoyuan Aerotropolis life circle zoning. The total benefit from the same action taken for the other three agricultural development life circles leads to a reduction in the transaction price by 1.89–134.89%. This result indicates that the cancellation of the high price level for contaminated farmlands is not anticipatec they are planned for agricultural purposes. Under the specific marginal benefit, the positi total benefit indicates that there is a longer period of time either from the announceme to the cancellation or from the cancellation to the transaction, so the transacted farmla price increases because it tends to gain enough time to earn the trust of people regardi the non-toxicity and cleanliness of the farmland for all purposes.

Furthermore, there are various factors that can either be characterized by the NIM or YIMBY attribute, such as a piece of transacted farmland surrounded by a controlled s which has existed there longer than the specific transacted farmland. Likewise, a transact piece of farmland surrounded by other cancelled contaminated controlled sites has a simi impact on the price of a piece of transacted farmland. Although the contaminated site h been cancelled, its existence around the transacted farmland might have a stigma attach to it. The impact on the price of the transacted farmland arises not only from the exister of the surrounding potential NIMBY or YIMBY sites, but also from the distance betwe the designated transacted farmland and these sites. As with the distance between t surrounding sites and the designated transacted farmland, there is a minor impact on t price of the farmland.

When both the distances between the sites and the transacted farmland are shorter, t total transacted farmland price for the Taoyuan Aerotropolis life circle is higher, resulti in an increase in the price of the transacted farmland of 0.14–0.45%. It is obvious that t distance attributes of other controlled sites or cancelled contaminated sites are deemed YIMBY in the Taoyuan Aerotropolis life circle. This indicates that once the cancellati of contaminated farmland is not mainly for agricultural purposes, it does not matter the transacted farmland is close to the surrounding controlled or cancelled contaminat sites. These distance attributes, however, reduce the price of transacted farmlands 1.12–5.24% for different price levels in the other three life circles. This means that the contaminated transacted farmland is cancelled for agricultural purposes, the farth the transacted farmland is from its nearby surrounding controlled sites and cancell contaminated site, the higher the transacted farmland price is.

The impact of the existence of the nearby controlled sites and cancelled contaminat sites on the transacted farmland price is much higher than the distance between the sites and the transacted farmland. For the transacted farmlands located in the Taoyu Aerotropolis life circles, the existence of these two types of sites depletes all levels farmland by 0.56–12.97%, except for the price level of the 25% quantile. On the other har the existence of these sites increases all levels of farmland by 10.86–138.64%, except f the price level of the 25% quantile. The total effect of the existence of the controlled si and the cancelled contaminated site surrounding the designated transacted farmland a the distance of these sites from the transacted farmlands is shown in Figure 5. The figu shows that there is a highly negative impact on the highest farmland price for both li circles, i.e., the Taoyuan Aerotropolis life circle planned for non-agricultural purposes a the other three life circles zoned for agricultural purposes under the *Spatial Planning Act Taoyuan City*.

As there are various factors with different impacts on the price of all transacted far lands, it is worth computing and comparing the percentage of each factor that contribu to the total benefit measured by the price change in the transacted farmlands. According the total benefit computed, as shown in Table 6 above, all factors are classified into ni types. A comparison is made for the total benefit of all transacted farmlands in the Taoyu Aerotropolis life circle zoned for non-agricultural purposes under the *Spatial Planning A of Taoyuan City* and for the Zhongli metropolitan, Taoyuan metropolitan urban, and ru development life circles planned for agricultural development, respectively.

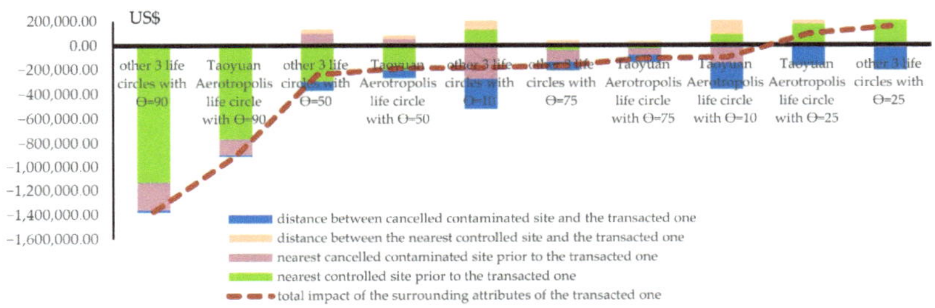

Figure 5. The impact of surrounding contaminated sites with different attributes on the price of transacted farmland.

The specific benefit share of each factor to the total benefit, either positive or negative, is presented in Table 6. Figures 6 and 7 are the composition and share of the total benefit through the change in the transacted farmland price for each factor. Both types of life circles have similar factors that contribute a larger share to the total benefit. These factors are the size of the transacted farmland, the distance between the main traffic artery and the transacted farmland, and the price of construction sites in both life circles for all price levels. The rankings of the benefit share for each quantile price level or for both life circles have slight differences. Moreover, it is not necessary for the higher quantile of the farmland price to contribute to the total benefit. The total benefit can be computed either in monetary terms or as a share of different combinations of benefit contributed by the attributes of the related contaminated farmlands.

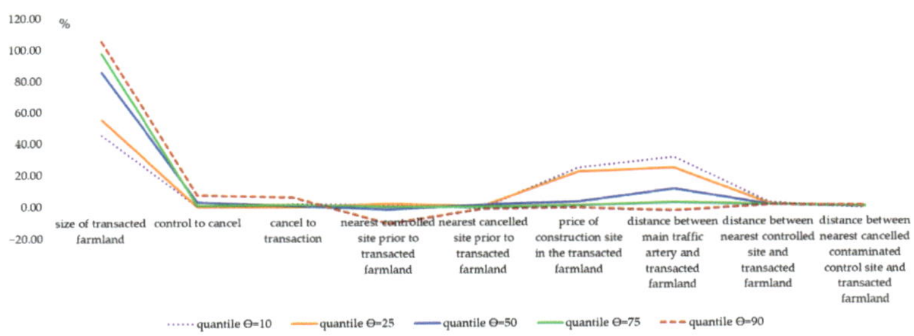

Figure 6. The share of the impact for each factor on the change in the total benefit for the Taoyuan Aerotropolis life circle.

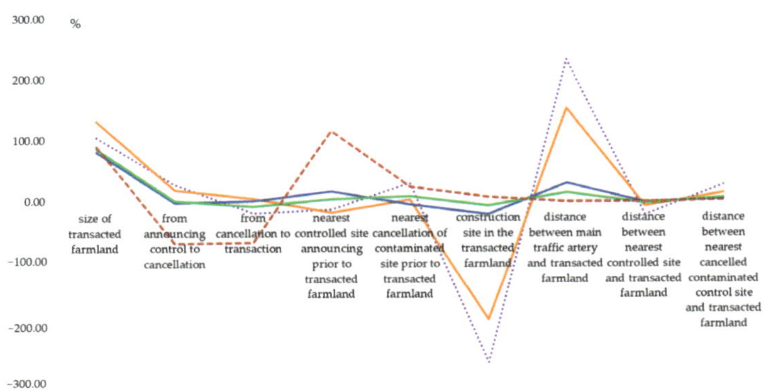

Figure 7. The share of the impact for each factor on the change in the total benefit for all life circles other than the Taoyuan Aerotropolis life circle.

6. Conclusions

This study developed a two-stage quantile spatial Durbin model to evaluate one the municipalities in Taiwan with the most cancelled contaminated controlled farmlan namely, Taoyuan city. The changes in the characteristics with significant impacts on t changes in the transacted farmland price were determined. Their corresponding benef via the changes in the farmland price were computed accordingly. The benefit contribut by the length of time from the cancellation of the contaminated land to the transacti involving the farmland per se and the most related surrounding attributes such as t duration of the nearest announced controlled site and/or the cancelled controlled s prior to the specific transacted farmland does not have the impact anticipated by t remediation mission. The significant attributes for the price impacts are the size of t transacted farmland, the distance between the transacted farmland and the main traf artery, and the price of the construction site. These attributes indicate that the cancellati of contaminated controlled farmlands apparently leads to the supply of more un-pollut lands for non-agricultural purposes based on the planning act drawn up by the city.

The impact of each attribute on the total benefit of farmland remediation was eva ated in this study. The magnitude of the evaluated benefit of each factor can be used accordance with the *Spatial Planning Act of Taoyuan City* for the zone planning of differe life circles. As the life circle is zoned to be mainly used for agricultural production purpos the farmland transaction price may not increase once the contaminated land is cancell However, retaining a certain amount of good-quality space for agricultural producti is essential not only for the city, but also for the whole country. Thus, the benefit eva ated in this study can contribute a guideline to selecting the appropriate attribute that mainly related to the contamination remediation of the transacted site per se and/or simi attributes surrounding the transacted site. Whether the compensation is implement depends not only on the evaluated benefit of each attribute accomplished here but al relies highly on understanding and interpreting the evaluated benefits by decision make From the farmland owners' viewpoint, once the contamination of farmland is remediat but causes a decline in the price of the farmland, compensation is an unavoidable and f action to encourage farmland owners to continue to cultivate farmland. Compensati for farmers remaining in agricultural production is contradictory to the viewpoint of t land speculator.

This study develops a comprehensive spatial and quantile model to evaluate t monetary-term benefit via the change in the transacted farmland price for each attribu that potentially has an impact on the remediation of contaminated farmland sites. T accomplishment of such a mission largely relies upon the accuracy of the investigati detection, elimination, and cancellation of the contaminated sites. However, uncertain in any process stated above might occur. This causes the elimination of the contaminat sites to lose momentum. On the contrary, the investigation or detection process might too stringent. This leads to too many controlled sites. This study does not account for a type of uncertainty arising from the confirmation of farmland contamination. Thus, futu work can go beyond the study accomplished here and account for either type of uncertai in determining the contaminated farmlands. Moreover, the study accomplished here c be applied to other cities, counties, or areas where the spatial zoning of land is plann Further analyses can also be conducted by classifying the planned zones and spaces ranking the importance of farmlands for different agricultural purposes.

Author Contributions: Conceptualization, P.-I.W., C.-E.S., J.-L.L. and T.-K.H.; methodology, P.-I. and C.-E.S.; software, C.-E.S., J.-L.L. and T.-K.H.; validation, P.-I.W., J.-L.L. and T.-K.H.; form analysis, P.-I.W. and C.-E.S.; investigation, J.-L.L. and T.-K.H.; resources, P.-I.W. and J.-L.L.; da curation, C.-E.S., J.-L.L. and T.-K.H.; writing—original draft preparation, P.-I.W. and C.-E.S.; writing review and editing, J.-L.L. and T.-K.H.; visualization, J.-L.L. and T.-K.H.; supervision, P.-I.W.; proj administration, P.-I.W. and C.-E.S.; funding acquisition, P.-I.W. and J.-L.L. All authors have read a agreed to the published version of the manuscript.

Funding: This research was funded by the Ministry of Science and Technology Council, Taiwan, grant number MOST 108-2410-H-002-191-MY2.

Data Availability Statement: All data used in this study are available upon request.

Acknowledgments: We sincerely appreciate the assistance of technical support from Pei-Ching Cheng.

Conflicts of Interest: The authors declare no conflict of interest.

ferences

Padilla, F.M.; Gallardo, M.; Manzano-Agugliaro, F. Global trends in nitrate leaching research in the 1960–2017 period. *Sci. Total Environ.* **2018**, *643*, 400–413. [CrossRef] [PubMed]

Ahmed, A.A.; Diab, M.S. Hydrologic analysis of the challenges facing water resources and sustainable development of Wadi Feiran basin, Southern Sinai, Egypt. *Hydrogeol. J.* **2018**, *26*, 2475–2493. [CrossRef]

Zhang, L.; Zhu, G.; Ge, X.; Xu, G.; Guan, Y. Novel insights into heavy metal pollution of farmland based on reactive heavy metals (RHMs): Pollution characteristics, predictive models, and quantitative source apportionment. *J. Hazard. Mater.* **2018**, *360*, 32–42. [CrossRef] [PubMed]

He, M.; Shen, H.; Li, Z.; Wang, L.; Wang, F.; Zhao, K.; Liu, X.; Wendroth, O.; Xu, J. Ten-year regional monitoring of soil-rice grain contamination by heavy metals with implications for target remediation and food safety. *Environ. Pollut.* **2019**, *244*, 431–439. [CrossRef]

Li, F.; Li, X.; Hou, L.; Shao, A. Impact of the coal mining on the spatial distribution of potentially toxic metals in farmland tillage soil. *Sci. Rep.* **2018**, *8*, 14925. [CrossRef]

Zhan, J.; Twardowska, I.; Wang, S.; Wei, S.; Chen, Y.; Ljupco, M. Prospective sustainable production of safe food for growing population based on the soybean (*Glycine max* L. Merr.) crops under Cd soil contamination stress. *J. Clean. Prod.* **2019**, *212*, 22–36. [CrossRef]

Chung, K.H.Y.; Adriaens, P. Financial exposure to environmental liabilities in Lake Huron drainage area farmlands: A GIS and hedonic pricing approach. *Agric. Financ. Rev.* **2023**, *83*, 144–167. [CrossRef]

Weber, E.F.; Wilson, J.; Kirk, M.; Tochko, S. Selected remedy at the Queen City farms superfund site: A risk management approach. In *Hydrogeology, Waste Disposal, Science and Politics, Proceedings of the 30th Symposium on Engineering Geology and Geotechnical Engineering, Boise, ID, USA, 23–25 March 1994*; Link, P.K., Ed.; Idaho State Government: Boise, ID, USA, 1994; pp. 413–427.

Soil and Groundwater Remediation Fund Management Board. *Soil and Groundwater Pollution Sites Searching*; Environmental Protection Administration: Taipei, Taiwan, 2022. Available online: https://sgw.epa.gov.tw/ContaminatedSitesMap/Default.aspx (accessed on 2 January 2023).

Directorate-General of Budget, Accounting and Statistics. *Population Housing and Census: General Statistical Analysis Report*; Directorate-General of Budget, Accounting and Statistics: Taipei, Taiwan, 2022. Available online: https://www.stat.gov.tw/ct.asp?xItem=47818&ctNode=6679&mp=4 (accessed on 12 February 2023).

Council of Agriculture. *Agricultural Statistics Yearbook 2020*; Council of Agriculture: Taipei, Taiwan, 2022. Available online: http://agrstat.coa.gov.tw/sdweb/public/book/Book.aspx (accessed on 20 January 2023).

Construction and Planning Agency. *Implementation of the Spatial Planning Act Spatial Planning in Municipalities, Counties (Cities) in Progress*; Construction and Planning Agency, Ministry of the Interior: Taipei, Taiwan, 2018. Available online: https://www.cpami.gov.tw/news/realtime-news.html?unid=24&sort=1 (accessed on 12 February 2023).

Deaton, B.J.; Vyn, R.J. The effect of strict agricultural zoning on agricultural land values: The case of Ontario's greenbelt. *Am. J. Agric. Econ.* **2010**, *92*, 941–955. [CrossRef]

Vitaliano, D.F.; Hill, C. Agricultural districts and farmland prices. *J. Real Estate Financ. Econ.* **1994**, *8*, 213–223. [CrossRef]

Choumert, J.; Phélinas, P. Determinants of agricultural land values in Argentina. *Ecol. Econ.* **2015**, *110*, 134–140. [CrossRef]

Cotteleer, G.; Gardebroek, C.; Luijt, J. Market power in a GIS-based hedonic price model of local farmland markets. *Land Econ.* **2008**, *84*, 573–592. [CrossRef]

Gracia, A.; Pérez y Pérez, L. Hedonic analysis of farmland prices: The case of Aragón. *Int. J. Agric. Resour. Gov. Ecol.* **2007**, *6*, 96–110. [CrossRef]

Hanson, E.D.; Sherrick, B.J.; Kuethe, T.H. The changing roles of urban influence and agricultural productivity in farmland price determination. *Land Econ.* **2018**, *94*, 199–205. [CrossRef]

Sheng, Y.; Jackson, T.; Lawson, K. Evaluating the benefits from transport infrastructure in agriculture: A hedonic analysis of farmland prices. *Aust. J. Agric. Resour. Econ.* **2018**, *62*, 237–255. [CrossRef]

Roka, F.M.; Palmquist, R.B. Examining the use of national databases in a hedonic analysis of regional farmland values. *Am. J. Agric. Econ.* **1997**, *79*, 1651–1656. [CrossRef]

Isgin, T.; Forster, D.L. A hedonic price analysis of farmland option premiums under urban influences. *Can. J. Agric. Econ.* **2006**, *54*, 327–340. [CrossRef]

Chen, M.-T. An empirical study of farmland prices in the agricultural-based area: The case of Taipei county in Yulin prefecture. *J. Law Commer.* **1997**, *33*, 159–190.

Ministry of the Interior. *Web Service of Actual Real Transaction of Real Estates*; Ministry of the Interior: Taipei, Taiwan, 2022. Available online: https://lvr.land.moi.gov.tw (accessed on 20 January 2023).

24. Lai, M.-C.; Wu, P.-I.; Liou, J.-L.; Chen, Y.; Chen, H. The impact of promoting renewable energy in Taiwan: How much hai added to snow in farmland prices? *J. Clean. Prod.* **2019**, *241*, 118519. [CrossRef]
25. Wu, W.-J.; Wu, P.-I.; Liou, J.-L. Boon or bane: Effect of adjacent YIMBY or NIMBY facilities on the benefit evaluation of op spaces or cropland. *Sustainability* **2020**, *13*, 3998. [CrossRef]
26. Huang, H.; Miller, G.Y.; Sherrick, B.J.; Gómez, M.I. Factors influencing Illinois farmland values. *Am. J. Agric. Econ.* **2006**, 458–470. [CrossRef]
27. Eagle, A.J.; Eagle, D.E.; Stobbe, T.E.; van Kooten, G.C. Farmland protection and agricultural land value at the urban-rural frir British Columbia's agricultural and reserve. *Am. J. Agric. Econ.* **2014**, *97*, 282–298. [CrossRef]
28. Bell, K.P.; Dalton, T.J. Spatial economic analysis in data-rich environments. *J. Agric. Econ.* **2007**, *58*, 487–501. [CrossRef]
29. Curtiss, J.; Jelínek, L.; Hruška, M.; Medonos, T.; Vilhelm, V. The effect of heterogeneous buyers on agricultural land prices: T case of the Czech land market. *Ger. J. Agric. Econ.* **2013**, *62*, 1–18.
30. Lehn, F.; Bahrs, E. Quantile regression of German standard farmland values: Do the impacts of determinants vary across conditional distribution? *J. Agric. Appl. Econ.* **2018**, *54*, 453–477. [CrossRef]
31. Mishra, A.K.; Moss, C.B. Modeling the effect of off-farm income on farmland values: A quantile regression approach. *Econ. Mo* **2013**, *32*, 361–368. [CrossRef]
32. Peeters, L.; Schreurs, E.; van Passe, S. Heterogeneous impact of soil contamination on farmland prices in the Belgian Camp Region: Evidence from unconditional quantile regressions. *Environ. Resour. Econ.* **2017**, *66*, 135–168. [CrossRef]
33. Uematsu, H.; Khanal, A.R.; Mishra, A.K. The impact of natural amenity on farmland values: A quantile regression appro *Land Use Policy* **2013**, *33*, 151–160. [CrossRef]
34. Koenker, R.; Bassett, G., Jr. Regression quantiles. *Econometrica* **1978**, *46*, 33–50. [CrossRef]
35. Anselin, L.; Syabri, I.; Kho, Y. GeoDa: An introduction to spatial data analysis. *Geogr. Anal.* **2006**, *38*, 5–22. [CrossRef]
36. LeSage, J.; Pace, R.K. *Introduction to Spatial Econometrics*; Chapman and Hall/CRC: Boca Raton, FL, USA, 2009; ISBN 978-042-913-80
37. Bekti, R.D.; Rahayu, A.; Sutikno. Maximum likelihood estimation for spatial Durbin model. *J. Math. Stat.* **2013**, *9*, 169–1 [CrossRef]
38. Kim, T.-H.; Muller, C. Two-stage quantile regression when the first stage is based on quantile regression. *Econom. J.* **2004** 218–231. [CrossRef]
39. Donoso, G.; Cancino, J.; Foster, W. Farmland values and agricultural growth: The case of Chile. *Econ. Agrar. Y Recur. Nat.* **20** *13*, 33–52.
40. Von Witzke, H.; Noleppa, S. The high value to society of modern agriculture: Global food security, climate protection, a preservation of the environment—Evidence from the European Union. In *Food Security in a Food Abundant World*; Schmitz, Kennedy, P.L., Schmitz, T.G., Eds.; Emerald Group Publishing Limited: Bingley, UK, 2016; pp. 55–65. ISBN 978-1-78560-215-
41. Awasthi, M.K. Socioeconomic determinants of farmland value in India. *Land Use Policy* **2014**, *39*, 78–83. [CrossRef]
42. Rosen, S. Hedonic prices and implicit markets: Product differentiation in pure competition. *J. Polit. Econ.* **1974**, *82*, 34– [CrossRef]
43. Feng, Z.; Chen, W. Environmental regulation, green innovation, and industrial green development: An empirical analysis bas on the spatial Durbin model. *Sustainability* **2018**, *10*, 223. [CrossRef]
44. Li, B.; Wu, S. Effects of local and civil environmental regulation on green total factor productivity in China: A spatial Durl econometric analysis. *J. Clean. Prod.* **2017**, *153*, 342–353. [CrossRef]
45. Tsai, M.-k.; Wu, P.-I.; Liou, J.-L. Spatial hedonic quantile Durbin model for benefit evaluation of diversified open spaces in urb planning division in Taoyuan City. *J. City Plan.* **2019**, *46*, 297–341.
46. Liou, J.-L.; Randall, A.; Wu, P.-I.; Chen, H.-H. Monetarizing spillover effects of soil- and groundwater- contaminated sites Taiwan: How much higher housing price people pay to avert? *Asian Econ. J.* **2019**, *33*, 67–86. [CrossRef]
47. Wu, P.-I.; Chen, Y.; Liou, J.-L. Housing property along riverbanks in Taipei, Taiwan: A spatial quantile modelling of landsca benefits and flooding losses. *Environ. Dev. Sustain.* **2020**, *23*, 2402–2438. [CrossRef]
48. Chernozhukov, V.; Hansen, C. Instrumental quantile regression inference for structural and treatment effect models. *J. Econ* **2006**, *132*, 491–525. [CrossRef]
49. Directorate-General of Budget, Accounting and Statistics. *Price: Consumer Price Indices*; Directorate-General of Budget, Accou ing and Statistics: Taipei, Taiwan, 2022. Available online: https://eng.stat.gov.tw/ct.asp?xItem=12092&ctNode=1558&mp (accessed on 16 January 2023).
50. Government of Taoyuan City. *The Draft of the Spatial Planning Act of Taoyuan City*; Government of Taoyuan City: Taoyuan C Taiwan, 2020. Available online: https://spatialplan.tycg.gov.tw/upload/news (accessed on 15 January 2023).
51. Ministry of the Interior. *State Road and Highway (Including Expressway above Grade) Road Centerline*; Ministry of the Interior: Tai Taiwan, 2021. Available online: https://data.moi.gov.tw/MoiOD/System/DownloadFile.aspx?DATA=1B5E71F3-C599-47 -ACBC-086B08C62BB9 (accessed on 24 January 2023).
52. Cavailhès, J.; Wavresky, P. Urban Influences on periurban farmland prices. *Eur. Rev. Agric. Econ.* **2003**, *30*, 333–357. [CrossRef]

land

icle

ustainable Management of Land Resources: The Case of hina's Forestry Carbon Sink Mechanism

ngyi Liu [1] and Tianyu He [2,*]

[1] Guizhou Qiannan College of Science and Technology, Guiyang 550600, China
[2] School of Business Administration, Zhejiang Gongshang University, Hangzhou 310018, China
* Correspondence: 20020181111@pop.zjgsu.edu.cn

Abstract: Compared to more developed countries, the use of land resources is less efficient in China. China's vast forest land area gives it a rich and underutilized carbon sink. This is an important way for China to achieve the goals of "carbon peaking" and "carbon neutrality", which is of great significance to China's sustainable development. In the past 20 years, China has designed a series of policies to serve the development of forestry carbon sinks, namely the forestry carbon sink mechanism (FCSM). However, the questions of which policy is the most important, and what is the socio-economic value it generates, have not been fully investigated. Accordingly, this paper studied 30 provincial-level regions in China from 2005 to 2020 using the difference-in-differences (DID) model. The conclusions show that: (1) the FCSM does increase the socio-economic value of land resources, thus improving the sustainability of land resources; (2) the FCSM helps to increase forest coverage, forest stock volume and the forest coverage rate, which increases the social value of land resources from the greening path; (3) the FCSM helps to increase the gross forestry product, which increases the economic value of land resources through the path of increasing production value.

Keywords: difference-in-differences model; forestry carbon sink mechanism; governance and policy of land resources; socio-economic value of land resource management

ation: Liu, H.; He, T. Sustainable nagement of Land Resources: The e of China's Forestry Carbon Sink chanism. *Land* **2023**, *12*, 1188. ps://doi.org/10.3390/ d12061188

ademic Editors: Matjaž Glavan Jianjun Zhang

eived: 5 April 2023
ised: 17 May 2023
epted: 3 June 2023
olished: 6 June 2023

1. Introduction

The *World Conservation Strategy* of 1980 called on humankind to recognize the need to consider the resource and ecological needs of future generations in the pursuit of economic development and the enjoyment of natural wealth. The *Rio Declaration* of 1992 emphasized that ecological protection is central to achieving sustainable development and called on countries around the world to take action according to their own circumstances. In 2015, 17 Sustainable Development Goals (SDGs) were presented in *Transforming our World: The 2030 Agenda for Sustainable Development*, which included the three dimensions of sustainable development: social, economic and environmental, as well as important aspects related to peace, justice and effective institutions [1]. The SDG15 emphasized the sustainable management of forests [1]. Today, sustainable development has become an important goal in the common quest for humanity. Researchers have been carrying out rich work on sustainable development from many perspectives, such as environmental protection [2–4], economic development [5–7] and human quality of life [8,9]. Greenhouse gas emissions are an issue that must be taken into account in the process of sustainable development. As shown in Figure 1, annual global CO2 emissions are increasing year by year and will be almost seven times higher in 2020 than in 1920. According to the IPCC, 1.5 °C of global warming above pre-industrial levels is due to greenhouse gases [10]. In addition, it has been shown that the increase in GHG emissions is associated with infectious and non-infectious diseases, negative effects on nutrition, water insecurity and other social disruptions [11]. Therefore, controlling greenhouse gases, especially carbon dioxide emissions, has become an important issue for humanity as it seeks to achieve sustainable development. China's total CO_2 emissions have been among the highest in the world for more than a decade [Undoubtedly, China's practices and exploration of low-carbon development has import social value for the world's carbon reduction and environmental protection, as well being an important inspiration for other developing countries to plan their own carb reduction path.

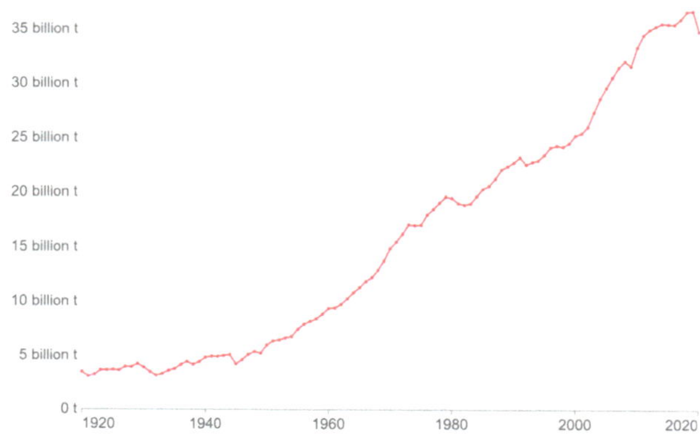

Figure 1. Annual global CO_2 emissions over the last 100 years. Source: Global Carbon Project [1

On the other hand, after decades of rapid development, China's insufficient efficien in the use of land resources has become increasingly evident [13]. Along with land deg dation and increased urban pollution, the past extensive mode of growth can no long meet China's demand for high-quality economic development. China urgently needs improve the way it uses its land resources. A better way is to implement land cont policies, such as what was described in the World Bank's "Land Reform Policy Paper" 1975 [14], China's "Requisition-Compensation Balance of Arable Land" [15], etc. Althou land management policies may also have negative effects, such as economic conflicts, soc injustice and government corruption [16], overall, they have more or less increased economic output of land resources.

In order to achieve low carbon development, the Chinese government has enact a number of environmental regulatory policies in the last decade or so. One of the m promising and effective approaches is the emissions trading scheme (ETS) [17,18]. The E allows companies to purchase carbon sinks to partially offset emissions in excess of th allowances, creating the basis for the forestry carbon sink mechanism (FCSM). China ha potential market of nearly USD 31.35 billion in Chinese Certified Emission Reduction [19] well as abundant forestry resources [20]. Therefore, against the background of record-hi world carbon sink prices [21], the study of forestry carbon sinks in China is of practi value. The FCSM therefore not only contributes to environmental protection, but al generates economic value and thus improves the use of land resources.

Through reviewing the literature, this study found that the studies on forestry carb sinks can be divided into the following categories.

The first category involves research on carbon sink forest construction technolo Technical barriers are one of the challenges in building carbon sink forests [22]. Exa ples include the depth of the oxic peat layer [23], landscape type [24,25], ground-lev ozone concentration [26], atmospheric CO_2 concentration [27], tree age [28], blockcha technology [29], etc. This type of research is usually in the scientific and technical fie and requires certain technical tools and analytical instruments to complete the resear work. A few scholars have also conducted research from a management perspective, su as reasonable deforestation patterns [30] and the investment models applicable to fores carbon sink projects [31].

In the second category, the main focus is on the measurement of the productivity forestry carbon sinks and carbon storage. The common measures are the Slacks-Based M sure (SBM) method [32], the DEA model [33], and the Malmquist–Luenberger index [34]. In addition, many researchers have furthered the analysis and discussion based on the measurement results. Chen et al. pointed out the contribution of forestry carbon sinks in reducing greenhouse gases and achieving carbon neutrality [35]. Shi et al. pointed out that there are deficiencies in the carbon trading market and imperfect accounting of forestry carbon sinks in China [36]. Wei and Shen analyzed the factors (urbanization rate, forestry financial allocation, etc.) affecting forestry carbon sinks in China [37]. Montagnini and Porras pointed out the shortcomings of the FCSM in China from the perspective of economic, social and environmental considerations [38].

The third category, which focuses on the role and drivers of forestry carbon sinks, is also a literature that is of interest in this study. As an investment project, forestry carbon sinks are not a sure thing [39,40]. In the field of economics, researchers typically use econometric models to examine the role and drivers of the FCSM. In terms of its role, the FCSM helps to upgrade the industrial structure at the county level and has a pro-poor effect [41]; the FCSM improves the environment [42]. In terms of drivers, factors such as climate change have a huge impact on the forestry carbon sink [43,44]. Meanwhile, human intervention is also an important factor that cannot be ignored, such as government subsidies [45], investment of fixed assets and FCSM projects [46].

It can be found that economists' studies in the field of forestry carbon sinks are focused on two effects: economic and social effects. The social effects are mainly in the promotion of greening and the reduction of greenhouse gases. However, from an economic standpoint, these studies are relatively crude in the use of econometric models, failing to better overcome the endogeneity problem. Therefore, this paper aims to improve the existing literature by using the DID model. On the one hand, several FCSM policies have been published in China since 2011, but no study has yet answered the question of which policy is most important for the development of the forest carbon sink. On the other hand, DID models have been used to identify causal relationships between variables. This helps to avoid the problem of model endogeneity, and thus can better identify whether the FCSM is delivering social and economic value. Figure 2 shows the process of proving the above questions.

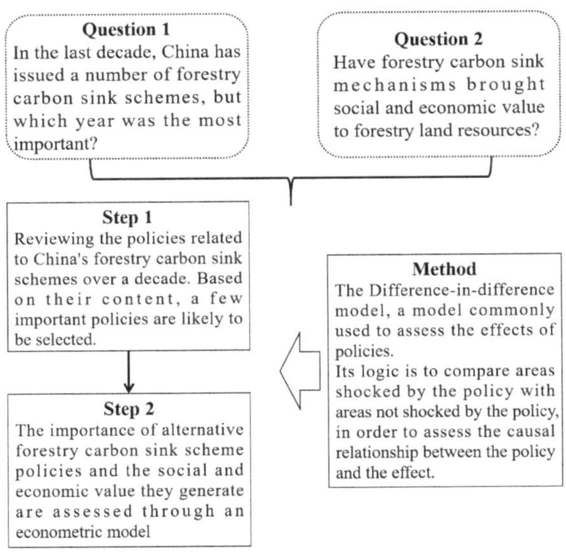

Figure 2. Guidance on research idea.

The remainder of this paper is structured as follows: Section 2 describes the materi needed for the study, including the details of the FCSM, the strategy for selecting the obj of study (treatment group area), the basis for selecting variables, and the sources. Sectio also shows the construction process of the research method (DID model). Section 3 repc the results of the study and the corresponding analysis, mainly including an analysis of regression results of the DID model, an analysis of the social and economic value genera by the FCSM, and an extended discussion. Section 4 summarizes the previous findings a proposes policy recommendations.

2. Materials and Methods

2.1. Materials

2.1.1. Introduction to the Forestry Carbon Sink Mechanism

The forestry carbon sink is an important element in the process of "carbon peaki and "carbon neutrality". The forestry carbon sink refers to the market-based means participate in forestry resources trading, thus generating additional economic value (c bon sink). It can be divided into carbon sink afforestation, bamboo afforestation, for management, bamboo forest management and grassland carbon sink, according to t type of subject matter. Carbon sink forests are forests created and managed accordi to certain methodologies and technical standards, and common methods include: t Climate-Community-Biodiversity Standard (CCB standard), the Verified Carbon Standa (VCS), the Kyoto Protocol Clean Development Mechanism (CDM), etc. Therefore, c bon sink forests are different from ordinary plantation forests in that they ensure that project brings real, measurable and long-term carbon sequestration and environmen optimization effects for climate change mitigation [47].

The 1992 United Nations Framework Convention on Climate Change (UNFCCC) i basic framework for international cooperation to address global climate change and t cornerstone of forestry carbon sink implementation. The 1997 Kyoto Protocol further est lished three cooperative mechanisms for reducing greenhouse gas emissions: Internatio Emissions Trading (IET), the Joint Implementation Mechanism (JI) and the Clean Dev opment Mechanism (CDM). This was the prototype of the international carbon tradi market, and also laid the foundation for the circulation of forestry carbon sinks. In Chi the year 2011 is generally taken as the initial point for the rapid development of fores carbon sinks. During the period from 2011 to 2021, many policies related to forestry carb sinks were issued by relevant parts of China (see Table 1).

The above analysis shows that the FCSM is dependent on the ETS for its realization. other words, the Chinese carbon emission trading market is the key to the implementati of the FCSM. Therefore, policies that contribute to the integration of the FCSM into t carbon emission trading markets are potentially important. From Table 1, there are tv policies that have a turning point: the *Interim Measures for the Management of Volunta Greenhouse Gas Emission Reduction Trading*[1] in 2012 (FCSM2012), and the *Guidance of the St Forestry Administration on Promoting Forestry Carbon Sink Trading*[2] in 2014 (FCSM2014).

FCSM2012 aims to promote ecological civilization, facilitate innovation in institutio mechanisms and give full play to the decisive role of the market in the allocation resources for greenhouse gas emissions. FCSM2012 emphasizes that Chinese Certifi Emission Reduction (CCER) projects can participate in carbon trading after applicati and certification. In terms of allowance management, free allocation will be the ma focus initially, with paid allocation introduced in due course and the proportion of pa allocation gradually increased. For emission units that fail to fulfil their obligations to cle their allowances on time, the carbon trading authorities of the relevant regions will impc administrative penalties in accordance with the law. As a result, enterprises with excessi emission allowances will purchase forestry carbon sink transactions for offsetting.

Table 1. Policies related to the forestry carbon sink scheme in China.

Date	Policy Name	Release Department	Key Content
2011-10	Notice on the Pilot Project of Carbon Emission Trading	National Development and Reform Commission	Agreed to launch carbon emissions trading pilot in 6 regions.
2012-06	Interim Measures for the Management of Voluntary Greenhouse Gas Emission Reduction Trading	National Development and Reform Commission	Recorded and certified CCER projects can participate in carbon trading.
2014-04	Guidance of the State Forestry Administration on Promoting Forestry Carbon Sink Trading	State Forestry Administration	Improved and promoted CCER forestry carbon sink projects.
2018-01	Opinions on the Implementation of Rural Revitalization Strategy	State Council	Promoted exploring market-based compensation systems such as forest carbon sinks.
2018-05	Opinions on further liberalization of collective management rights	State Forestry and Grassland Administration	Promoted actively developing forest carbon sinks and further exploring market-based compensation systems such as forest carbon sinks.
2018-12	Action Plan for Establishing Market-oriented and Diversified Ecological Protection Compensation Mechanism	National Development and Reform Commission and other nine departments	Sound carbon offset mechanism were based on CCER; encouraged the development of forestry carbon sinks through carbon-neutral, carbon-inclusive and other forms of support.
2020-12	Interim Measures for the Management of Carbon Emissions Trading (for Trial Implementation)	Ministry of Ecology and Environment	Provisions on the proportion of carbon emissions that could be offset by forestry carbon sinks.
2021-03	Interim Regulations on the Management of Carbon Emission Trading (Draft Revision) (Draft for Comments)	Ministry of Ecology and Environment	Encouraged enterprises and institutions to carry out forestry carbon sink projects.
2021-04	Opinions on Establishing a Sound Mechanism for Realizing the Value of Ecological Products	State Council	Sound carbon emission trading mechanism.

FCSM2014 aims to accelerate the construction of ecological forestry and forestry for people's livelihoods, strives to increase forestry carbon sinks, and actively promotes forestry carbon sink trading. In terms of basic principles, FCSM2014 emphasizes the need to adhere to and improve the trading of forestry carbon sink projects for CDM. In terms of a carbon trading market, China's National Development and Reform Commission has identified Beijing, Tianjin, Shanghai, Chongqing, Hubei, Guangdong and Shenzhen as the national pilot regions for carbon emissions trading, requiring the pilot regions to study and consider relevant measures, including forestry, in light of the actual local situation. The FCSM2014 encourages research on forestry carbon sink laws and regulations, implementation schemes, trading models and regulatory systems. Overall, FCSM2014 brings forestry carbon sinks to the forefront on its own, and is a guiding policy to promote the development of forestry carbon sinks.

These two policies play a key role in the construction of the forestry carbon sink market. Therefore, this study adopts these two years (2012 and 2014) as the FCSM impact times of the DID model, respectively.

2.1.2. Treatment Group

Since the DID model was used to conduct this study, it was necessary to distingu treatment and control areas. Here, treatment areas refer to areas where the FCSM been effectively implemented; the remaining areas are control areas. Forestry carbon si projects exist in most provinces in China. Taking CCER as an example, by 2021 th will be more than 23 provinces in China with CCER carbon sink projects. Further tak CCER as an example, the incomplete 2021 statistics show that 23 provinces in China h implemented CCER carbon sink projects. However, the scale of forestry carbon sinks most provinces is small and cannot be considered as an effective impact due to their o forestry location factors (growing environment, plantable forest area, etc.). It is theref necessary to select representative areas from these regions as treatment group areas.

In this study, the larger the number of treatment group areas, the more difficul is to accurately capture the social and economic value generated by the FCSM; howe the smaller the number of treatment group areas, the more it affects the accuracy of DID model from a statistical perspective. In this case, it is common practice to select ar that are more affected by the FCSM as treatment group areas, according to the act situation. Because forest carbon sink projects can be implemented in any area in China, project sponsor will prioritize the benefits as well as the feasibility of the project. Theref factors such as a more suitable environment for carbon sink forestry and a larger a of forested land were the primary considerations for project sponsors. Given the act conditions in China, this study decided to select eight typical areas as treatment group ar including Heilongjiang, Jilin, Hubei, Zhejiang, Jiangxi, Fujian, Guangdong and Guan The remaining 22 provincial areas (excluding Hong Kong, Macau, Taiwan and Tibet) w taken as the control group. Figure 3 shows the distribution of the treatment group ar Table 2 shows the basic information on the treatment group areas, i.e., the reasons their selection.

Figure 3. The distribution of treatment areas.

Table 2. Basic information on treatment group areas.

Province	Total Area	Landscape Features	Climate Environment	Forestry Resources
Heilongjiang	473,000 km^2	Characterized by "five mountains, one water, one grass and three fields".	(1) Continental monsoon climate; (2) Precipitation: 400–650 mm, concentrated in summer.	(1) One of the largest forestry provinces in China; (2) Forested land: 232,451 km^2.
Jilin	187,400 km^2	The eastern mountains and the central and western plains.	(1) Temperate continental monsoon climate; (2) Precipitation: 400–600 mm, with 80% concentrated in summer.	(1) Asia's largest planted forest: Jingyuetan National Forest Park; (2) Forested land: 87,690 km^2.
Hubei	185,900 km^2	56% mountains, 24% hills, 20% plains and lakes.	(1) Subtropical monsoonal humid climate; (2) Precipitation: 860–2100 mm.	(1) Forestry land: 92,801 km^2.
Zhejiang	105,500 km^2	High in the southwest and low in the northeast, dominated by mountains.	(1) Monsoonal humid climate; (2) Precipitation: 1600 mm.	(1) Known as the "southeast plant treasure house"; (2) Forest land: 66,797 km^2.
Jiangxi	166,900 km^2	Mainly hilly and mountainous.	(1) Subtropical warm and humid monsoon climate; (2) Precipitation: 1341–1943 mm.	(1) With 139.66 million acres of red soil, suitable for plant growth. (2) Forest land: 106,667 km^2.
Fujian	124,000 km^2	90% mountainous and hilly areas.	(1) Subtropical maritime monsoon climate; (2) Precipitation: 1400–2000 mm.	(1) Highest forest cover in China; (2) Forest land: 76,667 km^2.
Guangdong	179,725 km^2	Mountains (33.7%), hills (24.9%), tablelands (14.2%), plains (21.7%).	(1) Central subtropical, southern subtropical and tropical climate; (2) Precipitation: 1300–2500 mm.	(1) Forestry land: 100,342 km^2; (2) Fertile soil.
Guangxi	237,600 km^2	Mainly mountainous, hilly, terrace and plain types of landforms.	(1) Subtropical monsoon climate zone and tropical monsoon climate; (2) Precipitation: 1653 mm.	(1) Forest land: 148,667 km^2, and the mountains account for 69.7%; (2) High-quality plant growing environment.

Source: information from provincial government websites, summarized by the authors.

2.1.3. Variables Design

The explained variables aim to reflect the social and economic value of the output from the use of forestry resources. Therefore, this study designed the explained variables based on two dimensions, social value and economic value, respectively.

As mentioned above, the social value of the FCSM is mainly in the form of greening, which, if implemented effectively, will contribute to the construction of carbon sink forests, thereby increasing the area of green vegetation in a sustainable and long-term manner. The increase in green vegetation will help to reduce carbon dioxide in the atmosphere and release oxygen. At the same time, the increase in green vegetation helps to restore ecosystems, stop desertification and mitigate soil erosion. This study uses forest coverage (100 km^2), forest cover (%), forest stock volume (10,000 m^3) and area of nature reserve (100 km^2), respectively, to reflect the increase in green vegetation, indirectly showing its social value.

1. Forest coverage (FC) includes the area of coniferous forests of natural origin and artificial origin, the area of broad-leaved forests, the area of mixed coniferous forests

and the area of bamboo forests, excluding the area of shrub woodlands and the area open woodlands. In addition, this indicator also does not include the area of gre land for residents and the area of road and bank protection woodland of railroa roads, rivers and ditches in the statistics. Since this indicator covers the subject of m forestry carbon sink projects, it can better reflect growth in the area of green vegetati

2. The forest coverage rate (FCR) is the ratio of forest area to total land area, and is important indicator of the actual level of forest resources and forest land occupati in a country (or region). Since its denominator is the total land area, the indicator w show significant changes only when the forest resources increase significantly.
3. Forest stock volume (FSV), also known as wood storage, is the total volume of standi timber in a forest. It is a measure of the abundance of forest resources in a country region. General forestry production can be harvested from mature forests. Althou most carbon sink forests cannot be harvested, China still classifies carbon sink fore as forest stock for statistical purposes.
4. Area of nature reserve (NR) is the area of a typical territory designated for the purp of protecting various important ecosystems and their environments, saving end gered species, and preserving natural historical heritage. Nature reserves have high-quality ecological environment and are very suitable for carbon sink proje Although natural forests in nature reserves do not include forestry carbon sinks, th superior geographical locations and ecological environments are suitable for fores carbon sink projects. However, at present, China does not allow forestry carbon si projects in nature reserves. It is worth mentioning that the construction of natu reserves and the establishment of forestry carbon sink projects may develop syner tically. Therefore, the aim of this study is to investigate whether the FCSM promo the construction of nature reserves.

According to the previous section, land management policies such as the FCSM c increase the economic output of forest land resources. The FCSM converts the amount CO_2 absorbed by forests into a share of the carbon sink, which is traded and profital via the carbon sink market. Although the economic output of forest land is reflect in the primary, secondary and tertiary industries, the base of these three is too large reflect the increase in output due to the FCSM. Therefore, this paper only uses the grc forestry product indicator (RMB million) to reflect the development of forestry from economic perspective.

In the literature that uses the DID model, the explanatory variable is an interacti term between a time dummy variable and an individual dummy variable. This explanatc variable captures whether a region is shocked by an event (policy) in a specific period. this study, it is expressed as whether region i has implemented an FCSM policy in year This study sets this dummy variable to $D_{it'}$. Details of the construction of $D_{it'}$ are shown Section 2.2. In addition, as the study needs to compare the regression results of differe explained variables, they are required to have common control variables. However, th requirement is clearly difficult to meet. At the same time, the spatial fixed effects varial (μ_i) used in this paper also functions as a control variable, controlling for the effe from provincial characteristics. Therefore, the inclusion of other control variables is n considered in this study.

2.1.4. Data Sources

The raw data of this paper are mainly from the *China Statistical Yearbook* (https://da cnki.net/, accessed on 15 October 2022.) and the *China Research Database Service Platfo* (https://www.ceads.net/, accessed on 15 October 2022.). Table 3 shows the results descriptive statistics for these raw data. Due to significant order-of-magnitude differenc between the data, their logarithmic form was used for the regression analysis.

Table 3. Descriptive statistics of raw data.

Variable	Units	Mean	Std. Dev.	Min	Max	Obs.
Forest coverage (FC)	100 km^2	685.4464	597.9148	1.89	2614.85	480
Forest coverage rate (FCR)	%	32.30269	18.3175	2.94	66.8	480
Forest stock volume (FSV)	10,000 m^3	40,185.84	49,560.69	33.24	197,265.8	480
Area of nature reserve (NR)	100 km^2	355.4916	558.4313	9	2183.49	453
Gross forestry product (GFP)	RMB million	14,600,000	18,100,000	16,987.5	96,800,000	419
$D_{i,2012}$	-	0.1500	0.3574	0	1	480
$D_{i,2014}$	-	0.4375	0.4966	0	1	480

2.2. Method: Difference-in-Difference Model

The difference-in-differences (DID) model is one of the more popular econometric models, and is often used in economics to assess whether there is a causal link between a shock generated by an event and an outcome. Furthermore, DID naturally alleviates the endogeneity problem of econometric models. The purpose of this paper is to examine the social and economic value of China's FCSM. Since FCSM implementations are event shocks, they meet the basic requirements of the DID model. The basic construction of the DID model used in this study is shown in Equation (1).

$$Y_{i,t} = \alpha + \beta * D_{it'} + \mu_i + \varepsilon_{i,t} \tag{1}$$

where, i denotes the serial number of a province ($1 \leq i \leq 30$), and t denotes the serial number of a year ($2005 \leq i \leq 2020$), t' is the time point of the event impact (2012 or 2014, Section 2.1.1 for detail). $Y_{i,t}$ are the set of explained variables. $D_{it'}$ is the key explanatory variable, whose coefficient β is our interest value. μ_i is a spatial fixed effects variable, also known as provincial fixed effects in this study. Because the 30 provincial areas each have their unique characteristics, it is necessary to use the variable μ_i to control for possible interference from these characteristics. $\varepsilon_{i,t}$ is the error term.

Assume that $D_{it'} = Treat_i \times Post_{t'}$. The $Treat_i$ term is the treatment group dummy variable. If i region belongs to the treatment group, then i = 1; otherwise, i = 0. The $Post_{t'}$ term is the time dummy variable. If t' is in and after the policy shock year, then $Post_{t'} = 1$; otherwise, $Post_{t'} = 0$. For $Post_{t'} = 1$ in this study, there are two situations, i.e., $t' \geq 2012$ and $t' \geq 2014$.

3. Results and Discussion

3.1. Results and Analysis of Baseline Regression

Introduction to the Forestry Carbon Sink Mechanism

Table 4 shows the regression results based on Equation (1). Among them, the $D_{i,2012}$ term is the result of using 2012 as the initial year of FCSM shock, whereas the $D_{i,2014}$ term is the result of using 2014 as the initial year of FCSM shock.

Table 4. Results of FCSM identifying forestry land use variables.

Explained Variables	ln (FC)	FCR	ln (FSV)	ln (NR)	ln (GFP)
$D_{i,2012}$	0.1142 ***	4.2681 ***	0.3202 ***	−0.0361	1.3507 ***
	(3.67)	(6.60)	(6.08)	(−1.00)	(13.44)
$D_{i,2014}$	0.2096 ***	4.7496 ***	0.3961 ***	−0.0360	1.2529 ***
	(7.19)	(8.44)	(8.10)	(−0.55)	(17.56)
Provincial fixed effect	YES	YES	YES	YES	YES
R^2 ($D_{i,2012}$)	0.2382	0.3995	0.2527	0.1055	0.3245
R^2 ($D_{i,2014}$)	0.3367	0.3803	0.4425	0.1055	0.5848

Note: t-value in parentheses; *** $p < 0.01$.

It can be found that the regression results with FC, FCR, FSV and GFP as the explai variables all passed the 99% confidence test, and the corresponding coefficients are posit The regression results for the explained variable of area of nature reserve did not pass 90% confidence level test. The above results indicate that the implementation of the FC significantly increased the forest area in treatment group areas, increased their forest co and forest resource, and increased their forestry income. Meanwhile, comparing differ FCSM shock years, it shows that the regression results with 2014 as the FCSM shock y were better than those of 2012, overall. The possible explanations are as follows: (I) ther a time lag due to the development of forestry carbon sink projects and the constructior carbon sink forests, which may be 1–3 years, according to the basic regulations of CC projects; (II) the FCSM policies of 2012 and 2014 complement each other and their pol effects are superimposed. The proof towards these possibilities will need to be develor in further research in the future.

When the explained variable is FC, the coefficient of the $D_{i,2012}$ term is 0.1142 a the *t*-value is 3.67; the coefficient of the $D_{i,2014}$ term is 0.2096 and the *t*-value is 7 Obviously, the effect of FCSM2014 on the corresponding regional forest coverage is mu more improved than that of FCSM2012. In other words, the eight treatment group ar had an overall enhancement of about 20.96% in forest coverage after the implementatior FCSM2014 compared to the control group areas.

When the explained variable is FCR, the coefficient of the $D_{i,2012}$ term is 4.2681 w a *t*-value of 6.60, and the coefficient of the $D_{i,2014}$ term is 4.7496 with a *t*-value of 8 Similarly to above, the stronger policy effect was generated by FCSM2014. This indica is similar to the indicator for FC, but it is a relative indicator, as it takes into account effect of the area of the region's own land area on the growth of the forest area. The resu show that the forest coverage rate in treatment group areas increased by 4.7496 percent points overall compared to control group areas after the implementation of FCSM2C which also indicates that the FCSM has a greater impact on forest areas and has been a to significantly increase the level of forest coverage rate.

When the explained variable is FSV, the coefficient of the $D_{i,2012}$ term is 0.3202 a the t-value is 6.08; the coefficient of the $D_{i,2014}$ term is 0.3961 and the *t*-value is 8.10. can be found that when 2014 was the impact year of the FCSM, the forest stock volu is elevated more than that of 2012, but the difference was limited. Of course, compar with the increase of the FCSM on forest coverage, its contribution to forest stock volu was greater, i.e., the treatment group areas had a greater increase of 39.61% in forest st volume than the other areas, as a whole.

When the explained variable is GFP, the coefficient of the $D_{i,2012}$ term is 1.3507 w a *t*-value of 13.44, and the coefficient of the $D_{i,2014}$ term is 1.2529 with a *t*-value of 17. Unlike the previous variables, the enhancement of the gross forestry product by FCSM2 was smaller than that of that of 2012, but the difference is not significant. It shows overall increase of 135.07% in gross forestry output values in the treatment group ar compared to other areas after the implementation of FCSM2012. One possible reason is t the capital inflow of forestry carbon sink investment after the FCSM2012 led to an incre in gross forestry product, but the growth of carbon sink forests and further output ga take a certain amount of time, thus leading to a relatively smaller corresponding coeffici in 2014.

3.2. Analysis of the Social and Economic Value Generated by the FCSM

The results of Section 3.1 argue for the social and economic value of the FCSM, i.e., creasing the area of green vegetation and improving the economic output of forestry la resources. This subsection will combine the mapped data to show the impact effect of FCSM more visually. The specific strategies are as follows.

After the implementation of the FCSM, it generated social and economic value. The fore, with the policy shock as the node, there should be significant differences in fores land use in the two sample intervals before and after. This study used maps to show these changes and agglomeration characteristics.

3.2.1. Social Value Generated by the FCSM

The social value generated by the FCSM is reflected in the construction of the greening, i.e., the increase in forestry coverage (FC), forestry coverage rate (FCR) and forest stock volume (FSV). Figure 4 shows the changes in FC and FCR between 2014 and 2020, respectively. The "Top ten", "Middle ten" and "Last ten" represent the range of positive change from high to low.

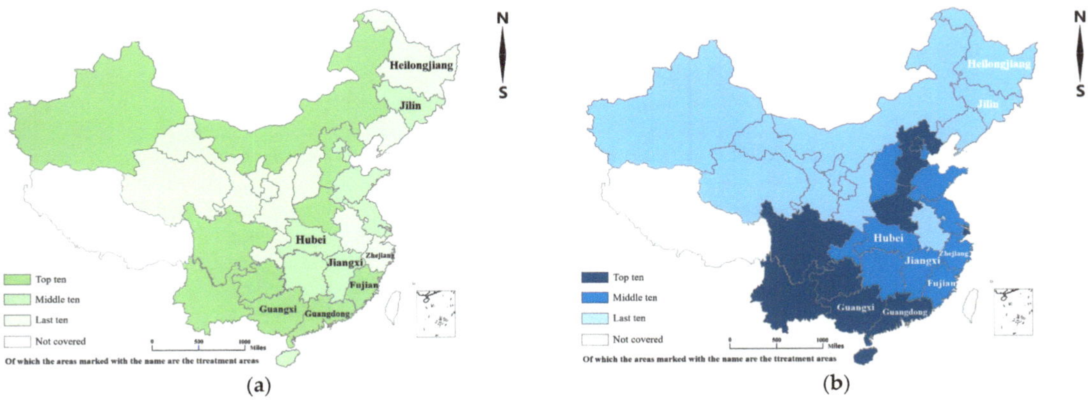

Figure 4. Distribution of forestry land use indicators. (**a**) Change of forestry coverage (FC); (**b**) Change of forestry coverage rate (FCR).

From Figure 4a, the "Top ten" regions are mainly located in southern and northwestern China, indicating that the FC in these regions has increased more after 2014. One important reason for the high growth of FC in southern China is the suitable climate and environment for plant growth; the large variation of FC in northwest China is related to its vast land area and abundant land resources, in addition to the climatic and environmental factors. However, among the eight treatment areas set in this paper, only Guangxi, Guangdong and Fujian are in the "Top ten". This indicates that the impact of FCSM policy on FC goes beyond the treatment group areas set in this paper, which is consistent with the previous section. Of course, this does not affect the conclusions of the econometric model, because the growth of FC in the remaining five areas is also above the average.

From Figure 4b, the "Top ten" regions are mainly located in southern China, indicating that the FCR in these regions has been increased to a higher extent after 2014. At the same time, Figure 4a,b show that the "Top ten" areas overlap to a greater extent, mainly in southern China. Combined with the analysis in the previous paragraph, it can be found that the increase of FC and FCR in southern China was higher after 2014, which indirectly reflects that the FCSM had a higher promotion effect on forest area in southern China than in northwestern China. An important reason for this is that northwestern China has a large volume of land and forest land, which provides the basis for a substantial increase in its forest area. This explanation also applies to Fujian, which has a larger increase in forest area, but has the highest forest cover in the country. Therefore, the increase in FCR in Fujian is limited and it is not a "Top ten" region in FCR.

Overall, the FC and FCR of 8 treatment areas improved more under the impact of FCSM, while some of the 22 control areas also gained significant positive improvem This is mainly concentrated in the southwestern part of China. This indicates that, on one hand, these areas are rich in forestry resources and suitable ecological environme on the other hand, there may be a spillover effect of the FCSM, which triggers the learn imitation behavior of neighboring regions. Finally, both Figure 4a and b show spa clustering characteristics, which indicates that the impact of the FCSM is closely relate the locational characteristics of regions.

From Figure 5, the "Top ten" areas are mainly located in southern and northeast China, indicating that the improvement of forest stock volume (FSV) in these areas is h after 2014. The reasons for the concentration in southern China are similar to the previc section. However, an important reason for the concentration in northeastern Chin that the FCSM does not yet involve nature reserves, while northeastern China has ma nature reserves with extremely rich forest stock volume. Among the "Top ten" areas, of them are treatment group areas, namely, Guangxi, Guangdong, Fujian, Jiangxi, Ji and Heilongjiang, which indicate that the FCSM is more prominent in improving F Combined with the findings in Figure 5, it can be concluded that the FCSM has a signific policy impact on the southern region of China. Of course, due to data limitations, this stu only analyzes and discusses the effect of the FCSM on forestry land use in the green dimension based on four aspects: forest coverage, forest coverage rate, forest stock volu and area of nature reserve. It is believed that when more abundant data are available, m detailed and richer studies can be conducted using the methods of this study.

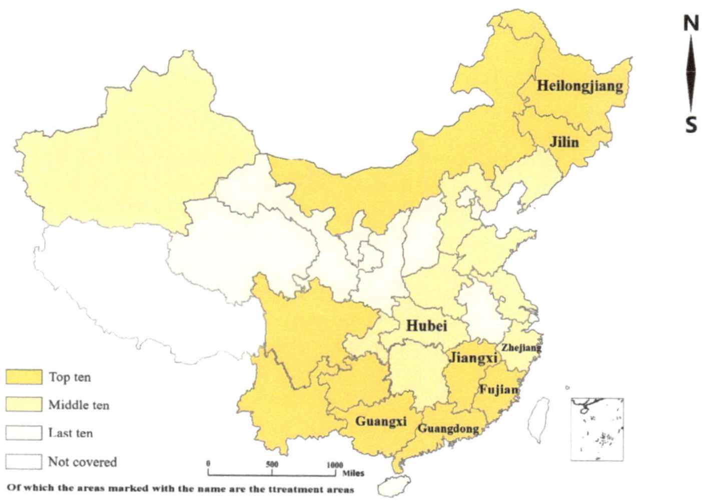

Figure 5. Change of forest stock volume (FSV).

3.2.2. Economic Value Generated by the FCSM

The economic value generated by the FCSM is reflected in the gross forestry producti (GFP). Figure 6 shows the overall change in GFP between 2014 and 2020. The "Top te "Middle ten" and "Last ten" represent the range of positive change from high to low.

As shown in Figure 6, the "Top ten" regions are clearly clustered in southern Chi indicating that the GFP growth in these regions was significant after 2014.

The possible reasons for the high change in GFP in the southern provinces are as follows: (1) since the currently opened carbon emission trading markets (Guangdong, Shenzhen, Fujian, Shanghai, Hube, and Chongqing) are mainly located in the south of China, it is easier for the neighboring provinces to learn about carbon emission trading and start related project activities; (2) many southern provinces have favorable ecological environments and richer land resources, and the plant growth cycle is significantly shorter than that in the north, which facilitates the successful implementation of forestry carbon sink projects; (3) the FCSM policy may have spillover effects. When treatment areas carry out forestry carbon sink activities and obtain objective benefits from them, their neighboring areas can be informed of the relevant information faster than other areas, so that they can imitate and learn from them.

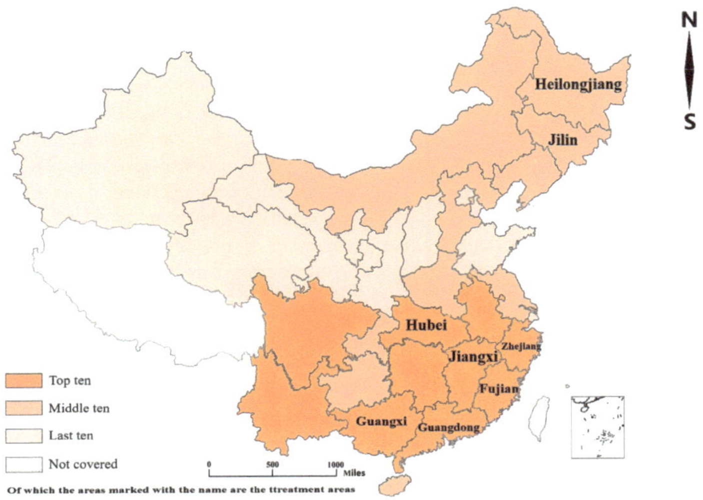

Figure 6. Change of gross forestry production (GFP).

Among the eight treatment areas in this paper, only Heilongjiang and Jilin in northeastern China are not among the "Top ten" areas, suggesting that FCSM policy has a significant policy impact on the southern region of China from the economic development dimension. Combined with the findings in Figure 4, it can be concluded that the FCSM has had a significant policy impact on the southern region of China. Of course, due to data limitations, this paper only uses a comprehensive variable, GFP, as a measure. It is believed that more interesting conclusions will be obtained in future studies using more microscopic data and richer indicators.

3.3. Discussion

The social and economic value of the FCSM as a land resource management tool has been discussed in this study. This subsection discusses the argumentation ideas and methods for the effectiveness of the FCSM, respectively. This section also analyses the shortcomings of the study and looks forward to future work.

3.3.1. Discussion on the Argumentation Ideas for the Effectiveness of the FCSM

The argument for the effectiveness of the FCSM is made by examining and comparing the social and economic value it generates. Kooten and Sohngen start from a cost–benefit perspective and show that not all forestry carbon sink projects are worthwhile. They find that the costs of sequestering carbon are some USD 3–280 per tCO_2, with variations from country to country. For example, in Canada and the USA, carbon sequestration costs range from a low of about USD 2 to nearly 80 per tCO_2 [39]. Bjørnstad and Skonhoft compare the carbon reduction effects of forests as a substitute for fossil fuels and as carbon sequestration tool, based on the substitutability of commodities. The former reduc fossil fuel consumption and indirectly reduces carbon emissions, whereas the latter direc absorbs CO_2 from the atmosphere and thereby reduces carbon emissions. In comparis the emission reduction effect of forests as a carbon sequestration tool is superior [4 This view is supported by Ma et al., who found that replacing fossil energy with for bioenergy was effective in reducing emissions in mature commercial forests where it alrea existed [42]. The above studies demonstrate the effectiveness of the FCSM in terms of co benefit ratio and commodity substitution. This paper argues for the effectiveness of FCS policies from the perspective of factual evidence. This study used provincial panel d from 2005 to 2020 to verify the social and economic value of the FCSM using a DID moc As initiating a forestry carbon sink project is a commercial activity, if the project did r bring positive benefits to the project initiator, then such a project is not said to have hac large-scale impact, i.e., the social and economic benefits mentioned in this paper. Therefc if the FCSM had a significant impact, then forestry carbon sink projects under the influer of the FCSM policy are worthwhile overall. In addition, Song and Peng point out th government subsidies, especially indirect subsidies to insurance companies, are benefic to the development of forestry carbon sink projects [45]. With the support of the FCS policy, China's forestry carbon sink projects have received support from local governmer which has also led to a reduction in their costs. Similarly, government support has al helped to strengthen investment in forestry fixed assets, which is important for developi forestry carbon sinks and increasing forestry GDP [46,48].

3.3.2. Discussion on the Argumentation Methods for the Effectiveness of the FCSM

The methods of examining the effectiveness of the FCSM are diverse. Lin and Ge us the SBM-DEA model to assess the role of carbon sink forests in mitigating the greenhou effect and increasing regional production [34]; Sun et al. used the differential game moc to derive the ecological utility (abatement mechanism) of forestry carbon sink projects [2 Tian et al. used structural dynamic methods to predict the role of carbon sink forests mitigating the greenhouse effect in the USA over the next century [49]; similarly, Kaipain et al. studied the issue using the CO_2FIX model [50]. In contrast to these studies, this stu is characterized by the use of an econometric model (DID model). This approach allevia the problem of pseudo-causality. For example, when assessing the impact of the FCS using methods such as the DEA model, the biggest struggle is the inability to identify range of impacts as originating from FCSM policies. Similarly, Shi et al. used the differen in-differences with variation in treatment timing (VTT-DID) model for their study. Th examined the pro-poor role and ecological value of the FCSM from an industrial structu provincial perspective, using county-level data from Sichuan Province as an example [4 However, their study took the implementation of forestry carbon sink projects as the no of DID, and there are significant differences between different forestry carbon sink proje Therefore, its method has some degree of shortcomings. This study was better able to avc this problem by taking the implementation of policy as the nodal point. It also avoided t problem that the findings of this study may only be applicable to one region by using provincial regions in China as the target population.

3.3.3. Research Shortcomings and Future Work

The shortcomings of this study are as follows: (1) This paper lists nine major FCS policies in China for the period 2011–2021. However, due to data length constrai (model requirements), the other seven FCSM policies are not examined in this paper. Th of course, does not affect the conclusions drawn in this paper. (2) In Section 2.1.2, t screening strategy for treatment group areas is not data dependent.

Accordingly, future work could be extended in two ways: (1) To use the methodolo of this paper to compare the impact of FCSM policies across countries. (2) To investiga various other FCSM policies in China, especially those that are not yet supported by sufficient data.

4. Conclusions and Recommendations

In attempting to explore the economic and social value generated by sustainable management for land resources, this paper uses the example of the FCSM in China. Using panel data for 30 provincial regions in China from 2005 to 2020, and with the help of a DID model, the paper addresses two questions: (1) in which year the most effective FCSM policies were implemented; (2) whether the FCSM generates economic and social value. The main marginal contributions of this paper are that (1) it enriches the study of FCSM policies by drawing on the methodology of economics; (2) it designs a strategy that can effectively identify the social and economic value generated by the FCSM, compared to existing studies, by drawing on the DID model.

The conclusions show that:

1. The FCSM policies for 2012 and 2014 are all in effect. However, overall, FCSM2014 has had a stronger effect than FCSM2012.
2. In terms of social value, the implementation of the FCSM has significantly increased forest coverage (FC), the forest coverage rate (FCR) and the forest stock volume (FSV) in the eight treatment group provinces. Among them, the largest increase in forest stock volume was observed. However, the implementation of the FCSM has not contributed to the construction of nature reserves (NR). On the one hand, this is due to the fact that natural forests in nature reserves are not the target of FCSM; on the other hand, it may be due to the fact that there are no relevant incentive policies and regulations, so that investors do not include nature reserves in their investment visits. These promotion effects are characterized by spatial agglomeration. The spatial agglomeration characteristics of FC, FCR and FSV differed when the FCSM generated policy shocks, but all of them had the characteristics of "southern agglomeration".
3. In terms of economic value, the implementation of the FCSM has significantly increased the gross forestry product of the eight treatment group provinces. This promotion effect has spatial agglomeration characteristics. When the FCSM produced a policy shock, the increase in gross forestry production (GFP) was higher in southern China than in other regions.

Based on the above, it can be concluded that the FCSM contributes to social and economic value. Therefore, this paper makes the following recommendations on how to optimize the effectiveness of the FCSM.

1. The FCSM makes forestry land resources sustainable in terms of social and economic value. However, the development of the forestry carbon sink business in China is lagging behind. This paper finds that the 2014 FCSM policy is the most efficient forestry carbon sink policy, which verifies this point. Therefore, the current scale of forestry carbon sinks in China is relatively small and the system is still not perfect. In this regard, this paper puts forward recommendations to expand the scale of forestry carbon sinks and improve the related systems in order to give full play to the effectiveness of the FCSM policy. (1) Increase the proportion of forestry carbon sinks to offset corporate carbon emissions. According to Section 2.1.1, the value of China's forestry carbon sinks is that they can offset the carbon emission amount of polluting industries, but this percentage is 5% or 10%, which cannot meet the demand of many enterprises to offset carbon emissions. Therefore, the initial number as well as the allocation ratio of allowances issued in national and local carbon markets can be improved to increase the proportion of forestry carbon sinks offsetting carbon allowances. This pathway can directly increase market demand for carbon sinks, thereby stimulating the establishment of forestry carbon sink projects. (2) Simplify the management process of forestry carbon sink projects. Forestry carbon sink projects are characterized by large areas, multiple elements, and complex measurement and monitoring methods, which constrain the efficiency of their development and implementation. In order for enter-

prises to actively participate in promoting forestry carbon sink projects, it is necessa to simplify the relevant processes and clear the project obstacles for participati entities, for example, through the use of innovative accounting methods, such as usi remote sensing, big data and other means. This pathway is an important eleme in improving the forestry carbon sink system, which can speed up and smooth o forestry carbon sink projects and lower the entry threshold. (3) Developing regiona differentiated FCSM policies. The FCSM has the most obvious positive impact on t southern regions of China, but there is variability in the impact on northern Chi (northeast and northwest). This is related to the locational factors in the northe region. Therefore, policymakers can better leverage and incentivize the impact of t FCSM on different regions by developing policies with regional differentiation.

2. The FCSM is unable to act on the expansion of nature reserves. At present, FCS policies are not allowed to be carried out in nature reserves, public welfare fore and other forest stands. However, it cannot be denied that the areas around natu reserves have rich carbon sink potential. In the future, we can study new developme strategies, state-sponsored development and other means to release the carbon si potential of related areas.

Author Contributions: Conceptualization, H.L. and T.H.; methodology, H.L.; software, T.H.; form analysis, H.L.; writing—original draft preparation, H.L. and T.H.; writing—review and editing, T All authors have read and agreed to the published version of the manuscript.

Funding: This research received no external funding.

Data Availability Statement: Not applicable.

Conflicts of Interest: The authors declare no conflict of interest.

Notes

1 Original information from: http://www.gov.cn/gongbao/content/2015/content_2818456.htm, accessed on 5 May 2022.

2 Original information from: http://www.forestry.gov.cn/sites/main/main/gov/content.jsp?TID=2088, accessed on 5 May 20

References

1. Foundational Primer on the 2030 Agenda for Sustainable Development. Available online: https://unsdg.un.org/SDGPrim (accessed on 26 March 2023).
2. Liu, L.; Chen, C.; Zhao, Y.; Zhao, E. China's carbon-emissions trading: Overview, challenges and future. *Renew. Sustain. Ener Rev.* **2015**, *49*, 254–266. [CrossRef]
3. Liang, W.; Yang, M. Urbanization, economic growth and environmental pollution: Evidence from China. *Sustain. Comput. Info Syst.* **2019**, *21*, 1–9. [CrossRef]
4. Tang, Y.; Wang, K.; Ji, X.; Xu, H.; Xiao, Y. Assessment and Spatial-Temporal Evolution Analysis of Urban Land Use Efficien under Green Development Orientation: Case of the Yangtze River Delta Urban Agglomerations. *Land* **2021**, *10*, 715. [CrossRe
5. Evangeliou, N.; Grythe, H.; Klimont, Z.; Heyes, C.; Eckhardt, S.; Lopez-Aparicio, S.; Stohl, A. Atmospheric transport is a ma pathway of microplastics to remote regions. *Nat. Commun.* **2020**, *11*, 3381. [CrossRef]
6. Liu, X.; Zou, B.; Feng, H.; Liu, N.; Zhang, H. Anthropogenic factors of PM2.5 distributions in China's major urban agglomeratio A spatial-temporal analysis. *J. Clean. Prod.* **2020**, *264*, 121709. [CrossRef]
7. Nizalov, D.; Thornsbury, S.; Loveridge, S.; Woods, M.; Zadorozhna, O. Security of property rights and transition in land u *J. Comp. Econ.* **2016**, *44*, 76–91. [CrossRef]
8. Desforges JP, W.; Galbraith, M.; Dangerfield, N.; Ross, P.S. Widespread distribution of microplastics in subsurface seawater in NE Pacific Ocean. *Mar. Pollut. Bull.* **2014**, *79*, 94–99. [CrossRef]
9. Ethan, C.J.; Mokoena, K.K.; Yu, Y.; Shale, K.; Fan, Y.; Rong, J.; Liu, F. Association between PM2.5 and mortality of stomach a colorectal cancer in Xi'an: A time-series study. *Environ. Sci. Pollut. R* **2020**, *27*, 22353–22363. [CrossRef]
10. The Intergovernmental Panel on Climate Change. Available online: https://www.ipcc.ch/sr15/ (accessed on 23 March 2023
11. Mikhaylov, A.; Moiseev, N.; Aleshin, K.; Burkhardt, T. Global climate change and greenhouse effect. *Entrep. Sustain. Issues* **20** *7*, 2897. [CrossRef]
12. Ritchie, H.; Roser, M.; Rosado, P. CO_2 and greenhouse gas emissions. *Our World Data* **2022**.
13. Wu, C.; Wei, Y.D.; Huang, X.; Chen, B. Economic transition, spatial development and urban land use efficiency in the Yang River Delta, China. *Habitat Int.* **2017**, *63*, 67–78. [CrossRef]

Deinlnger, K.; Binswanger, H. The evolution of the World Bank's land policy: Principles, experience, and future challenges. *World Bank Res. Obs.* **1999**, *14*, 247–276. [CrossRef]

Liu, Y.; Fang, F.; Li, Y. Key issues of land use in China and implications for policy making. *Land Use Policy* **2014**, *40*, 6–12. [CrossRef]

Ding, C. Land policy reform in China: Assessment and prospects. *Land Use Policy* **2003**, *20*, 109–120. [CrossRef]

van Meijl, H.; Van Rheenen, T.; Tabeau, A.; Eickhout, B. The impact of different policy environments on agricultural land use in Europe. *Agric. Ecosyst. Environ.* **2006**, *114*, 21–38. [CrossRef]

Tang, Y.; Yang, Y.; Xu, H. The Impact of China Carbon Emission Trading System on Land Use Transition: A Macroscopic Economic Perspective. *Land* **2022**, *11*, 41. [CrossRef]

2021 China Carbon Price Survey. Available online: http://www.tanpaifang.com/tanguwen/2022/0312/83391.html (accessed on 15 November 2022).

14th Five Year Plan for the Protection and Development of Forestry and Grassland. Available online: http://www.forestry.gov.cn/main/5461/20210819/091113145233764.html (accessed on 15 November 2022).

State and Trends of Carbon Pricing 2022. Available online: http://hdl.handle.net/10986/37455 (accessed on 15 November 2022).

Mitchell, C.D.; Harper, R.J.; Keenan, R.J. Current status and future prospects for carbon forestry in Australia. *Aust. For.* **2012**, *75*, 200–212. [CrossRef]

Lohila, A.; Minkkinen, K.; Aurela, M.; Tuovinen, J.P.; Penttilä, T.; Ojanen, P.; Laurila, T. Greenhouse gas flux measurements in a forestry-drained peatland indicate a large carbon sink. *Biogeosciences* **2011**, *8*, 3203–3218. [CrossRef]

Hommeltenberg, J.; Schmid, H.P.; Drösler, M.; Werle, P. Can a bog drained for forestry be a stronger carbon sink than a natural bog forest? *Biogeosciences* **2014**, *11*, 3477–3493. [CrossRef]

Zhang, C. Carbonate rock dissolution rates in different landuses and their carbon sink effect. *Chin. Sci. Bull.* **2011**, *56*, 3759–3765. [CrossRef]

Matyssek, R.; Wieser, G.; Ceulemans, R.; Rennenberg, H.; Pretzsch, H.; Haberer, K.; Löw, M.; Nunn, A.J.; Werner, H.; Wipfler, P.; et al. Enhanced ozone strongly reduces carbon sink strength of adult beech (*Fagus sylvatica*)–Resume from the free-air fumigation study at Kranzberg Forest. *Environ. Pollut.* **2010**, *158*, 2527–2532. [CrossRef] [PubMed]

Launiainen, S.; Katul, G.G.; Leppä, K.; Kolari, P.; Aslan, T.; Grönholm, T.; Korhonen, L.; Mammarella, I.; Vesala, T. Does growing atmospheric CO2 explain increasing carbon sink in a boreal coniferous forest? *Glob. Change Biol.* **2022**, *28*, 2910–2929. [CrossRef] [PubMed]

Zhou, T.; Shi, P.; Jia, G.; Dai, Y.; Zhao, X.; Shangguan, W.; Du, L.; Wu, H.; Luo, Y. Age-dependent forest carbon sink: Estimation via inverse modeling. *J. Geophys. Res. Biogeosciences* **2015**, *120*, 2473–2492. [CrossRef]

Sun, R.; He, D.; Yan, J.; Tao, L. Mechanism Analysis of Applying Blockchain Technology to Forestry Carbon Sink Projects Based on the Differential Game Model. *Sustainability* **2021**, *13*, 11697. [CrossRef]

Pukkala, T. Does management improve the carbon balance of forestry? *For. Int. J. For. Res.* **2017**, *90*, 125–135. [CrossRef]

Li, X.; Hu, W.; Zhang, F.; Zhang, J.; Sheng, F.; Xu, X. Carbon Sink Cost and Influence Factors Analysis in a National Afforestation Project under Different Investment Modes. *Int. J. Environ. Res. Public Health* **2022**, *19*, 7738. [CrossRef]

Yin, S.; Gong, Z.; Gu, L.; Deng, Y.; Niu, Y. Driving forces of the efficiency of forest carbon sequestration production: Spatial panel data from the national forest inventory in China. *J. Clean. Prod.* **2022**, *330*, 129776. [CrossRef]

He, X.; Chen, L.; Huang, Y. A Study of Forest Carbon Sink Increment from the Perspective of Efficiency Evaluation Based on an Inverse DEA Model. *Forests* **2022**, *13*, 1563. [CrossRef]

Lin, B.; Ge, J. Carbon sinks and output of China's forestry sector: An ecological economic development perspective. *Sci. Total Environ.* **2019**, *655*, 1169–1180. [CrossRef]

Keren, C.; Qiyuan, C.; Nan, Z.; Yinan, L.; Changyong, L.; Yuanfei, L. Forest carbon sink evaluation—An important contribution for carbon neutrality. *IOP Conf. Ser. Earth Environ. Sci.* **2021**, *811*, 012009.

Shi, X.; Wang, T.; Lu, S.; Chen, K.; He, D.; Xu, Z. Evaluation of China's forest carbon sink service value. *Environ. Sci. Pollut. Res.* **2022**, *29*, 44668–44677. [CrossRef] [PubMed]

Wei, J.; Shen, M. Analysis of the Efficiency of Forest Carbon Sinks and Its Influencing Factors—Evidence from China. *Sustainability* **2022**, *14*, 11155. [CrossRef]

Montagnini, F.; Porras, C. Evaluating the role of plantations as carbon sinks: An example of an integrative approach from the humid tropics. *Environ. Manag.* **1998**, *22*, 459–470. [CrossRef] [PubMed]

Van Kooten, G.C.; Sohngen, B. Economics of forest ecosystem carbon sinks: A review. *AgEcon Search* **2007**. *submitted*. [CrossRef]

Bjørnstad, E.; Skonhoft, A. Wood fuel or carbon sink? Aspects of forestry in the climate question. *Environ. Resour. Econ.* **2002**, *23*, 447–465. [CrossRef]

Shi, Y.; Guan, Y.; Li, L.; Huang, M. Empirical analysis on the impacts of carbon sink afforestation project on county industrial structural upgrading. *Alex. Eng. J.* **2022**, *61*, 207–216. [CrossRef]

Ma, X.; Wu, L.; Zhu, Y.; Wu, J.; Qin, Y. Simulation of Vegetation Carbon Sink of Arbor Forest and Carbon Mitigation of Forestry Bioenergy in China. *Int. J. Environ. Res. Public Health* **2022**, *19*, 13507. [CrossRef]

Pugh, T.A.; Arneth, A.; Kautz, M.; Poulter, B.; Smith, B. Important role of forest disturbances in the global biomass turnover and carbon sinks. *Nat. Geosci.* **2019**, *12*, 730–735. [CrossRef]

44. Duque, A.; Peña, M.A.; Cuesta, F.; González-Caro, S.; Kennedy, P.; Phillips, O.L.; Calderón-Loor, M.; Blundo, C.; Carilla, Cayola, L.; et al. Mature Andean forests as globally important carbon sinks and future carbon refuges. *Nat. Commun.* **20** *12*, 2138. [CrossRef]
45. Song, Y.; Peng, H. Strategies of Forestry Carbon Sink under Forest Insurance and Subsidies. *Sustainability* **2019**, *11*, 4607. [CrossR
46. Zhang, Y.; Fang, L.; Jing, P. Analysis of broad leaved forest carbon sinks changes and forest economics and management in Chi *Environ. Sci. Pollut. Res.* **2020**, *27*, 12922–12931. [CrossRef] [PubMed]
47. Carbon Trading Network. Available online: http://www.tanjiaoyi.com/article-2542-1.html (accessed on 11 November 2022)
48. Zhang, Y.; Song, W.; Irland, L.C. Forest carbon stocks change and silvicultural investment in China–the case of coniferous fore *Int. J. Environ. Stud.* **2015**, *72*, 166–178. [CrossRef]
49. Tian, X.; Sohngen, B.; Baker, J.; Ohrel, S.; Fawcett, A.A. Will US forests continue to be a carbon sink? *Land Econ.* **2018**, *94*, 97–1 [CrossRef] [PubMed]
50. Kaipainen, T.; Liski, J.; Pussinen, A.; Karjalainen, T. Managing carbon sinks by changing rotation length in European fore *Environ. Sci. Policy* **2004**, *7*, 205–219. [CrossRef]

land

ticle

'he Distributional Effects Associated with Land Finance in :hina: A Perspective Based on the Urban–Rural Income Gap

xing Wang [1] and Meirong Zhang [2,*]

[1] Information and Management School, Guilin Shanshui Vocational College, Guilin 541199, China
[2] Department of Public Finance and Taxation, School of Finance and Economics, Jimei University, Xiamen 361021, China
* Correspondence: mrzhang@jmu.edu.cn

Abstract: Land finance has become an important way of generating fiscal incomes in developing countries, while the urban–rural income gap (URIG) in developing countries remains high. However, existing research has not paid much attention to the connection between land finance and the URIG. Therefore, this study used a fixed-effects model to test this relationship for 275 prefecture-level cities in China from 2014 to 2017. To identify the effects of the potential omitted variables, this study conducted additional robustness checks using placebo tests. The results showed that land finance significantly widened the URIG, and this finding was maintained after a set of tests. Further study found that the effect of land finance on the URIG showed significant heterogeneity. Land grants by tender, listing, and auction significantly widened the URIG, while land grants by agreement did not affect the URIG; the effect of land finance on the URIG was more significant in Eastern and Middle regions, but not marked in Western regions; and land finance had no impact on the URIG in large and medium-sized cities, while it had a significant impact in small cities. Based on the above results, this study offers recommendations to improve land fiscal policy and urban-biased development strategies, which aim to promote the equalization of the basic rights and interests of urban and rural residents and reduce the URIG.

Keywords: land finance; income distribution; urban–rural income gap; urban preference

ation: Wang, Z.; Zhang, M. The tributional Effects Associated h Land Finance in China: A spective Based on the an–Rural Income Gap. *Land* **2023**, 1771. https://doi.org/10.3390/ d12091771

ademic Editor: Matjaž Glavan

eived: 26 July 2023
ised: 3 September 2023
epted: 5 September 2023
lished: 13 September 2023

1. Introduction

With the rapid progress of global urbanization, it is a remarkable fiscal phenomenon to use land assets to raise extra-budgetary funds for the authorities in many developing countries [1]. Some examples include China, India, Viet Nam, and sub-Saharan Africa [2–5]. Land grant incomes and land taxes have turned into one of the main sources of income for local authorities, and this revenue strategy is often referred to as "land finance" [6]. The phenomenon of land finance is typical in China as the largest developing country. Although there are differences between other developing countries regarding the system and the institutional context in which land finance is established, there are many commonalities among them [2]. There is no doubt that the experience and exploration of land finance in China provide a valuable lesson for other developing countries.

Since the reform of the marketization of land transactions in 2004, land finance has been growing rapidly at an average annual rate of 20.7% over the past ten years, based on the Data of the Ministry of Finance [7]. In 2021, the scale of land finance reached USD 1.35 trillion, which accounted for 78% of local fiscal incomes. Land finance plays an essential role in regional socioeconomic development. On one hand, land finance increases the local authorities' income, accelerates the process of regional urbanization and industrialization [2,8], promotes the rapid development of infrastructure [9], and improves the regional economy [10]. On the other hand, the negative effects produced by land finance are also increasingly obvious. For example, it boosts the rapid growth of property prices [11], exacerbates the risk of regional debts [12], induces illegal land use [13], bree corruption [14], and increases macroeconomic fluctuations, etc. [15].

A high dependence on land finance has kept income gaps at a high level for a lo time in China. In the 21st century, the Gini coefficient in China has remained above the hi level of 0.4 [16]. One study found that excessive income inequality may be harmful to soc stability, hinder economic development, and weaken residents' sense of well-being [1 Income inequality in China mainly stems from the URIG. For example, based on War estimation, the URIG in China accounts for 70–80% of the total income gap [18]. In a la study, Chen et al. concluded that this ratio was 58% [19]. Figure 1 indicates that the p capita disposable income ratio of urban and rural residents stayed above 2.5 from 2013 2021, although it slightly decreased. Considering the welfare difference between urban a rural residents, this gap will continue to increase [20]. The huge URIG has also attract the criticism that "the welfare of urban areas is built on the backs of farmers" [21].

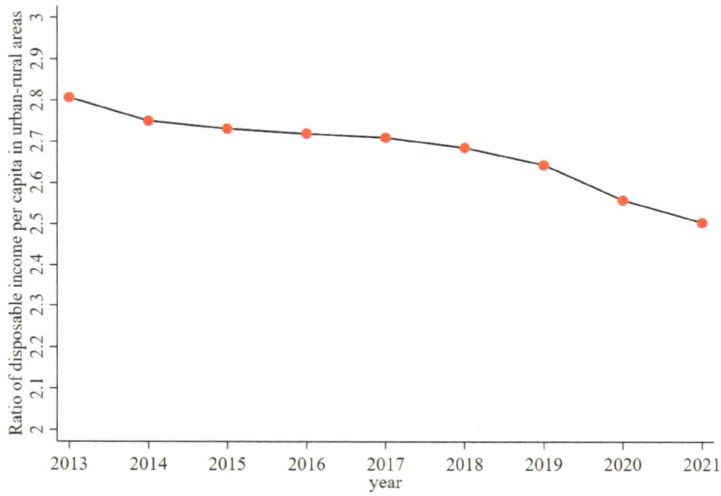

Figure 1. Ratio of disposable income per capita in urban–rural areas from 2013 to 2021.

The urban preference theory is often used to explain the URIG in developing countri which attributes its cause to policies that favor the development of the urban sector [2 Yang's study showed that urban preference policies, which include urban–rural lab market segregation, price control of agricultural commodities, discriminatory social be efits, and differentiated security systems, widened the URIG [23]. Sicular et al. fou that the household registration system explained more than 50% of the middle-body come gap [24]. The household registration system distorts the labor market and hel urban residents obtain greater job opportunities with higher labor remuneration [25,2 Moreover, the URIG, induced by a series of urban-preferred fiscal policies, including frastructure investment, financial development, and education budgets, has been wide discussed [27–29]. Although the existing studies have analyzed the relationship betwe urban preferences, fiscal systems, and the URIG, there are few studies on the impact land finance on the URIG. Land finance is dominated by the government with an obvio urban preference, and the government is probably contributing to the unequal distributi of factors between urban and rural areas through its intervention in land policy, which turn affects income distribution [30]. Therefore, the relationship between land finance a the URIG needs to be further analyzed.

Based on the above background, this study analyzed the relationship between la finance and the URIG and tested this relationship using a fixed-effects model for 2 prefecture-level cities in China over the period 2014–2017, as well as an additional robu ness test using a placebo test in order to identify the effects caused by the potential omitted variables. The results indicated that land finance significantly widened the urban–rural income gap, and the conclusion still stood after a set of tests. The results showed that land finance significantly widened the URIG. Further study found that the effect of land finance on the URIG showed significant heterogeneity. Land grants by tender, listing, and auction significantly widened the URIG, while land grants by agreement did not affect the URIG; the effect of land finance on the URIG was more significant in Eastern and Middle regions, but not in Western regions; and land finance had no impact on the URIG in large and medium-sized cities, while it had a significant impact in small cities.

The potential contributions of this study are as follows. First, it enriches the literature on the socioeconomic consequences of land finance, expanding the framework of land finance-related studies. Second, it integrates land finance and the URIG into a unified analytical framework, exploring the causes and mechanisms of the continuing widening of the URIG, and providing a new explanation for the understanding of China's URIG. Finally, it examines in detail the effects of land finance on the URIG, which deepens the understanding of the relationship between land finance and the URIG, proving that the exploration of policies can effectively promote the reform of the land system and reduce the URIG. This study also provides an empirical reference for other developing countries.

The remainder of the study is structured as follows. Section 2 describes the institutional background and theory analysis. Section 3 constructs the econometric regression model and describes the measurement of the date and variables. Section 4 shows the results of the empirical analysis. Section 5 discusses the implications of the empirical evidence and points out the shortcomings. Section 6 draws the conclusions and proposes relevant policy recommendations.

2. Institutional Background and Theoretical Analysis

2.1. Institutional Background

2.1.1. Land Finance in China

In the 1994 tax-sharing reform, taxes were classified into three categories: central taxes, local taxes, and shared taxes. While the largest of these, VAT (value added tax), was shared between the central and local governments at the ratio of 75% to 25%, the consumption tax of enterprises was included in the central tax. The income tax reform implemented in 2002 even changed the corporate income tax and the personal income tax from a local tax to a shared tax, which was shared 50%:50% between the central government and the local government. Since 2003, it has shifted to 60% to the central government and 40% to the local governments [31]. In 2016, China fully implemented the business tax compared to VAT and further reduced local government income by reducing VAT and double taxation [32]. However, the cost of public services and infrastructure borne by the localities was not reduced at the same time [6]. Local governments must find new ways of generating income to make up for the fiscal gap. The reform of the tax system also has a significant impact on the extra-budgetary and non-budgetary funds of local governments. In particular, agricultural coordination and land transfer income related to land development are not in the budget. There is no effective management system after tax reform, and the land transfer income is fully owned by the local government. Therefore, non-budgetary funds, especially land transfer income, have become the main source of financial growth for the local government [31].

In 2007, land transfer income and expenditure were fully incorporated into the budget management of local government funds. Compared to the general budget, it obviously reflected "weak constraints". The management of land grant revenues is still irregular, and expenditures are not transparent. Meanwhile, government officials, under the pressures of promotion and performance appraisal, must not only maintain a high GDP growth rate but also achieve higher rankings than the other candidates. As a result, they tend to adopt approaches that are inconsistent with the central government's policies and increase urban infrastructure investment through land finance to promote regional economic growth [32,33]. In addition, after the global financial crisis in 2008, local governments hanced the role of local government financing platforms (LGFPs) to support infrastruct development [34] and used future land grant income as an invisible guarantee for d financing. The Ministry of Finance announced in 2009 that "local governments are allow to finance investment projects from essentially all sources of funds, including land inco budgetary income, and funds borrowed from local financing instruments," which a provided policy support for future invisible guarantees of land transfer income [35]. 2014, to curb the risky rise of local financing platforms, the Chinese government amenc its budget law to allow "provincial governments to raise capital for local investmen part by issuing local government bonds within the limits prescribed by the State Counc The Ministry of Finance (MOF) issued guidelines for local government bonds in 2C which allowed local governments issuing special bonds to use cash inflows from fut infrastructure projects and the government's future land sale income to repay the bon which also provided a formal legal link between local government land financing and d financing [2], while strengthening the local government's reliance on land-based finan to maintain land finance at a high level.

2.1.2. Land Acquisition Strategies in China

China has adopted a dualistic land system divided into urban and rural lands, wh urban land is owned by the state, and rural land is owned collectively by peasants. The La Management Law (unrevised until 2020) provides that the state may acquire collect land in the "collective interest" by law [36]. In practice, most urban land needs to acquired by the local government. Only after the change of ownership to state-owr land can the land be used for industrial, commercial, or infrastructure construction. Th mechanisms strengthened the right of the government to manage and control land resour and achieve the monopolistic position of the government in the supply of urban land [There are two ways for the government to obtain newly developed urban land. C is to transform the land owned by rural collectives into new developable urban la whereby the government exchanges its rural land ownership by paying compensati to rural collectives and farmers. The second is to obtain the right to use urban la through compensation [2]. However, in terms of compensation standards based on Land Management Law, land expropriation only reflects the value of the land bef expropriation and does not take into account the value of the land after expropriation [For example, based on the Land Management Law, compensation for expropriation arable land consists of three components: land compensation (the monetary value the average value of agricultural output over the past three years); compensation resettlement (4–6 times the productivity of the derived land); and compensation for ancill assets of the land [36]. The Ministry of Land and Resources (MLR) further stipulates t the maximum compensation for land expropriation cannot exceed 30 times the deriv land's productivity. While the central government requires localities to dynamically upd compensation standards to match the rate of economic growth, only a few provinces raising compensation standards as recommended. The vast difference in the cost of la expropriation has led local governments to prefer expropriating rural collective land, wh is cheaper. Local governments made huge profits by buying land at low prices and selli it at high prices. This mechanism is an important reason why local governments are able implement land finance as well.

2.1.3. Local Government Income–Expenditure Structure of Land Concessions

Under the current financial framework of China, local governments have significa autonomy over the use of land concessions, and this autonomy of local governments has been significantly limited, even though the full amount of income and expenditures fr land concessions was incorporated into the budgetary management of local governm funds in 2007. In 2008–2010, the average portion of land grant proceeds spent on cit was 69.56%, and the average portion spent on rural areas was 9.52% [38]. In 2010–20 the cost expenditures for land acquisition compensation, and subsidies for expropriated farmers decreased from 80% to 20%, and the proportion of expenditures for rural development was smaller and has been decreasing constantly, creating a widening gap between expenditures on rural and urban development [30]. Overall, urban residents gain more than rural residents in land grant income expenditures, which highlights the urban bias in these expenditures.

2.2. Theoretical Analysis

Under government domination, the land finance development strategy has an obvious urban preference in terms of income and expenditure, which will inevitably have a negative impact on urban–rural income distribution. On the one hand, in the land expropriation procedure, collective landowners (rural residents) do not have the ability to negotiate directly with urban land users, nor can they directly transfer land ownership to urban land users, and the amount of compensation for land transfer is entirely unilaterally decided by the local authorities [36]. This directly leads to land compensation for farmers being far lower than the market price of the land [30]. In the meantime, urban residents are receiving higher compensation in the process of land expropriation. Local authorities have gained funds for urban development, promoted urban expansion, and increased the income of urban residents by selling expropriated collective land at high prices [39,40]. This huge gap between low compensation prices and high land premiums can widen the URIG [41].

On the other hand, the main source of the local authority's tax income has changed from an enterprise tax to a business tax, which is mainly collected from tertiary industries and the construction industry. With the construction industry being the main target of the business tax, it is logical that the local authorities allocate land grant income to the urban areas where the secondary and tertiary industries are concentrated [31]. This strategy of urban-oriented development significantly increases the provision of infrastructure and public services in urban areas [42]. Improvements in public services and infrastructure enhance the value of housing for urban residents, which increases the wealth of urban residents and increases their personal incomes through higher rates of return on capital [21,30]. However, rural residents are unable to trade their own houses and land freely due to institutional constraints, thus making it difficult for them to access the benefits of rising property prices, which in turn raises the cost of urban labor [43] and inevitably widens the URIG.

In addition, land is not only a factor in agricultural production, but it is also the basis of farmers' survival [44]. The expropriation of rural land means that rural residents will no longer own their production factors or housing, resulting in many rural residents migrating between urban and rural areas or choosing to settle in urban areas. This not only reduces the productivity of agriculture but also inhibits an increase in the income of rural residents who work at home [45]. At the same time, due to the poor education and skill levels of urban migrant workers, they are at an obvious disadvantage in the job market. Therefore, it is difficult for them to find employment, and they can only take up low-paid and high-risk jobs with insufficient labor protection, which results in a lower growth in wages than that of urban residents [36,46]. Moreover, rural residents who choose to settle in urban areas must often pay a mortgage from their labor income [47], and the rising prices of urban housing exacerbate this function, which is also not conducive to their employment choices and wage income [48].

As a result of the above analysis, this study puts forward the following hypothesis:

The dependence of local authorities on land finance will significantly widen the URIG.

3. Model, Variable and Data

To identify the relationship between land finance and the URIG and taking into account the type of data and the features of the variables, this study adopted the fixed-effects model, which is commonly used in economics and management to conduct such a test. The remainder of this section is arranged as follows. It first introduces the fixed-effects mo then describes the construction of the relevant variables, and finally explains the sou and treatment of the data.

3.1. Fixed-Effects Model

The fixed-effects model is a panel data analysis method that is widely used in fields of economics, sociology, and management by introducing dummy variables in ordinary least squares regression model or using within-group de-meanings to prev the endogeneity problem caused by omitted variables [49,50]. Usually, omitted variab affect the explanatory variables as well as the explained variables, leading to inaccur estimation coefficients of the explanatory variables. Therefore, it is necessary to add m control variables. However, some of the control variables are unobservable, whereas fixed-effects model can control the unobservable non-time-varying only-individual-vary omitted variables and the non-individual-varying only-time-varying omitted variables. T advantage is that it can largely mitigate the endogeneity problem caused by specific omit variables, as well as eliminate inter-individual heterogeneity and improve the estimati efficiency and accuracy. The disadvantages are that it is unable to estimate the impact eff of the observable variables that do not vary over time, it cannot recognize inter-individ variability, and it may miss important differences between the individuals [51]. Taking i account the data features and variable features of this study, the fixed-effects model w used to test the relationship between land finance and the URIG. The specific form of fixed-effects model is shown in Equation (1).

$$IG_{it} = \alpha_0 + \alpha_1 land\ fin_{it} + \alpha_2 X_{it} + \theta_i + \rho_t + u_{it},$$

where IG indicates the URIG; $land\ fin$ indicates land finance; X indicates a set of cont variables; θ indicates the city fixed effects used to control the omitted variables that va only with the city and not with time, such as a city's geographic location; ρ indicates time fixed effects used to control the omitted variables that vary only with time and n with cities, such as macro-level covariation; u indicates a random error term; and i an indicate the city and year, respectively. The focus of this study was on the core index of which should be positive if the hypothesis is true. The coefficient estimation was carri out using the Stata 15.1 software.

3.2. Variable Description

3.2.1. Explained Variables

There are two main indicators commonly used to measure the URIG. One is the ra of disposable income of urban and rural residents as a direct measure of the URIG [30, The second is the Thiel index [20,52]. Compared to the Thiel index, the ratio of disposa income of urban and rural residents is more intuitive due to it being simple to meas However, it ignores the proportion of urban and rural populations and fails to reflect changes in mobility between urban and rural populations [20]. Therefore, we used t Thiel index, which better reflected the flow of urban and rural populations, to measure URIG. Its formula is shown in Equation (2). In addition, the ratio of disposable income urban and rural residents was used for the additional robustness test.

$$TL_{it} = \sum_{j=1}^{2} \left(\frac{P_{ijt}}{P_{it}} \right) \ln \left(\frac{P_{ijt}}{P_{it}} \Big/ \frac{Z_{ijt}}{Z_{it}} \right)$$

where TL_{it} indicates the Thiel index of city i in year t; $j = 1,2$ indicates the urban a rural areas, respectively; P_{ijt} indicates the per capita disposable income of city i in yea P_{it} indicates the per capita disposable income of city i in year t; Z_{ijt} indicates the number urban and rural resident populations of city i in year t; and Z_{it} indicates the total reside population of city i in year t. The larger the Thiel index, the larger the URIG.

3.2.2. Main Explanatory Variables

Land finance: Land finance income primarily includes land grant, land-related tax income, and land mortgage, where the land grant counts as the main part of land finance [53]. Therefore, this study took the ratio of land grant income to public finance income as a proxy variable for land finance. Moreover, this study constructed the ratio of land grant income to GDP and per capita land grant income to measure land finance for the robustness test.

3.2.3. Control Variables

To avoid the possible omitted variables, the following control variables were selected in this study. The level of economic development (econ): measured the level of economic development by using GDP per capita with natural logarithmic treatment; industrial structure (indus): measured the industrial structure by using the ratio of the gross domestic product of the secondary and tertiary industries to the regional GDP; financial development (fina): measured the financial development by using the ratio of the balance of deposits and loans of financial institutions to the GDP at the end of the year; the level of opening up: measured opening by using the ratio of the total amount of regional imports and exports to the regional gross domestic product; authorities' intervention (gove): measured the authorities' intervention by using the ratio of public fiscal expenditure to the regional GDP; education level (educ): measured the education level by using the number of students enrolled in general higher education institutions per 100,000 people with a natural logarithmic treatment. Since some cities did not have any higher education institutions, one was added to the original data before taking the logarithmic value.

3.3. Data Sources

The data used in this study were mainly from the *China Land and Resources Statistical Yearbook* and *China Urban Statistical Yearbook*, and some of the missing data were supplemented by local statistical yearbooks or statistical bulletins. In 2013, the National Bureau of Statistics launched a survey on the income, expenditure, and living conditions of urban and rural residents, and the survey methodology and caliber of urban and rural incomes were changed. Thus, the data after 2013 were not comparable with the previous data, and the *China Statistical Yearbook of Land and Resources* had only counted the data up to 2017. Therefore, this study selected the sample interval as 2014–2017. In addition, to avoid the influence of the price factor, the variables in the nominal value of money statistics were deflated using the GDP deflator and using 2003 for the base period. The descriptive statistics for each variable are shown in Table 1.

Table 1. Descriptive statistics.

Variable	Obs.	Min.	Max.	Mean	Std. Dev.
TL	1100	0.0045	0.252	0.074	0.0373
$land\ fin$	1100	0.0086	3.141	0.494	0.361
econ	1100	8.737	11.78	10.21	0.512
indus	1100	0.495	0.997	0.878	0.077
fina	1100	0.717	22.32	2.485	1.369
open	1100	0	10.61	0.178	0.47
gove	1100	0.0439	2.06	0.207	0.127
educ	1100	0	4.752	2.471	0.928

4. Results

4.1. Baseline Regression Results

The results of the base regression results of this study are shown in Table 2. Where column (1) only controlled for the time fixed effects and city fixed effects, the regression coefficient of land finance was 0.0024, and it was significantly positive at the 5% level. Column (2) added two control variables, which were the economic development level and industrial structure, and the results showed that the regression coefficient of land finance was 0.0028, and it was significantly positive at the 1% level. Column (3) added the finan development level and opening-up level based on column (2), and it was shown that regression coefficient and significance level of land finance were not significantly chang compared to column (2). Column (4) added authorities' intervention and education le to column (3), and Table 2 shows that the regression coefficients and significance levels land finance did not change significantly. According to the above results, land finance v indeed an important reason for the widening of the URIG in China, which initially verif the hypothesis of this study.

Table 2. Base regression results.

Variable	**(1) TL**	**(2) TL**	**(3) TL**	**(4) TL**
land fin	0.0024 ** (2.3654)	0.0028 *** (2.7604)	0.0027 *** (2.7272)	0.0028 *** (2.7960)
econ		−0.0161 *** (−3.0017)	−0.0165 *** (−2.8612)	−0.0158 *** (−2.8486)
indus		−0.0154 (−0.5697)	−0.0151 (−0.5611)	−0.0153 (−0.5617)
fina			−0.0002 (−0.6717)	−0.0004 (−1.2928)
open			0.0001 (0.3308)	0.0001 (0.3527)
gove				0.0047 (1.4958)
educ				−0.0015 (−1.6437)
_cons	0.0784 *** (121.2969)	0.2550 *** (4.4574)	0.2595 *** (4.2166)	0.2551 *** (4.2378)
Year fixed	Yes	Yes	Yes	Yes
City fixed	Yes	Yes	Yes	Yes
N	1100	1100	1100	1100
Adj. R2	0.289	0.325	0.324	0.326
F	82.0712	53.8820	40.4140	32.9928

Note: *t* statistics in parentheses (cluster in cities level). *** $p < 0.01$, ** $p < 0.05$.

4.2. Robustness Tests

4.2.1. Replace Relevant Variables

To verify the robustness of the results of the basic regression, robustness tests we conducted by replacing the variables. The results of the robustness tests are shown Table 3. Column (1) was a measure of land finance using the per capita land transfer pri denoted by *land fin*2. The coefficient of *land fin*2 was 0.0037 and was significantly positi at the 5% level. Column (2) was a measure of land finance using the ratio of the la transfer price to GDP, and the coefficient of *land fin*3 was 0.0221 and was significan positive at the 10% level. Column (3) was a measure of the URIG using the ratio of t disposable income per urban resident to the disposable income per rural resident (IR), a the coefficient of *land fin* was 0.0316 and was significantly positive at the 5% level. T above results demonstrated that after replacing the main explanatory variables, the sig and significance of the coefficients were not significantly different from the basic regressi which further indicated that the empirical results were robust.

Table 3. Robustness test regression results.

Variable	**(1) TL**	**(2) TL**	**(3) IR**	**(4) TL**
*land fin*2	0.0037 ** (2.4686)			

Table 3. *Cont.*

Variable	(1) TL	(2) TL	(3) IR	(4) TL
$land\ fin3$		0.0221 * (1.9408)		
$land\ fin$			0.0316 ** (2.3153)	0.0014 (0.6646)
$land\ fin^2$				0.0007 (0.6863)
Control	Yes	Yes	Yes	Yes
Year fixed	Yes	Yes	Yes	Yes
City fixed	Yes	Yes	Yes	Yes
N	1100	1100	1088	1100
Adj. R2	0.323	0.322	0.228	0.326
F	34.4127	33.5594	15.6242	30.3025

Note: t statistics in parentheses (cluster in cities level). ** $p < 0.05$, * $p < 0.1$.

4.2.2. Non-Linear Regression

To verify the non-linear relationship between land finance and the URIG, the secondary term of land finance was also included in the regression equation in this section, and the regression results are shown in column (4) of Table 3. The coefficients of both the primary and secondary terms of land finance were positive and insignificant, and this result indicated that there was no non-linear relationship between land finance and the URIG.

4.2.3. Placebo Test

It is possible that, even after controlling for fixed effects and confounding variables, unobservable factors may still have affected the regression results. There may have been mechanisms that we could not observe that affected both land finance and the URIG, which could bias the estimated coefficients. Therefore, we used the placebo test to investigate whether the impact effect of land finance on the URIG stemmed from other unobservable factors. Referring to other placebo tests such as in La Ferrara et al. [54], the steps were as follows. First, the land finance data were randomly assigned to each city. Second, the regression was performed using Equation (1) to extract and preserve the regression coefficients of the land finance. Third, the above process was repeated 1000 times, and the regression coefficients extracted from these 1000 times were summarized and preserved. Finally, the 1000 regression coefficients were plotted according to the preserved kernel density distribution.

Figure 2 shows the distribution of kernel densities generated by the placebo test. The regression coefficients were approximated to a normal distribution with a mean of 0 and were mainly focused on the wings, while the absolute values of the 1000 regression coefficients were smaller than the base regression coefficients, which indicated that there was no effect related to virtual land finance. By inversion, the unobservable variables hardly affected the regression results.

4.3. Analysis of the Heterogeneity

4.3.1. The Heterogeneity of Land Grants for Industrial and Commercial/Residential Sites

In the process of land transfer, the authorities have adopted different modes of transfer for industrial land and commercial/residential land. In particular, the authorities usually grant industrial land at a low price, mainly by agreement, while they grant commercial and residential land at a higher and restrictive price, mainly by tender and auction. It is worthwhile to examine in detail whether this difference in the mode of land transfer may have different impacts on the URIG. Therefore, we used the agreement transfer transaction price as a proxy variable for the returns of industrial land, further exploring the impact of different land transfer returns on the URIG. We also used the tender and auction transfer transaction price as a proxy variable for the returns of commercial and residential land.

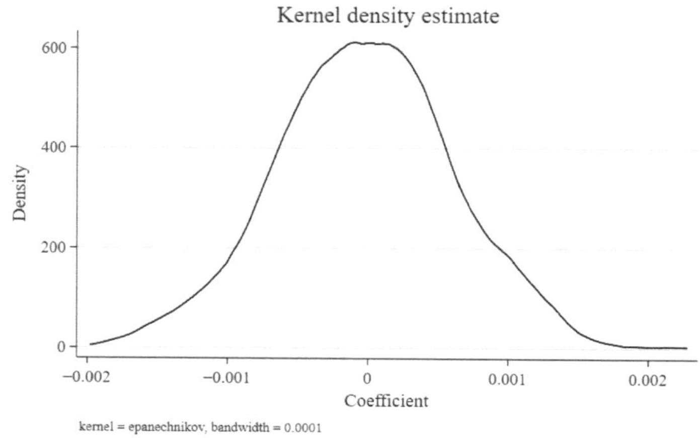

Figure 2. Placebo testing.

The regression results are shown in Table 3. Column (1) shows the regression resu of the agreement to let, and column (2) shows the results of tender and auction. In colu (1), the coefficient of agreement to transfer was −0.0017 but was not significant. In colu (2), the regression coefficient of bidding, listing, and auctioning was 0.0034 and it w significantly positive at the 1% level. The above results showed that the proceeds fr industrial land concessions by local authorities in the form of agreements did not have impact on the URIG, while the proceeds from commercial and residential land concessi by means of tenders and auctions could aggravate the widening of the URIG.

4.3.2. Regional Heterogeneity

Due to the sheer size of China, there are large differences among regions. The la finance dependence varies from region to region, and the effect of land finance on t URIG may differ greatly. Therefore, this study further examined the effect on the URIG different regions. The entire sample was classified into East, Middle, and West, according the region, the regression tests were separately run, and the results are shown in colum (3–5) of Table 4. Column (3) shows the results for the East region, where the regressi coefficient of land finance was 0.0037 and was significantly positive at the 1% level. Colur (4) shows the results for the Middle region, where the coefficient of land finance was 0.0 and was significantly positive at the 10% level. Column (5) shows the results for the W region, where the land finance coefficient was 0.0024 but was not significant. All the results indicated that in the Eastern and Middle regions, land finance widened the UR while in the Western region, it had no significant effect.

Table 4. Heterogeneity analysis regression results for industrial and commercial/residential lanc

Variables	(1) TL	(2) TL	(3) East	(4) Middle	(5) West
Agreement	−0.0017 (−0.9475)				
Tender and auction		0.0034 *** (3.4550)			
land fin			0.0037 *** (2.9845)	0.0024 * (1.8334)	0.0024 (0.6465)
Control	Yes	Yes	Yes	Yes	Yes
Year fixed	Yes	Yes	Yes	Yes	Yes
City fixed	Yes	Yes	Yes	Yes	Yes
N	1100	1100	384	400	316

Table 4. *Cont.*

Variables	(1) TL	(2) TL	(3) East	(4) Middle	(5) West
Adj. R2	0.320	0.328	0.295	0.531	0.439
F	33.8263	32.7332	20.6137	13.8762	16.5488

Note: *t* statistics in parentheses (cluster in cities level). *** $p < 0.01$, * $p < 0.1$.

4.3.3. Heterogeneity of City Size

There are large differences among Chinese cities in terms of functional positioning and resource concentration. Local authorities in different-sized cities differ markedly in their resource allocation rights, leading to differences in the impact of land finance on the URIG. Therefore, we further tested the effect of land finance on the URIG in different-sized cities. The Circular of the State Council on the Adjustment of the Standard for the Division of City Size classified city size into five types and seven levels. Considering that such a classification will lead to a reduction in the sample size, the city size was classified into three types in this study: cities with 500,000 or fewer residents were small cities, cities with 500,000 to 1,000,000 residents were medium cities, and cities with more than 1,000,000 residents were large cities. The city size was classified according to the 2014 resident population.

Table 5 shows the results for the different city sizes. In column (1), the coefficient of land finance was 0.0033 but was not significant in large cities. In column (2), the coefficient of land finance was 0.0011 and was not significant in medium cities. In column (3), the coefficient of land finance was 0.0035 and was significantly positive at the 5% level in small cities. All the results showed that land finance had no effect on the URIG in large and medium-sized cities, while in small cities, land finance had a widening effect on the URIG.

Table 5. Heterogeneity analysis regression results for city size.

Variables	(1) Large Cities	(2) Medium Cities	(3) Small Cities
land fin	0.0033 (1.1119)	0.0010 (0.5455)	0.0035 ** (2.5718)
Control	Yes	Yes	Yes
Year fixed	Yes	Yes	Yes
City fixed	Yes	Yes	Yes
N	252	372	476
Adj. R2	0.200	0.381	0.388
F	6.3603	12.7303	21.0853

Note: *t* statistics in parentheses (cluster in cities level). ** $p < 0.05$.

5. Discussion

This study theoretically analyzed and empirically tested the impact of land finance on the URIG in China, and we found evidence that land finance significantly widens the URIG. This finding not only helps us to understand the causes of the URIG in China and deepens our awareness of the income distribution effects of land finance, but also provides guidance for other developing countries. Nowadays, many developing countries are moving toward modernization and share many similarities with China in terms of the URIG [55,56]. Meanwhile, intense land expropriation is widespread [57]. This finding is a cautionary tale for other developing countries, who should be aware that rural residents whose land is expropriated bear the costs of urbanization, which is an important cause of frequent land conflicts [58]. Therefore, these countries should balance equity with the pursuit of efficiency in the future land development process to avoid infringing on the interests of rural residents, and they need to actively improve the current inequality.

This study found differences in the effect of land finance on the URIG depending on the way in which the land was granted. Specifically, it was found that land granting by tender and auction significantly widened the URIG, while land granting by agreement h no effect on the URIG. This was due to the local authorities granting industrial land a low price by way of agreement, which decreased the gap between the land appreciat and the land compensation price. Meanwhile, industrial expansion also absorbed ma rural laborers and increased their income [59]. On the other hand, the local authori granted commercial and residential land at high and restrictive prices, which increased cost and reduced the supply of housing, thus pushing up the price of houses. It was eas to increase the income of urban residents through the wealth effect and increase the cos living of rural residents who work in the city [21,60].

This study found that in the Eastern and the Middle regions, land finance ha widening effect on the URIG, and the widening effect was more significant in the East region, while it had no effect in the Western region. This ranking was consistent with level of economic development of each region, where the Eastern region had the best le of economic development, the Middle of the country was second, and the Western reg was the worst. Additionally, this ranking was consistent with the degree of dependence local governments on land finance, which was also highest in the East, next highest in Middle, and lowest in the West. The reason for this was that the higher the level of regio economic development, the higher the urban land premium, and that the larger the g with the land compensation price, the more the dependence of the local government on la finance is intensified, which increases the urban–rural income gap. In turn, the higher dependence of the local government on land finance, the more it could promote regio economic development. Thus, the urban–rural income gap gradually increased un this bidirectional feedback mechanism. In addition, the Western region tended to ben from the central government due to its lower economic development, thus alleviating financial pressure, which explained its lower dependence on land finance [46,61].

Furthermore, this study found that the effect of land finance on the widening of URIG was more significant in small cities, while it had no effect in large and mediu sized cities. A larger urban population meant a higher level of local urbanization, wh could absorb more rural surplus labor and increase farmers' income, thereby reducing URIG [62,63]. The reduction in the urban–rural income gap as a result of urbanizati weakened the role of land finance in widening the urban–rural income gap. Compar to large cities, small cities had fewer sources of fiscal income, which made them mo dependent on land finance, thus leading to the widening effect of land finance on urban–rural income gap being more significant. Another possible explanation is that sample sizes of large cities were relatively smaller, which led to insignificant regressi coefficients, which could be verified in future research by increasing the sample size.

6. Conclusions and Policy Recommendations

The phenomenon of land finance has become more widespread in developing co tries, with profound impacts on local economic and social development. However, th are few studies focusing on the income distribution effects it induces. This study examin the impact of land finance on the URIG from the perspective of the URIG, based on pa data from 275 prefecture-level cities in China from 2014 to 2017.

In general, the results indicated that land finance significantly widened the UR Further study found that the effect of land finance on the URIG showed significant h erogeneity. From a methodology perspective, land grants by tender, listing, and aucti significantly widened the URIG, while land grants by agreement did not affect the UR From a regional perspective, the effect of land finance on the URIG was more significa in Eastern and Middle regions, but not in Western regions. From the city size perspecti land finance had no impact on the URIG in large and medium-sized cities, while the imp was significant in small cities.

These findings have important policy consequences for the reform of the land regi and the promotion of fairness in income distribution. Firstly, the authorities should celerate the market-oriented reform of land supply to remove the monopolistic stat of local authorities in the primary land market, offer fair rights to rural residents in the land market, optimize the land income distribution system, ensure that rural residents receive fair and reasonable compensation for their landlessness under the market system, and rectify the uneven distribution of land premiums. Secondly, the authorities should change the present development strategy, accelerate the integration of urban and rural areas, and use the income from land concessions to enhance the public services provided in rural areas, providing the same opportunities for rural residents to obtain employment as urban residents and giving them the rights to the same public services. Lastly, authorities should actively promote the reform of the fiscal and taxation systems and change the local income structure to convert the income of local authorities from land transfer to stable and sustainable tax income. The dependence of local authorities on land finance must be removed at the source.

This study incorporated land finance and the URIG into a unified analytical framework, which deepened our comprehension of the relationship between land finance and the URIG, enriched the literature on the socioeconomic consequences of land finance, and provided empirical references for other developing countries. However, some shortcomings remained in the study that can be addressed in future research. First, this study only analyzed data from 2014 to 2017 due to the increasingly frequent adjustments in land finance policies in recent years with the continued advancement of urban–rural integration construction. In particular, with the revision of the Land Management Law in 2020, the government has strengthened the protection of rural residents' interests, and the effect of land finance on the urban–rural income gap may be transformed. Due to limitations on the availability of the data, this study did not discuss and test the change. However, future research can use this as an entry point and compare the changes in the impact of the policy before and after the revision. In addition, due to limitations on the available data, this study only analyzed the influence path of land finance on the URIG in theory. Lastly, this study did not explore the possible spatial spillover effects of land finance on the URIG, which will be the focus of our next phase of research. Further, the external validity of the results of this study has yet to be tested. The land acquisition and transfer mechanism has significant Chinese characteristics, and the institutional contexts of other developing countries will be markedly different from those of China, along with the resulting impacts. However, the development strategy of urban preferences is consistent, which is an important reason why other developing countries can learn from China's experience.

Author Contributions: Conceptualization, M.Z.; data curation, Z.W.; formal analysis, Z.W.; funding acquisition, M.Z.; methodology, Z.W.; software, Z.W.; supervision, M.Z.; writing—original draft, Z.W.; writing—review and editing, M.Z. All authors have read and agreed to the published version of the manuscript.

Funding: This research was funded by Jimei University, grant number "600008/HW22093".

Data Availability Statement: Some of the data were sourced from commercial databases. If necessary, please contact the corresponding authors.

Conflicts of Interest: The authors declare no conflict of interest.

References

1. Ye, F.; Wang, W. Determinants of Land Finance in China: A Study Based on Provincial-level Panel Data. *Aust. J. Publ. Admin.* **2013**, *72*, 293–303. [CrossRef]
2. Gyourko, J.; Shen, Y.; Wu, J.; Zhang, R. Land finance in China: Analysis and review. *China Econ. Rev.* **2022**, *76*, 101868. [CrossRef]
3. Vyas, I.; Vyas, H.N.; Mishra, A.K. Land-based financing of cities in India: A study of Bengaluru and Hyderabad and directions for reforms. *J. Public Aff.* **2022**, *22*, e2378. [CrossRef]
4. Nguyen, T.B.; van der Krabben, E.; Musil, C.; Le, D.A. Land for infrastructure' in Ho Chi Minh City: Land-based financing of transportation improvement. *Int. Plan. Stud.* **2018**, *23*, 310–326. [CrossRef]
5. Berrisford, S.; Cirolia, L.R.; Palmer, I. Land-based financing in sub-Saharan African cities. *Environ. Urban.* **2018**, *30*, 35–52. [CrossRef]
6. Qun, W.; Yongle, L.; Siqi, Y. The incentives of China's urban land finance. *Land Use Policy* **2015**, *42*, 432–442. [CrossRef]

7. Lu, Y.; Zhang, K.; Ouyang, J. Does Land Finance Hinder Regional Innovation? Based on the Data of 267 Prefectural—Level C in China. *J. Financ. Res.* **2018**, *5*, 101–119.
8. Liu, Y.; Chen, J. Land System, Financing Mode and Industrialization with Chinese Characteristics. *China Ind. Econ.* **2020**, *3*, 5 [CrossRef]
9. Guo, S.; Shi, Y.Y. Infrastructure investment in China: A model of local government choice under land financing. *J. Asian E* **2018**, *56*, 24–35. [CrossRef]
10. Mo, J. Land financing and economic growth: Evidence from Chinese counties. *China Econ. Rev.* **2018**, *50*, 218–239. [CrossRef
11. Du, H.; Ma, Y.; An, Y. The impact of land policy on the relation between housing and land prices: Evidence from China. *Q. Econ. Financ.* **2011**, *51*, 19–27. [CrossRef]
12. Pan, F.H.; Zhang, F.M.; Zhu, S.J.; Wojcik, D. Developing by borrowing? Inter-jurisdictional competition, land finance and lo debt accumulation in China. *Urban Stud.* **2017**, *54*, 897–916. [CrossRef]
13. Tang, P.; Feng, Y.; Li, M.; Zhang, Y. Can the performance evaluation change from central government suppress illegal land us local governments? A new interpretation of Chinese decentralisation. *Land Use Policy* **2021**, *108*, 105578. [CrossRef]
14. Cai, H.; Henderson, J.V.; Zhang, Q. China's land market auctions: Evidence of corruption? *Rand J. Econ.* **2013**, *44*, 488– [CrossRef]
15. Liu, Z.; Wang, P.F.; Zha, T. Land-Price Dynamics and Macroeconomic Fluctuations. *Econometrica* **2013**, *81*, 1147–1184. [CrossR
16. Wang, Q.; Feng, G.; Chang, C. Financial Structure Foundation of The Urban–Rural Income Gap in China: An Investigation fr the Perspective of the Double Dual Structure. *Bul. Ekon. Monet. Dan. Perbank.* **2019**, *22*, 177–194. [CrossRef]
17. Su, B.; Heshmati, A. Analysis of the determinants of income and income gap between urban and rural China. *China Econ. Po Rev.* **2013**, *02*, 1350002. [CrossRef]
18. Wan, G. Understanding Regional Poverty and Inequality Trends in China: Methodological Issues and Empirical Findings. *Income Wealth* **2007**, *53*, 25–34. [CrossRef]
19. Chen, J.; Fang, F.; Hou, W.; Li, F.; Pu, M.; Song, M.; Yuan, J. Chinese Gini Coefficient from 2005 to 2012, Based on 20 Grou Income Data Sets of Urban and Rural Residents. *J. Appl. Math.* **2015**, *2015*, 939020. [CrossRef]
20. Ji, X.; Wang, K.; Xu, H.; Li, M. Has Digital Financial Inclusion Narrowed the Urban-Rural Income Gap: The Role of Entrepren ship in China. *Sustainability* **2021**, *13*, 8292. [CrossRef]
21. Wang, Y.; Li, Y.; Huang, Y.; Yi, C.; Ren, J. Housing wealth inequality in China: An urban-rural comparison. *Cities* **2020**, *96*, 102 [CrossRef]
22. Bezemer, D.; Headey, D. Agriculture, Development, and Urban Bias. *World Dev.* **2008**, *36*, 1342–1364. [CrossRef]
23. Yang, D.T. Urban-Biased Policies and Rising Income Inequality in China. *Am. Econ. Rev.* **1999**, *89*, 306–310. [CrossRef]
24. Sicular, T.; Ximing, Y.; Gustafsson, B.; Shi, L. The Urban–Rural Income Gap and Inequality in China. *Rev. Income Wealth* **2007**, 93–126. [CrossRef]
25. Wang, C.; Wan, G.; Yang, D. Income Inequality in the People's Republic of China: Trends, Determinants, and Proposed Remed *J. Econ. Surv.* **2014**, *28*, 686–708. [CrossRef]
26. Wu, X.; Treiman, D.J. The household registration system and social stratification in China: 1955–1996. *Demography* **2004**, 363–384. [CrossRef]
27. Lei, G.; Cai, X. The distortion of primary income distribution, urban-biased fiscal expenditure policy and urban-rural inequa *J. Quant. Tech. Econ.* **2012**, *29*, 76–89. [CrossRef]
28. Lin, J.Y.; Chen, B. Urbanization and Urban-Rural Inequality in China: A New Perspective from the Government's Developm Strategy. *Front. Econ. China* **2011**, *6*, 1–21. [CrossRef]
29. Chen, B.; Zhang, P.; Yang, R. Government education budget, human capital in_vestment and urban-rural income disparity China. *Manag. World* **2010**, *1*, 36–43. [CrossRef]
30. Wang, S.; Tan, S.; Yang, S.; Lin, Q.; Zhang, L. Urban-biased land development policy and the urban-rural income gap: Evide from Hubei Province, China. *Land Use Policy* **2019**, *87*, 104066. [CrossRef]
31. Sun, X.; Zhou, F. Land Finance and the Tax-sharing System: An Empirical Interpretation. *Soc. Sci. China* **2014**, *35*, 47–64. [CrossR
32. Fan, X.; Qiu, S.N.; Sun, Y.K. Land finance dependence and urban land marketization in China: The perspective of strategic cho of local governments on land transfer. *Land Use Policy* **2020**, *99*, 105023. [CrossRef]
33. Li, H.; Zhou, L. Political turnover and economic performance: The incentive role of personnel control in China. *J. Public Ec* **2005**, *89*, 1743–1762. [CrossRef]
34. Chen, Z.; He, Z.; Liu, C. The financing of local government in China: Stimulus loan wanes and shadow banking waxes. *J. Fin Econ.* **2020**, *137*, 42–71. [CrossRef]
35. Bai, C.; Hsieh, C.; Song, Z. *The Long Shadow of a Fiscal Expansion*; University of Chicago: Chicago, IL, USA, 2016. [CrossRef]
36. Cao, G.; Feng, C.; Tao, R. Local "Land Finance" in China's Urban Expansion: Challenges and Solutions. *China World Econ.* **20** *16*, 19–30. [CrossRef]
37. Liu, S.; Xiong, X.; Zhang, Y.; Guo, G. Land System and China's Development Mode. *China Ind. Econ.* **2022**, 34–53. [CrossRef]
38. Liu, S. Characteristics, Problems and Reform of China's Dual Urban-Rural Land Tenure. *Int. Econ. Rev.* **2014**, *3*, 9–25. Availa online: https://kns.cnki.net/kcms2/article/abstract?v=3uoqIhG8C44YLTlOAiTRKgchrJ08w1e7M8Tu7YZds8_WqnnxqWBNR zkWJvSjcnCJND6GNXhoQIBu6-NQCurcdkA5JwVQO8&uniplatform=NZKPT (accessed on 5 September 2023).
39. Huang, Z.; Du, X. Government intervention and land misallocation: Evidence from China. *Cities* **2017**, *60*, 323–332. [CrossRe

Du, J.; Thill, J.; Feng, C. Wealth redistribution in urban land development under a dual land system: A case study of Beijing. *Pap. Reg. Sci.* **2014**, *93*, 501–517. [CrossRef]

Zhong, S.; Wang, M.; Zhu, Y.; Chen, Z.; Huang, X. Urban expansion and the urban–rural income gap: Empirical evidence from China. *Cities* **2022**, *129*, 103831. [CrossRef]

Zhu, C.; Zhang, X.; Wang, K.; Yuan, S.; Yang, L.; Skitmore, M. Urban–rural construction land transition and its coupling relationship with population flow in China's urban agglomeration region. *Cities* **2020**, *101*, 102701. [CrossRef]

Yu, A.T.W.; Wu, Y.; Shen, J.; Zhang, X.; Shen, L.; Shan, L. The key causes of urban-rural conflict in China. *Habitat Int.* **2015**, *49*, 65–73. [CrossRef]

Hui, E.C.M.; Bao, H.J.; Zhang, X.L. The policy and praxis of compensation for land expropriations in China: An appraisal from the perspective of social exclusion. *Land Use Policy* **2013**, *32*, 309–316. [CrossRef]

Xie, D. Rural-urban Income Distribution Effect of Land Supply: A Perspective of Imbalance Development of Urbanization. *Nankai Econ. Stud.* **2017**, *2*, 76–95. [CrossRef]

He, S.; Liu, Y.; Chris, W.; Wu, F. Property Rights Redistribution, Entitlement Failure and the Impoverishment of Landless Farmers in China. *Urban Stud.* **2009**, *46*, 1925–1949. [CrossRef]

Zhang, J.; Ge, Y. The Influence of Land Market Allocation on the Rural—Urban Income Distribution in China—A Perspective of Imbalance Development of Urbanization. *J. Huazhong Agric. Univ. Soc. Sci. Ed.* **2021**, *3*, 172–181. [CrossRef]

Pintér, G. House Prices and Job Losses. *Econ. J.* **2019**, *129*, 991–1013. [CrossRef]

Bai, J. Panel Data Models with Interactive Fixed Effects. *Econometrica* **2009**, *77*, 1229–1279. [CrossRef]

Flannery, M.J.; Rangan, K.P. Partial adjustment toward target capital structures. *J. Financ. Econ.* **2006**, *79*, 469–506. [CrossRef]

DeHaan, E. Using and interpreting fixed effects models. *Soc. Sci. Electron. Publ.* **2021**. [CrossRef]

Yuan, Y.; Wang, M.S.; Zhu, Y.; Huang, X.J.; Xiong, X.F. Urbanization's effects on the urban-rural income gap urban-rural income gap in China: A meta-regression analysis. *Land Use Policy* **2020**, *99*, 104995. [CrossRef]

Hou, S.; Song, L.; Wang, J.; Ali, S. How Land Finance Affects Green Economic Growth in Chinese Cities. *Land* **2021**, *10*, 819. [CrossRef]

La Ferrara, E.; Chong, A.; Duryea, S. Soap Operas and Fertility: Evidence from Brazil. *Am. Econ. J. Appl. Econ.* **2012**, *4*, 1–31. [CrossRef]

Cai, F. Rural urban income gap and critical point of institutional change. *Econ. Chang. Restruct.* **2007**, *40*, 189–206. [CrossRef]

Tan, Y.; Xu, H.; Zhang, X. Sustainable urbanization in China: A comprehensive literature review. *Cities* **2016**, *55*, 82–93. [CrossRef]

PETERS, P.E. Inequality and Social Conflict Over Land in Africa. *J. Agrar. Chang.* **2004**, *4*, 269–314. [CrossRef]

Lin, Q.; Tan, S.; Zhang, L.; Wang, S.; Wei, C.; Li, Y. Conflicts of land expropriation in China during 2006–2016: An overview and its spatio-temporal characteristics. *Land Use Policy* **2018**, *76*, 246–251. [CrossRef]

Lagakos, D. Urban-Rural Gaps in the Developing World: Does Internal Migration Offer Opportunities? *J. Econ. Perspect.* **2020**, *34*, 174–192. [CrossRef]

Wu, L.; Zhang, W. Rural migrants' homeownership in Chinese urban destinations: Do institutional arrangements still matter after Hukou reform? *Cities* **2018**, *79*, 151–158. [CrossRef]

Ji, F.; Ba, S. Local Government Competition, Transfer Payment and Land Finance. *China Soft Sci.* **2020**, *11*, 100–109. Available online: https://kns.cnki.net/kcms2/article/abstract?v=3uoqIhG8C44YLTlOAiTRKibYlV5Vjs7i8oRR1PAr7RxjuAJk4dHXouqIIUjyeQixOny5fc_PPlSibtvSj9RCTilkxao6g1uB&uniplatform=NZKPT (accessed on 5 September 2023).

Chen, S.; Jiang, J. The Relationship of Financial Development, Urbanization and Urban-Rural Income Gap: An Empirical Research Based on Provincial Panel Data in China. *SHS Web Conf.* **2014**, *6*, 1228–1233. [CrossRef]

Su, C.W.; Liu, T.Y.; Chang, H.L.; Jiang, X.Z. Is urbanization narrowing the urban-rural income gap? A cross-regional study of China. *Habitat Int.* **2015**, *48*, 79–86. [CrossRef]

land

Article

Spatio-Temporal Evolution and Multi-Scenario Simulation of Non-Grain Production on Cultivated Land in Jiangsu Province, China

Chengge Jiang [1], Lingzhi Wang [1,*], Wenhua Guo [2,3], Huiling Chen [4,5], Anqi Liang [1], Mingying Sun [1], Xinyao Li [1] and Hichem Omrani [6]

1 College of Earth Sciences, Jilin University, Changchun 130061, China; jiangcg22@mails.jlu.edu.cn (C.J.)
2 Technology Innovation Center for Territorial & Spatial Big Data, MNR, Beijing 100830, China
3 Information Center of Ministry of Natural Resources, Beijing 100830, China
4 Technology Innovation Center for Territorial & Spatial Big Data, MNR, Jiangsu Branch, Nanjing 210017, Ch
5 Information Centre of Land and Resources of Jiangsu Province, Nanjing 210017, China
6 Urban Development and Mobility Department, Luxembourg Institute of Socio-Economic Research, Univer of Luxembourg, 4366 Esch-sur-Alzette, Luxembourg
* Correspondence: wanglingz@jlu.edu.cn

Citation: Jiang, C.; Wang, L.; Guo, W.; Chen, H.; Liang, A.; Sun, M.; Li, X.; Omrani, H. Spatio-Temporal Evolution and Multi-Scenario Simulation of Non-Grain Production on Cultivated Land in Jiangsu Province, China. *Land* **2024**, *13*, 670. https://doi.org/10.3390/land13050670

Academic Editor: Matjaž Glavan

Received: 28 March 2024
Revised: 7 May 2024
Accepted: 10 May 2024
Published: 13 May 2024

Abstract: Cultivated land plays a crucial role as the basis of grain production, and it is essential effectively manage the unregulated expansion of non-grain production (NGP) on cultivated land order to safeguard food security. The study of NGP has garnered significant attention from schol but the prediction of NGP trends is relatively uncommon. Therefore, we focused on Jiangsu Provir a significant grain production region in China, as the study area. We extracted data on cultiva land for non-grain production (NGPCL) in 2000, 2005, 2010, 2015, and 2019, and calculated ratio of non-grain production (NGPR) for each county unit in the province. On this basis, Ker Density Estimation (KDE) and spatial autocorrelation analysis tools were utilized to uncover spatio-temporal evolution of NGP in Jiangsu Province. Finally, the Patch-Generating Land U Simulation (PLUS) model was utilized to predict the trend of NGP in Jiangsu Province in 2038 un the three development scenarios of natural development (NDS), cultivated land protection (CPS), a food security (FSS). After analyzing the results, we came to the following conclusions:(1) During period of 2000–2019, the NGPCL area and NGPR in Jiangsu Province exhibited a general decreas trend. (2) The level of NGP displayed a spatial distribution pattern of being "higher in the south a central and lower in the north". (3) The results of multi-scenario simulation show that under the N the area of NGPCL and cultivated land for grain production (GPCL) decreases significantly; un the CPS, the decrease in NGPCL and GPCL is smaller than that of the NDS. Under the FSS, NGP decreases, while GPCL increases. These results can provide reference for the implementation of la use planning, the delineation of the cultivated land protection bottom line, and the implementati of thee cultivated land use control system in the study area.

Keywords: cultivated land; non-grain production; spatio-temporal evolution; multi-scenario simulati Jiangsu Province

1. Introduction

Food security is a critical global issue that intersects with economies and peopl livelihoods. It serves as a fundamental pillar of national security [1]. The foundation food production lies in cultivated land, making it imperative to ensure the stability both its quantity and quality to guarantee food security. However, China has witness a decline in the quantity and quality of cultivated land in recent years, primarily due the increasing level of urbanization [2,3]. Meanwhile, various factors, such as the l efficiency of grain production, changes in dietary patterns, land property rights transfe and the influx of industrial and commercial capital into rural areas, have compelled certain agricultural entities to prioritize maximizing their own profits. Consequently, they selectively choose crops based on profitability, leading to a significant portion of cultivated land being allocated for non-grain production (NGP) [4]. This disorderly expansion of NGP not only threatens food security by causing shortages and imbalances in grain supply, but also contributes to the degradation of cultivated land quality and environmental issues resulting from changes in land utilization practices [5–8]. Recognizing the implications of NGP, the Chinese Government has acknowledged its significance. In November 2020, the General Office of the State Council issued the "Opinions on Preventing the Non-Grain Production of Cultivated Land and Stabilizing Grain Production". This document summarized the manifestations of the "non-grain" use of cultivated land into three categories: "some localities have unilaterally interpreted agricultural structural adjustment as a reduction in grain production", "business entities have illegally planted trees and dug ponds on basic farmland", and "industrial and commercial capital has transferred cultivated land on a large scale to cultivate non-grain crops". It also emphasized that the use of cultivated land should not be determined purely on the basis of economic benefits and that limited cultivated land resources must be prioritized for grain production. Specifically, the government aims to stabilize the area allocated for grain production, with particular attention to safeguarding the planting area of the three major grains: rice, wheat, and maize.

Research concerning NGP dates back to the beginning of the 21st century [9]. Scholars at the time raised concerns about this pursuit of economic efficiency at the expense of food security. In recent years, with the widespread concern for food security, scholars have been studying NGP in further depth. Currently, there is no universally accepted definition of "non-grain" within the academic community. Some scholars define "non-grain" as the utilization of cultivated land for non-grain purposes by agricultural operators, including the cultivation of cash crops such as fruit trees and flowers, as well as the development of livestock and poultry breeding [10,11]. Others consider "non-grain" as the process of converting cultivated land that previously grew grain crops into land that now produces non-grain crops [12,13]. Regarding the measurement of NGP, various approaches have been used by scholars. Some calculate the level of NGP based on statistical data or sectoral surveys, using indicators like the ratio of non-grain crop area to total agricultural crop area [14,15]. Alternatively, remote sensing and GIS technologies have been employed by other researchers to identify NGP phenomena [16–18].Studies about the trends in land use change to NGP [19,20] and the drivers [21,22] and driving mechanisms [21,22] of NGP provide valuable insights into understanding its causes and mitigating its effects. These studies found that factors such as natural resources [23,24], socio-economics [21], transportation location [23], and policy considerations [21,25,26] may contribute to NGP. Considering previous research and the current agricultural production landscape in this study area, we defined NGP as the cultivation of crops other than rice, wheat, and maize on cultivated land. NGPCL is identified through the analysis of land use data and the spatial distribution of the three major grain cultivars. The ratio of NGP (NGPR) was then utilized to assess the level of NGP in the region. Subsequently, we analyzed the spatio-temporal evolution of the NGP. Finally, the evolution of the NGP under multiple development scenarios was projected. This study not only contributes to the theoretical and methodological aspects of NGP research but also offers insights into future developments, providing recommendations and counter measures for the sustainable use of cultivated land and the assurance of food security.

In order to reveal the issue of NGP and its implications for food security, this study focused on Jiangsu Province, a significant grain-producing province in China, which has consistently ranked among the top provinces in China in terms of grain sown, total grain output, and total output value. Based on the analysis of the spatial and temporal evolution of NGP in Jiangsu Province from 2000 to 2019, the Patch-Generating Land Use Simulation (PLUS) model [27], which is widely followed by scholars in the field of land simulation and is used to conduct a multi-scenario simulation to reveal the pattern of NGP under different development scenarios in 2038. The results of the study will provide a ba and reference for land managers to formulate policies, scientifically set the bottom line NGP, and practically control the behavior of NGP in order to promote the rational use cultivated land resources and guarantee food security.

2. Materials and Methods

2.1. Study Area

Jiangsu Province is situated in the eastern coastal region of mainland China, wi latitude ranging from 30°45′ to 35°08′ N and longitude ranging from 116°21′ to 121°56 (Figure 1). As of 2020, the province had a total area of 102,521.20 km^2, with a cultivat land area of 62,773.09 km^2, accounting for 61.22% of the total area. Known for its f terrain, favorable climate, and abundant water resources, Jiangsu Province is well-suit for agricultural production. Between 2000 and 2022, the sown area of grain in Jiang Province varied from 53,043.1 km^2 to 54,444.4 km^2. During this period, the total gra output is expected to change from 31,066,300 tons to 37,691,300 tons, and the total outp value is anticipated to shift from USD 4379 million to USD 19,884 million. However, recent years, rapid socio-economic development and increasing urbanization have led the conversion of some of Jiangsu Province's cultivated land into artificial surfaces (A like urban and rural residential land [28]. This, along with the prominent issue of NGP, h posed a serious challenge to food security in the region. Protecting cultivated land resour and ensuring stable grain production in Jiangsu Province are of utmost importance national food security and the sustainable development of the national economy.

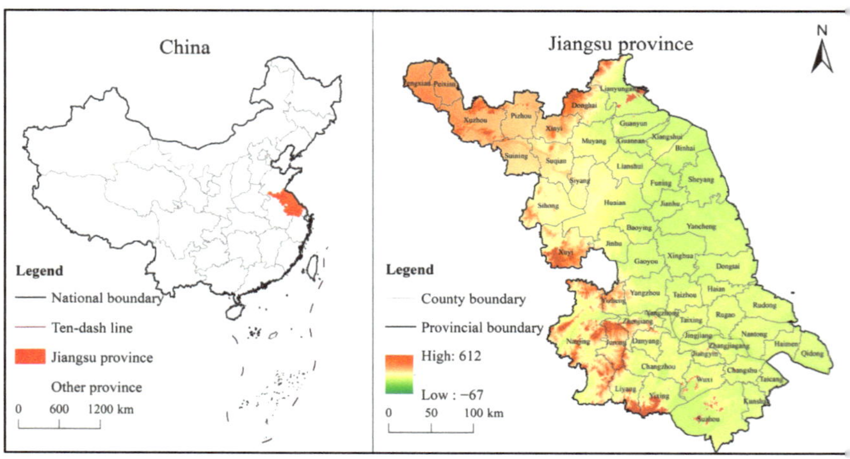

Figure 1. Location of the study area.

2.2. Data Sources and Pre-Processing

The data used in this paper include county-level administrative boundary data Jiangsu Province, land use/land cover (LULC) data, grain cultivation spatial distributi data, meteorological data, DEM data, soil type data, road data, POI data, population dens data, and GDP data (Table 1).

Given the extensive time span under consideration, the modifications to administrati divisions during the study period, and the presence of small-sized municipal districts, t study adopts the approach proposed by Li et al. [29,30] to merge these municipal distri and ultimately create 54 consolidated study units. Taking into account the situation the study area and data availability, land use data from the Resource and Environme Science and Data Center (RESDC) were selected for five time points: 2000, 2005, 2010, 20 and 2020; the spatial distribution data of grain cultivation at the five time points we obtained from the National Ecosystem Science Data Center (NESDC), which originated from Luo's study, which was based on cultivated land data and determined the spatial distribution of the cultivation of the three major grain crops in China according to the climatic characteristics of rice, maize, and wheat for the period of 2000–2019. The spatial distribution of cultivation of the three major grain crops in China was proved to be highly consistent with agricultural statistics ($R^2 > 0.8$) [31]; the meteorological dataset comprises annual precipitation (PRE), annual mean temperature (TEM), annual evaporation (EVP), and annual sunshine duration (SSD). This dataset was derived from daily observations collected at the meteorological element stations and utilizes the smoothing spline function integrated in the Anusplin 4.4 interpolation software to generate spatially interpolated data for each year's meteorological element [32]. We utilized Digital Elevation Model (DEM) data obtained from the US National Geological Survey (USGS). The slope and slope direction analysis functions in ArcGIS 10.7 software were employed to generate raster data representing slope and slope direction. Road data, including highways, primary roads, secondary roads, and railroad data, were sourced from Openmap. The Euclidean distance analysis function in ArcGIS was used to calculate distance data to all levels of roads. Government site data were derived from point-of-interest (POI) data collected by AMap, with distance data to government sites generated using the Euclidean distance analysis function in ArcGIS. Subsequently, all datasets were transformed into raster data and resampled to a resolution of 30 m.

Table 1. Data sources.

Name	Categories	Resolution	Source
Climate	raster	1 km	RESDC (https://www.resdc.cn (accessed on 1 April 2023))
DEM	raster	30 m	USGS (https://www.usgs.gov (accessed on 1 April 2023))
Soil types	vector plane	—	RESDC (https://www.resdc.cn (accessed on 1 April 2023))
Administrative boundary	vector line	—	RESDC (https://www.resdc.cn (accessed on 1 April 2023))
Land use	raster	30 m	RESDC (https://www.resdc.cn (accessed on 1 April 2023))
Grain cultivation spatial distribution	raster	1 km	NESDC (http://www.nesdc.org.cn (accessed on 1 April 2023))
GDP	raster	1 km	RESDC (https://www.resdc.cn (accessed on 1 April 2023))
Population density	raster	100 m	WorldPop (https://hub.WorldPop.org (accessed on 1 April 2023))
Roads	vector line	—	Openmap (https://openmaptiles.org (accessed on 1 April 2023))
Government sites	point	—	Amap (https://lbs.amap.com (accessed on 1 April 2023))

2.3. Methods

We analyzed the spatial and temporal evolution of NGP in Jiangsu Province from 2000 to 2019 and predicted the pattern of NGP under different scenarios for 2038. First, the spatial location of NGPCL was identified and the NGPR was calculated for each county. Then, land use transfer matrix, kernel density estimation (KDE), and spatial autocorrelation analysis were used to reveal the spatio-temporal evolution pattern of NGP. Finally, the PLUS model was used to simulate the different future scenarios of the NGP pattern. The general framework of the study is shown in Figure 2.

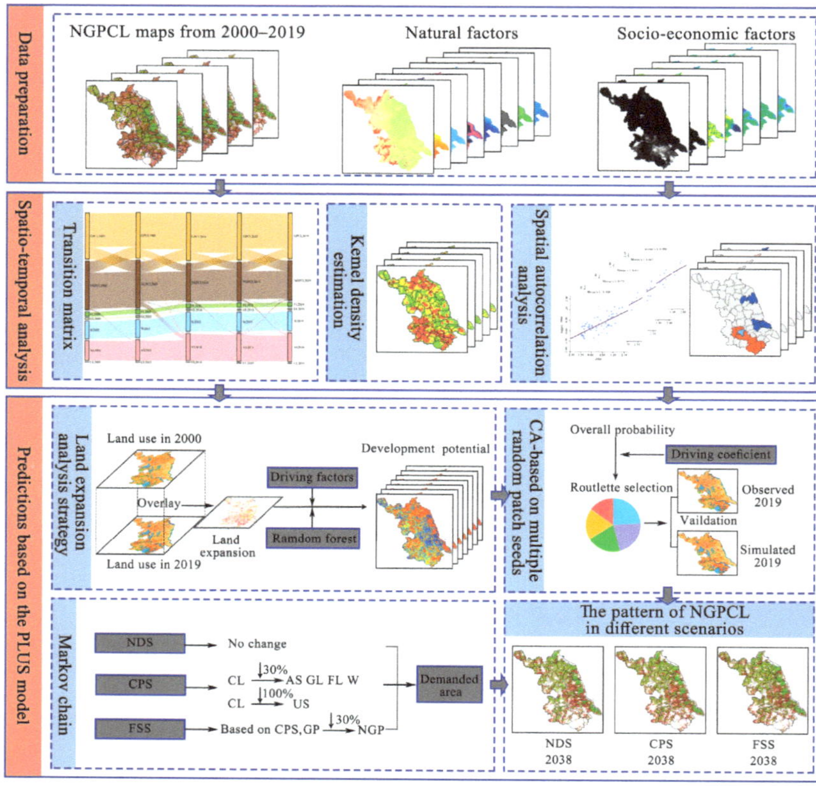

Figure 2. Study framework.

2.3.1. NGPR

The grain crops involved in this study include the three major categories of rice, whe and maize, and the cultivated land outside the three major categories of grain crops w defined as NGPCL. It should be noted that in identifying the 2019 NGPCL, we used 20 land use data as a substitute due to the lack of 2019 land use data. The degree of region non-grain was measured in terms of the NGPR with the following formula:

$$\mathrm{NGPR} = \left(1 - \frac{\mathrm{G}}{\mathrm{C}}\right) \times 100\%$$

where NGPR is the ratio of non-grain production; G is the sum of the area of wheat, co and rice; C is the total area of cultivated land.

2.3.2. Land Use Transfer Matrix

We categorized cultivated land into GPCL and NGPCL, and the land use maps different time periods were spatially superimposed by using ArcGIS to derive the trans matrix of the land use types in order to characterize the conversion between the land u types. The mathematical expression is as follows [33,34]:

$$\mathrm{P_{ij}} = \begin{bmatrix} \mathrm{P_{11}} & \mathrm{P_{12}} & \cdots & \mathrm{P_{1n}} \\ \mathrm{P_{21}} & \mathrm{P_{22}} & \cdots & \mathrm{P_{2n}} \\ \vdots & \vdots & \ddots & \vdots \\ \mathrm{P_{n1}} & \mathrm{P_{n2}} & \cdots & \mathrm{P_{nn}} \end{bmatrix}$$

where P denotes the area of each LULC type; i and j denote the area of LULC types at the beginning and end of the study period, respectively; and n denotes the total number of LULC types. Each row and column represents the outflow and inflow areas, respectively.

2.3.3. Kernel Density Estimate

Kernel Density Estimation (KDE) is a nonparametric method used to estimate the probability density function, which can effectively identify the spatial variability and continuity of the distribution of NGPCL and observe the agglomeration of their spatial distribution. In this paper, the raster data of NGPCL were converted to point data, and the kernel density value of NGPCL in Jiangsu Province was calculated with a kernel density bandwidth of 10 km [35]. The calculation formula is:

$$f(x) = \frac{1}{nh}\sum_{i=1}^{n} k\left(\frac{d(x,x_i)}{h}\right) \quad (3)$$

where f(x) is an estimate of the NGPCL kernel density, n is the number of NGPCL point data in the bandwidth range, h is the bandwidth, k () is the kernel density equation, and $d(x,x_i)$ is the distance from the NGPCL point x to the sample point x_i.

2.3.4. Spatial Autocorrelation Analysis

We employed global spatial autocorrelation and local spatial autocorrelation to explore the spatial correlation of NGP.

Global spatial autocorrelation analyzes and describes the overall spatial characteristics of geographic elements within a study area and can visually represent the degree of spatial interdependence of geographic variables. A commonly used measure of global spatial autocorrelation is the global Moran's I, which has a value range of [−1,1]. A positive value indicates the existence of positive spatial correlation, i.e., the same trend between the research object and the neighboring objects; a negative value indicates the existence of negative spatial correlation, i.e., the opposite trend between the research object and the neighboring objects; and a value equal to zero indicates no spatial correlation. We took the NGPR of the Jiangsu county-level administrative units as an observational variable and used the global Moran's I to determine its overall spatial characteristics. Its calculation formula is:

$$I = \frac{N\sum_{i=1}^{N}\sum_{j=1}^{N} w_{ij}\left(x_i - \bar{x}\right)\left(x_j - \bar{x}\right)}{\left(\sum_{i=1}^{N}\sum_{j=1}^{N} w_{ij}\right)\sum_{i=1}^{N}\left(x_i - \bar{x}\right)^2} \quad (4)$$

where I is the global spatial autocorrelation coefficient; x_i and x_j are the observed values of NGPR at the location of objects i and j in the study area; N is the number of county units; and w_{ij} is the spatial weight matrix.

Local spatial autocorrelation explores spatial correlation features from the local scale to identify spatial clustering or spatial anomalies of geographic elements. In this paper, local Moran's I was used to identify the spatial association patterns of NGPR, which is calculated by the expression:

$$I_i = \frac{x_i - \bar{x}}{\sigma^2}\sum_{j=1}^{N} w_{ij}\left(x_j - \bar{x}\right) \quad (5)$$

where I_i is the local spatial autocorrelation coefficient at location i in the study area; x_i and x_j are the observed values of NGPR at the locations of objects i and j in the study area; $\bar{x}$ is the mean value of x; σ^2 is the variance of x; w_{ij} is the spatial weighting matrix; and the range of values of I_i is the same as that of the global Moran' s I.

Local spatial autocorrelation involves four distinct patterns: high-high agglomeration (H-H), low-high agglomeration (L-H), low-low agglomeration (L-L), and high-low agglomeration (H-L). H-H and L-L signify positive spatial correlation, indicating clustering of similar high or low values, whereas H-L and L-H suggest negative spatial correlation, reflecting dissimilar neighboring values. We utilized Moran scatter plots and LISA cluster diagrams to provide a comprehensive analysis of the local spatial correlation attribute the context of NGP.

2.3.5. NGP Simulation

1. PLUS model. The PLUS model, introduced by the HPSCIL@CUG laboratory te at China University of Geosciences (Wuhan, China) in 2020 [27], is a patch-bas land-use change simulation model that operates on raster data. This model effectiv identifies the driving forces behind land expansion and landscape modificatio enabling a more accurate simulation of the progression of land-use patches. The PL model comprises two key components:
    - Land Expansion Analysis Strategy (LEAS): This strategy involves extracti various land use expansion areas between two stages of land use change. It th selects from the expanded areas, using the Random Forest algorithm to iden the influencing factors and driving forces behind each type of land use expans individually. This process allows for the determination of development proba ities and the contribution of driving factors for all types of land use expansi during the specified period.
    - Cellular Automata based on multiple random patch seeds (CARS) model: T PLUS model integrates random seed generation and a threshold-decreasi mechanism to dynamically simulate the automatic generation of patches in sp under specific development probability constraints.

In the study, the land use expansion between the years 2000 and 2019 was extract and the expansion areas were sampled in the LEAS module. The parameters of the LE module were set as follows: 50 decision trees, a sampling rate of 0.01, 15 training featu and the outputs include the probability of development for each class and the contributi of driving factors to the expansion of the site. We selected 15 driving factors from natu and socio-economic factors based on previous research (Table 2). These factors were us to determine the probability of development and the contribution of each factor to expansion of the land. The PLUS model requires that the interval between the projecti year and the end year should be the same as, or an integer multiple of, the interval betw the start year and the end year. The time interval for this study was 19 years, so we us CARS to predict the 2038 landscape based on the information and analysis obtained fr the LEAS module.

Table 2. Driving factors.

Variable Type	Name	Bibliography
Natural factors	DEM	[35,36]
	Slope	[35,36]
	Slope direction	[37]
	EVP	[35]
	PRE	[35]
	SSD	[38]
	TEM	[35]
	Soil types	[39]
Socio-economic factors	Population density	[36]
	GDP	[5,36]
	Distance to government departments	[39,40]
	Distance to highway	[35]
	Distance to primary road	[35,36]
	Distance to secondary road	[35,36]
	Distance to railroad	[35]

Pontious et al. [41,42] found that the Figure of Merit (FoM) index is superior to the Kappa coefficient in evaluating the accuracy of the simulated changes, so the FoM index was used for the accuracy evaluation of the simulations. Its calculation formula is:

$$\mathrm{FoM} = \frac{\mathrm{B}}{\mathrm{A}+\mathrm{B}+\mathrm{C}+\mathrm{D}} \tag{6}$$

where A represents the number of pixels that are predicted not to change and actually change; B represents the number of rasters that are accurately predicted; C represents the number of pixels that are predicted incorrectly for the land-use type; and D represents the number of pixels that are predicted to change and actually do not change. According to previous studies, when the calculated value of FoM is small, it indicates that the simulation accuracy is low. Conversely, when its value is larger, it suggests that the simulation accuracy is higher. An FoM value greater than 0.1 can be considered indicative of good simulation accuracy.

We set the neighborhood weights according to the proportion of the expansion area of different land types to the total area. After repeated debugging, the attenuation threshold was set to 0.9, the diffusion coefficient was set to 0.1, and the probability of random seeds was set to 0.001. At this time, the FoM index reached 0.29, indicating that the simulated 2019 NGPCL was highly consistent with the real situation and was able to meet the research needs.

2. Scenario analysis. Based on the documents "Jiangsu Province Land Spatial Planning (2021–2035)" and "Jiangsu Province 14th Five-Year Plan Grain Industry Development Plan", as well as drawing insights from previous studies [43–45], the study established three development scenarios as follows:
    - Natural development scenario (NDS): Based on the change rule of land use in Jiangsu Province from 2000 to 2019, the distribution of NGPCL under the natural growth scenario in 2038 was simulated without any constraints.
    - Cultivated land protection scenario (CPS): Considering the requirements of "stabilizing the amount of cultivated land, optimizing the layout of cultivated land, and strictly controlling the occupation of cultivated land by non-agricultural construction" mentioned in the land space planning of Jiangsu Province (2021–2035), this scenario reduced the probability of transferring cultivated land to construction land, forest land, grassland and water by 30% and 100% to unutilized land based on the probability of the natural development scenario to simulate the protection of cultivated land by land managers.
    - Food security scenario (FSS): Considering the requirements of "Strictly controlling non-grain production and guaranteeing food security" mentioned in the "Jiangsu Province Land Space Planning (2021–2035)" and the "14th Five-Year Plan for the Development of Grain Industry in Jiangsu Province", this scenario reduced the probability of transferring GPCL to NGPCL by 30% based on the probability of the CPS to simulate the control measures taken by the land managers on the NGP.

3. Results

3.1. Timing Evolution of NGP in Jiangsu Province

Between 2000 and 2019, the cultivated land area in Jiangsu Province decreased from 69,910.29 km^2 to 62,777.79 km^2. The NGPCL decreased by 18.67% from the initial year, while the NGPR decreased from 51.48% to 46.63% (Figure 3). The temporal evolution characteristics of NGP suggest two phases during the study period.

1. From 2000 to 2010, the province's NGPR declined by 6.59%. This decline can be attributed to China's active implementation of agricultural policies and reforms in the grain circulation system after 2003. In addition, in 2004, Jiangsu Province introduced supportive policies such as grain subsidy policies and agricultural tax reductions and

exemptions. This has boosted grain production and increased farmers' incentive cultivate grain. Consequently, the dominance of grain production increased, lead to a decline in the NGPR during this period.

2. In the period from 2010 to 2019, the NGPR in Jiangsu Province experienced a sli increase, primarily due to declining grain prices. The continuous decrease in gr prices, particularly rice prices, exhibited a downward trend from 2012, reaching C 2.4/kg in 2019. This decline in the profitability of grain cultivation prompted so farmers to shift towards planting cash crops with higher economic returns. T average annual increase in the NGPCL rate in Jiangsu Province during this per was 0.17%, with an average annual increment in the area of NGPCL amountin 51.32 km^2.

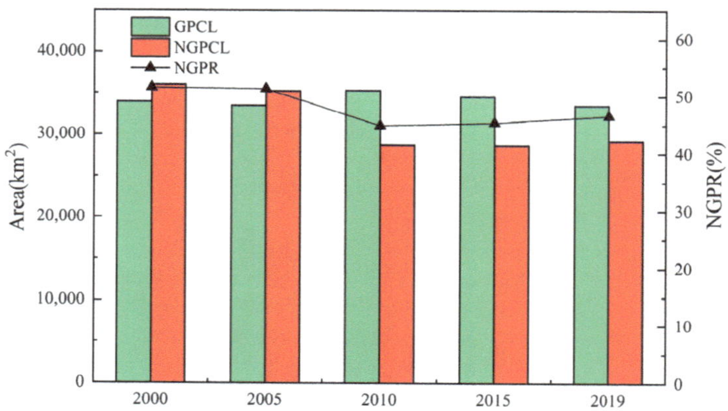

Figure 3. Temporal changes in NGPCL area and NGPR in Jiangsu Province.

During the study period, both the area of GPCL and NGPCL experienced a decrea with the reduction in NGPCL surpassing that of GPCL. This led to a decline in the NG aligning with the conclusions drawn by Wang et al. [46]. This observation indicated t despite the decrease in NGPR in Jiangsu Province, concerns regarding food security pers particularly given the rising demand for grain within the province [47,48].

The data presented in Figure 4 indicate that the mutual transformation betwe NGPCL and GPCL was a common occurrence across all time periods studied. The d from Table 3 reveal that during the period of 2000–2019, NGPCL accounted for 76.33% the transfer out of GPCL to other land categories, while the conversion of NGPCL to GP represented 63.18% of the outward transfers from NGPCL. Additionally, the conversion cultivated land to AS constituted a significant proportion, with 20.69% of GPCL outwa transfers and 30.86% of NGPCL outward transfers resulting in AS. These findings sugges that the expansion of AS during the study period led to a common occurrence of cultiva land being converted, with NGPCL being more prone to transformation into AS compa to GPCL.

Table 3. Land use transfer matrix for Jiangsu Province, 2000–2019.

Area in 2000/km^2	Area in 2019/km^2							
	GPCL	NGPCL	FL	GL	W	AS	UL	Transfers Ou
GPCL	—	8892.66	14.73	7.03	321.69	2410.33	3.82	11,650.26
NGPCL	10,681.86	—	105.15	47.07	831.04	5217.69	23.03	16,905.84
FL	60.73	191.45	—	6.46	21.30	170.44	30.09	480.47
GL	28.64	204.68	4.05	—	453.08	89.80	2.94	783.19
W	78.36	320.25	7.61	242.18	—	441.41	22.29	1112.09

Table 3. *Cont.*

Area in 000/km^2	Area in 2019/km^2							
	GPCL	NGPCL	FL	GL	W	AS	UL	Transfers Out
AS	389.27	576.18	22.97	16.42	829.22	—	7.02	1841.09
UL	0.02	0.15	1.57	0.06	2.20	1.59	—	5.59
ansfers in	11,238.88	10,185.39	156.08	319.22	2458.51	8331.26	89.19	—

"FL" indicates forest land; "GL" indicates grassland; "W" indicates water; "AS" indicates artificial surface; "UL" indicates unutilized land.

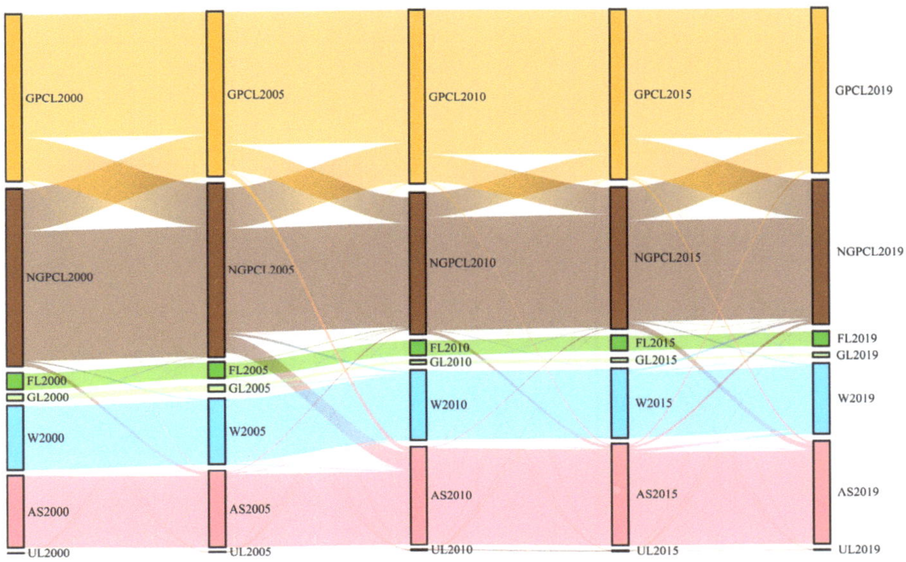

Figure 4. Land use transfer Sankey map of Jiangsu Province, 2000–2019.

3.2. *Spatial Evolution of NGP in Jiangsu Province*

The spatial distribution of NGPCL in Jiangsu Province is depicted in Figure 5. Across all counties in Jiangsu Province, NGPCL distribution exhibited clear spatial differentiation characteristics. Southern Jiangsu emerged as a focal point for NGPCL concentration throughout the study period, boasting the highest NGPR in the province. Central Jiangsu has experienced a notable increase in NGPCL from 2000 to 2019, accompanied by a rise in the NGPR. Conversely, northern Jiangsu has witnessed a decline in NGPCL and a corresponding decrease in NGPR, alongside a resurgence in grain cultivation activities. By 2019, Jiangsu Province had established a spatial pattern characterized by a "high prevalence of NGP in the central and southern regions, with lower levels in the northern areas".

As shown in Figure 6, the patches with decreasing NGPCL were mainly located in Lianyungang, Guanyun, and Gunnan in northeastern Jiangsu and Nanjing, Suzhou, and Wuxi in the south, while the areas with increasing NGPCL were mainly located in Rugao, Taixing, and Xinghua in central Jiangsu and Fengxian, Peixian, and Pizhou in the northwest.

3.2.1. Spatial Clustering Analysis of NGPCL

We utilized KDE to quantitatively illustrate the aggregation of NGPCL, as depicted in Figure 7. The results are segmented into three levels based on the natural breakpoint method: the low-value zone (0–22.84/km^2), the medium-value zone (22.84–43.78/km^2), and the high-value zone (43.78–97.08/km^2).

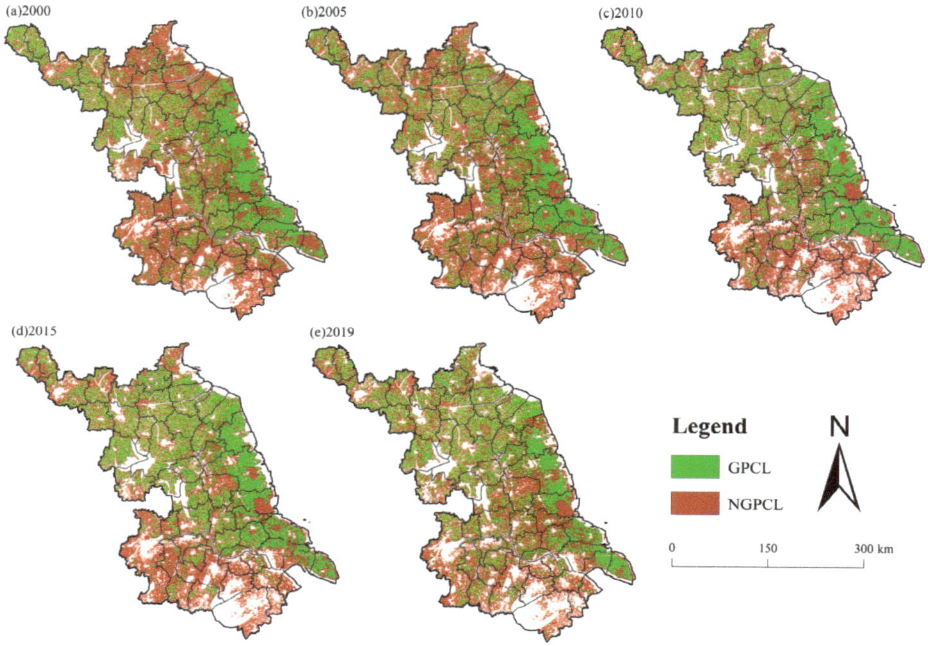

Figure 5. Spatial variation map of NGPCL in Jiangsu Province.

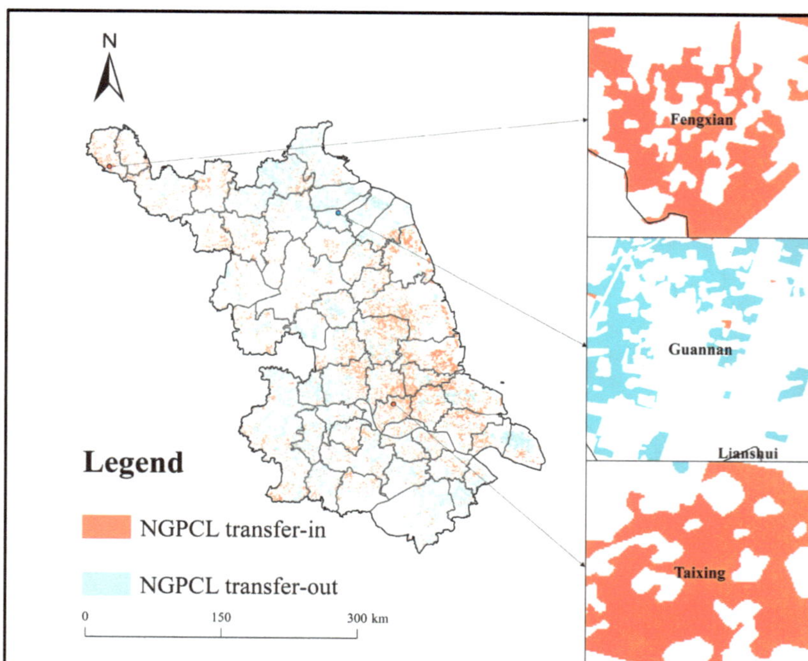

Figure 6. Distribution of transfer-in and transfer-out for NGPCL in Jiangsu Province.

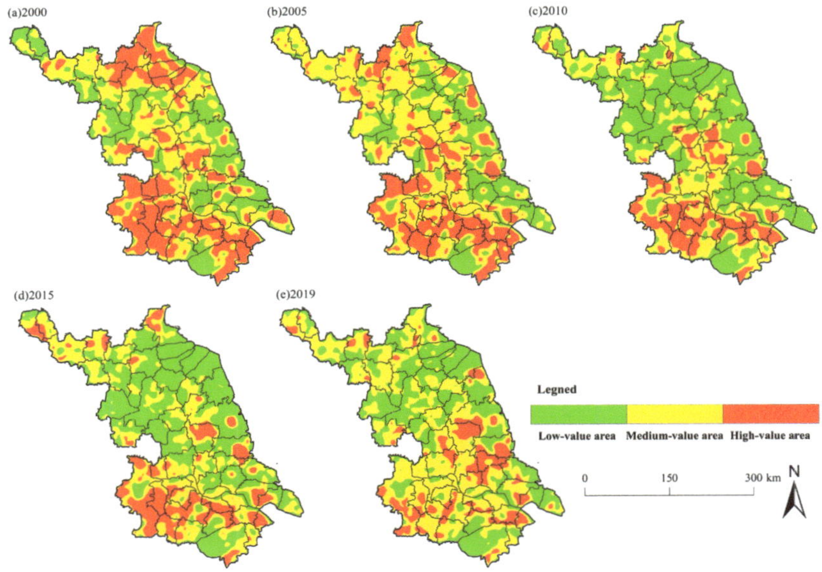

Figure 7. Spatial variation in KDE grade of NGPCL in Jiangsu Province.

Over the study period, the spatial distribution of NGPCL underwent a gradual transition from being "dense in the northeast and south and sparse in the center" to "dense in the middle of the south and sparse in the north". Initially, high-value areas were predominantly concentrated in Nanjing, Wuxi, and Suzhou in the south, as well as Lianyungang, Guanyun, and Gunnan in northeastern Jiangsu Province, indicating a high degree of NGPCL concentration in these regions.

Throughout the study period, the high-value zones in northeastern Jiangsu experienced a contraction, with low-value zones expanding and a decrease in NGPCL concentration. In central Jiangsu, high-value zones expanded, particularly in areas like Haian and Taizhou, indicating an increase in NGPCL concentration in the central region over time. In southern Jiangsu, high-value zones decreased, leading to the dominance of medium-density and high-density zones by the end of the study period, maintaining a relatively high level of NGPCL concentration.

3.2.2. Spatial Evolution of the NGPR

The county NGPR was used to classify the 54 county units into low-value areas (40% $\leq$ NGPR), medium-value areas (40% $\leq$ NGPR $\leq$ 58%), and high-value areas (NGPR $\geq$ 58%) based on the natural breakpoint method, as shown in Figure 8. Throughout the study period, the high-value zones decreased from 24 in 2000 to 15 in 2019, while the medium-value zones increased by 6, and the low-value zones increased by 3. In terms of spatial distribution, Jiangsu Province underwent a shift from "high NGPR in the north and south and lower in the center" to "high NGPR in the south and center and lower in the north". Across the entire study period, the southern part of Jiangsu exhibited a higher degree of NGP, primarily concentrated in the high-value areas. Cities like Nanjing and Suzhou remained in the high-value area for an extended period. In central Jiangsu, the NGPR was lower in 2000 but increased over time. Cities such as Taizhou and Taixing transitioned from the low-value area to the middle-value area, while Xinghua and Haian shifted from the middle-value area to the high-value area. In the northern part of Jiangsu, the NGPR was higher in 2000 but experienced a decline over time. Locations like Donghai, Guanyun, and Gunnan shifted from being high-value areas to low-value areas during this period.

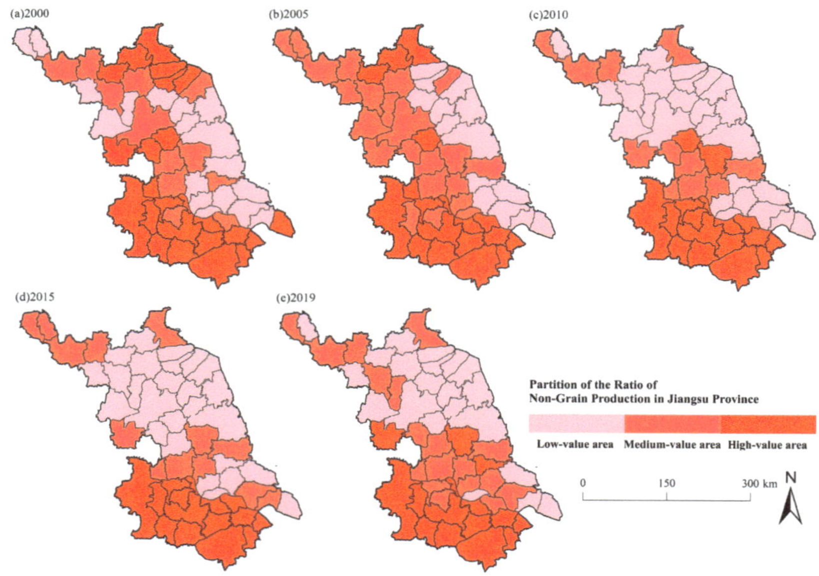

Figure 8. Spatial evolution of county NGPR in Jiangsu Province.

The spatial evolution of the NGPR in Jiangsu Province highlighted a clear developm trend where the NGP level was higher in the southern regions and lower in the northe areas, which is consistent with the findings of previous studies conducted by Li et al. [46, This pattern was influenced by the diverse geographic locations and variations in social a economic factors between the north and south of Jiangsu. The higher levels of urbanizati and economic development in southern Jiangsu drove a greater demand for NGP cro especially cash crops. Farmers in the southern region were more inclined towards N practices due to economic incentives. The process of urbanization in these areas creat more job opportunities in secondary and tertiary industries, reducing the dependence farmers solely on agriculture for their livelihoods and leading to a higher prevalence NGP. The central region of Jiangsu benefited from its proximity to the well-develop southern part of the province, underwent rapid urbanization and economic growth, a resulted in an increasing trend in NGP levels over the study period. On the contrary, t northeast, although it had a high degree of NGP in 2000, had a relatively low potent return on NGP due to relatively lagging social and economic development. As a resu the region experienced an increase in the status of grain cultivation and a decrease in t NGPR during the study period.

3.2.3. Spatial Correlation of NGP

The analysis presented in Figure 9 demonstrates that the global Moran's I for ea year in Jiangsu Province was consistently greater than 0 and passes the 99% confiden test. This indicates a strong positive correlation in the spatial distribution pattern of t NGPR over the five years under study. The small change in Moran's I suggests that t global characteristics exhibit a relatively smooth transition over time. The results of t global spatial autocorrelation analysis point towards a strong radiation effect in the NG distribution, indicating that the trend of NGP in one area drove NGP in neighboring regio

The spatial correlation pattern of the NGPR in Jiangsu Province, as illustrated Figure 10, was characterized by a prevalence of H-H agglomeration and L-L agglomerati High H-H agglomerations were predominantly situated in the southern region of t study area, while L-L agglomerations tend to concentrate gradually in the northeaste part of the province. The number of H-H agglomerations remained constant from 2000

2019, primarily clustered in the economically developed southern Jiangsu Province, where urbanization and agricultural diversity contribute to a high NGPR. In contrast, the number of L-L agglomerations increased and shifted towards the northeast of Jiangsu, an area marked by slower economic growth but favorable agricultural conditions conducive to grain crop production, resulting in a lower NGPR.

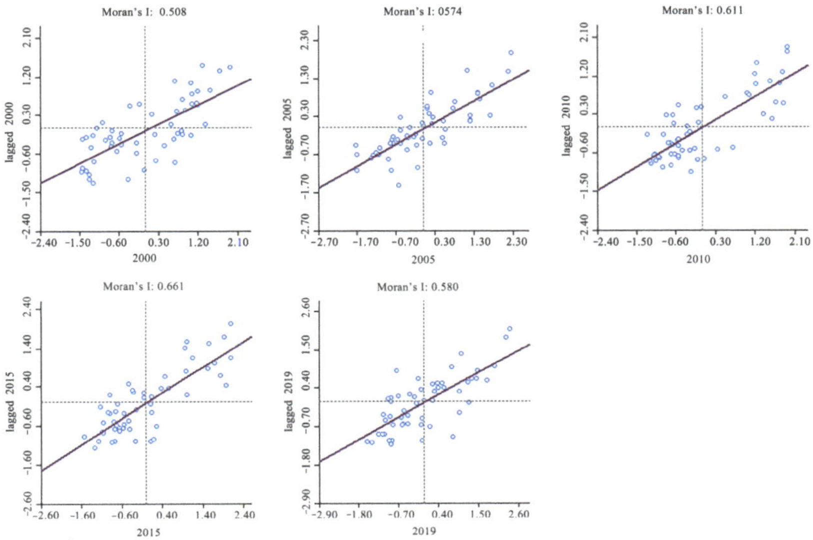

Figure 9. Moran scatterplot of NGPR in Jiangsu Province.

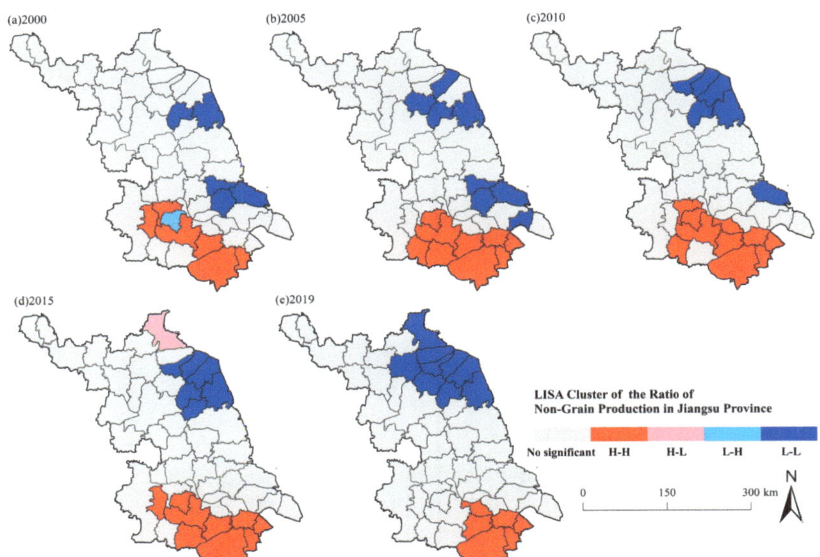

Figure 10. LISA clustering map of NGPR in Jiangsu Province.

3.3. *Multi-Scenario Simulation Results for NGP in Jiangsu Province*

The contribution of the drivers of NGPCL could be derived from LEAS (Figure 11). EVP, SSD, TEM, GDP, PRE, and population density had a high contribution to NGPCL, while distance to government departments, soil types and distance to roads at all lev also had a certain influence on NGPCL, and topographic factors had the lowest drivi force for NGPCL.

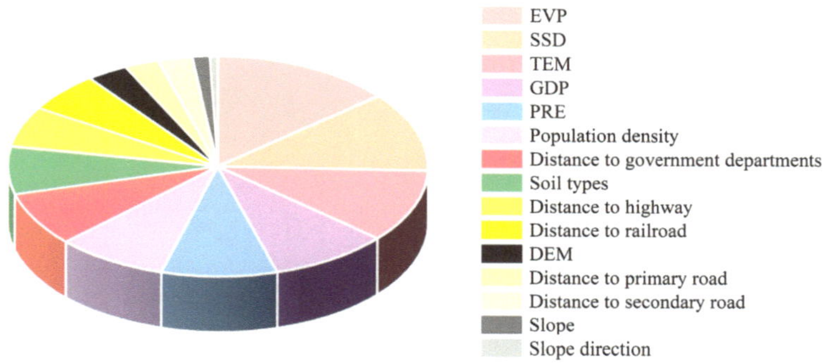

Figure 11. Distribution of contributions to driving factors of NGPCL.

The significant difference in the development trend of the NGP in Jiangsu Provir under different scenarios is depicted in Figures 12 and 13.

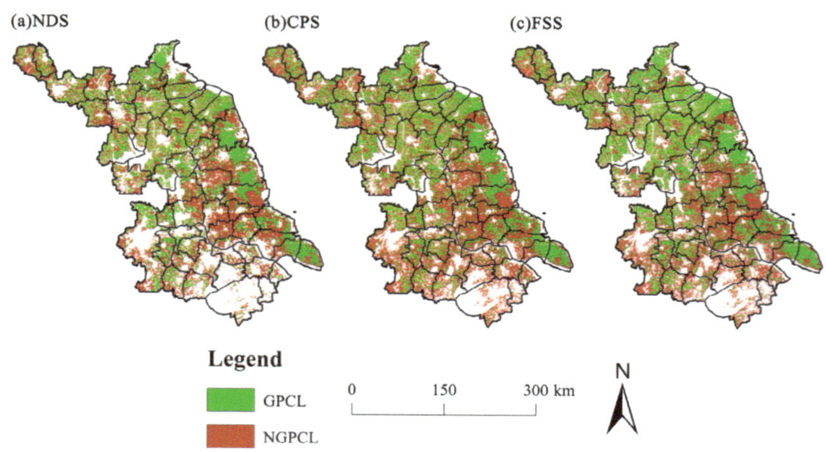

Figure 12. Spatial distribution of NGPCL under multi-scenario simulation.

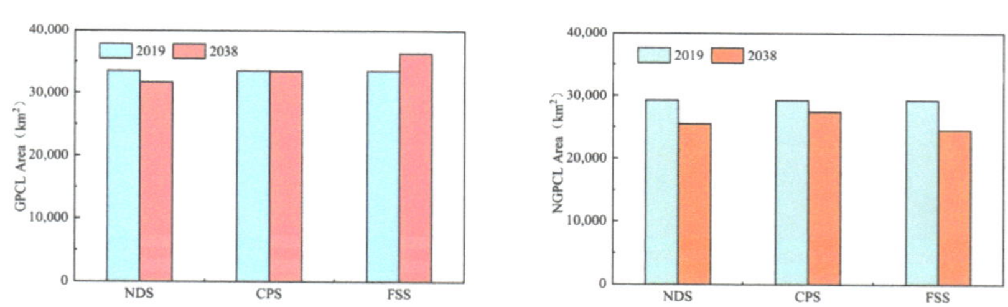

Figure 13. Area statistics of GPCL and NGPCL under multi-scenario simulation.

Under the NDS, the constraints imposed by land managers on the reduction in cr land and the NGP were not considered, and only the future development of NGPCL driv by a combination of socio-economic and natural factors was considered. Under this s nario, the area of cultivated land will decrease by 5569.77 km^2, with NGPCL decreasi by 12.96% and GPCL decreasing by 5.31% from 2019. Spatially, the distribution of NGPCL continued the characteristics of 2019, with more densely populated economically developed areas in central and southern Jiangsu and new NGPCL. Fengxian and Suining counties in northwestern Jiangsu have relatively poor conditions for grain production due to the limitations of precipitation, evaporation, and average temperature and also had some new NGPCL. In addition, there will be a clear mutual transformation between GPCL and NGPCL, which will be consistent with the trend of change from 2000 to 2019.

Under the CPS, land managers will rigorously enforce the compensation system for cultivated land occupation and the fundamental farmland protection system. They enhance cultivated land protection through the development of high-standard basic farmland and regulate the conversion of cultivated land to other land uses. In comparison to the NDS, the decline in cultivated land will diminish significantly, particularly as the encroachment of AS on cultivated land shows a notable decrease. In this scenario, the reduction in cultivated land amounts to 1915.27 km^2, with a 6.4% decrease in NGPCL and a 39.19 km^2 decrease in GPCL. Spatially, NGPCL will be predominantly concentrated in economically developed and urbanized areas such as Taicang, Changshu, Dongtai, and Haian in the southern and central regions of Jiangsu Province.

Under the FSS, land managers will not only impose restrictions on the conversion of cultivated land to other land types but also control the transformation of GPCL into NGPCL through various policies like cultivated land use control, NGP monitoring, and grain cultivation subsidies. This approach will lead to a decrease in the addition of new NGPCL and a reversal in the declining trend of GPCL. Cultivated land will diminish by 1931.08 km^2 in this scenario, with NGPCL decreasing by 16.40% and GPCL increasing by 2868.68 km^2. Notably, regions like Lianyungang, Guanyun County, and Xiangshui will witness substantial growth in new GPCL, aligning with local initiatives such as the grain industry development plan and the promotion of high-standard farmland construction. In addition, this is in line with the requirements of strengthening the construction of grain bases in the northern part of Jiangsu Province and steadily increasing the grain capacity, as mentioned in the Jiangsu Province Territorial Spatial Planning (2021–2035). Overall, this scenario will effectively maintain the quantity of GPCL, ensuring food security by slowing down the conversion of GPCL to NGPCL while safeguarding cultivated land.

The NGPR under the three scenarios was divided according to the previous grading (Figure 14). The distribution of the NGPR in Jiangsu Province under the three scenarios will maintain the overall distributional characteristics of 2019, with higher NGPR in the south and center and lower in the north.

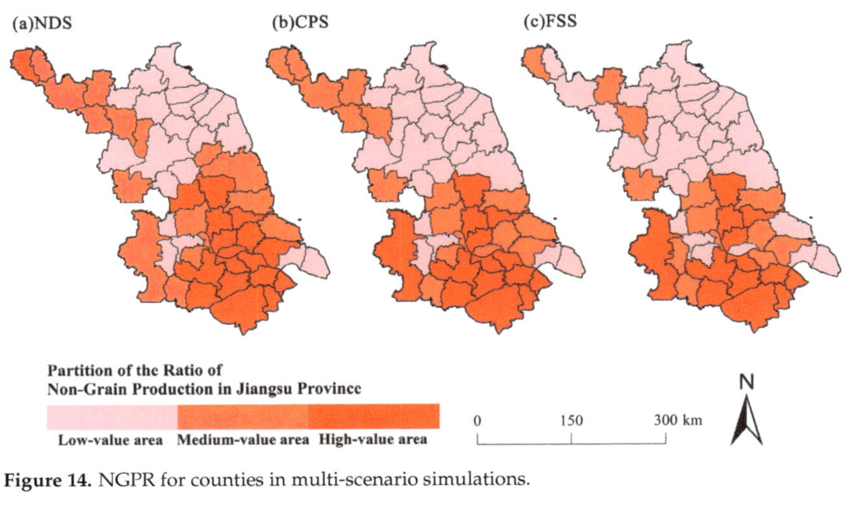

Figure 14. NGPR for counties in multi-scenario simulations.

Under the NDS, the overall NGPR will decrease, which is due to the fact that the r of decrease in NGPCL will be larger than that of GPCL. The high-value areas of the NG are Taizhou, Taixing, and Kunshan, while the medium-value areas will mainly be locat in Dongtai and Yangzhou, and the low-value area areas are Lianyungang and Guanyu

Under the CPS, the reduction in both GPCL and NGPCL is mitigated. In this scena the high NGPR zones will mainly be located in Changshu and Kunshan, the medium zo will mainly be located in Yangzhou and Dongtai, and the mild NGPR zones will mainly located in Guannan and Guanyun.

Under the FSS, the conversion of GPCL to NGPCL is limited and the NGPR will further reduced. The high-value areas of the NGPR in this scenario will be Suzhou a Taicang, the medium-value areas are mainly distributed in Yangzhou and Gaoyou, and t low-value areas will be Lianyungang and Guanyun.

4. Discussion

4.1. Research Significance

NGP has a negative impact on the stabilization of grain output. NGP takes away t cultivated land for grain production, which may lead to a shortage of grain planting are and thus trigger a decline in grain production and quality [14]. Furthermore, the sh to non-grain crops can alter the irrigation and drainage systems [6], making them mc suitable for the growth of non-grain crops or even causing complete destruction. So non-grain crops, like fast-growing poplar and eucalyptus, can contribute to soil infertil and acidification [10]. If there is a need to revert to planting grains in the future, significa economic investment may be required to restore the irrigation and drainage systems a address soil degradation issues caused by non-grain crop cultivation.

The decline in grain output caused by the NGP may lead to a decline in per cap grain availability, as well as an increase in grain prices, threatening human health and t stability of society [50]. While dietary adjustments, such as increasing the consumption meat and meat products, can alleviate some dependence on grains, the demand for gra crops for human consumption and livestock feed will persist over the long term [51,5 cross-border and inter-provincial grain trade can support the demand for grain up to certain extent, but it is unstable due to transportation costs, market price fluctuatio adjustments in policy factors, and changes in international relations [10]. Therefore, it crucial to control the unregulated expansion of NGP to stabilize grain yields.

The evolution of NGP is a complex and dynamic process, and it is difficult to comp hensively characterize the evolution of NGP by relying on the analysis of historical da only, and it lacks the consideration of multiple development scenarios in the future. T NGPCL prediction under multi-scenario simulation can complement the understanding the spatial and temporal evolution characteristics of NGPCL. Meanwhile, the simulation different future NGPCL distribution patterns based on different development goals such cultivated land protection and food security also provide an important reference for t current land management and cropland utilization in the region.

4.2. Exploration of Driving Mechanisms

We found that climatic factors contributed more to NGP, probably because gra production requires better water and heat conditions, and areas with poorer water and he conditions tend to favor NGP. Jiangyan, Rugao, and Taixing had higher EVPs and low PREs, and water resources were relatively scarce, leading to poorer conditions for gra production, so these counties experienced a significant trend of NGP change during t study period, with more newly added NGPCL. Soil types affects the suitability for gra production, which in turn affects the potential benefits of grain production and therefo also has an impact on the degree of NGP.

GDP and population density reveal the level of economic development and urbani tion in the counties, and central and southern Jiangsu have a higher level of developme than northern Jiangsu and therefore a higher degree of NGP. In developed regions, c sumption tends to be more diverse, with a higher demand for cash crops. At the same time, increased employment opportunities in secondary and tertiary industries have shifted farmers' focus away from viewing agricultural production as their sole livelihood. These reasons significantly reduced farmers' incentives to produce grain, leading to a higher likelihood of NGP. The proximity to various levels of roads reflects the convenience of transportation. The enhancement of the transportation system has accelerated the integration of urban and rural areas, fostering stronger inter-regional exchanges and communication. This development has also provided transportation facilities for the movement of cash crops, particularly those with shorter storage times like vegetables and fruits. As a result, farmers are more inclined towards pursuing NGP due to the comparative advantages and economic interests associated with these cash crops.

4.3. Policy Implications

Jiangsu Province holds a crucial position as a significant grain-producing region in China. It is imperative to address the uncontrolled expansion of non-grain crops and stabilize grain yields to ensure food security in Jiangsu Province and the nation as a whole. To achieve this goal, the following suggestions were proposed:

1. The AS has indeed resulted in the reduction of cultivated land in Jiangsu Province. Without effective restrictions, the trend of AS occupying a significant amount of cultivated land is likely to persist in the future. By enhancing the protection of cultivated land and imposing strict controls on the conversion of cultivated land to other land categories, such as, agricultural resources can be protected and a balance between urban development and agricultural production can be achieved. These measures are essential for ensuring food security, promoting sustainable land use practices, and preserving the long-term productivity of Jiangsu Province's agricultural sector.
2. The NGP is a result of farmers weighing various factors, including natural conditions, economic considerations, and other influences. To mitigate the impact of NGP, governmental intervention is crucial. Firstly, the government should implement financial policies such as subsidies and preferential loans for farmers and enhance the purchase price of grain crops to encourage their cultivation. Secondly, attention should be paid to guiding farming methods, and planting conditions should be improved by such means as strengthening the construction of water conservancy facilities. It should also build a production system that integrates grain processing, transportation, storage, and marketing in order to reduce production costs and increase income from grain production. Lastly, it is imperative to limit NGP through a comprehensive system of land use control.

4.4. Limitations and Future Prospects

1. The grain cultivation spatial distribution data were extracted from the 1 km resolution land use data provided by RESDC, which came from the same data source as the 30 m resolution land use data we used and had the same decoding method and classification method (we validated the two sets of data and obtained an overall accuracy of 94.57% and a kappa coefficient of 0.90 [53,54]). But when defining the NGPCL and the GPCL, there is still a certain degree of error, and subsequent studies should improve the resolution of the spatial distribution data of grain cultivation in order to identify NGPCL more accurately.
2. The evolution of NGPCL is affected by a variety of factors, including natural factors such as EVP and PRE, as well as socio-economic factors such as GDP and population density. In our study, we did not include factors that are difficult to quantify, especially policy factors, and due to the limitations of the PLUS model [27], we could only consider the driving factors as static factors. Future studies should aim to include policies as drivers and consider their dynamics.

5. Conclusions

Based on the spatio-temporal analysis of NGP in Jiangsu Province, we conducte multi-scenario simulation of the pattern of NGP in 2038, and came up with the followi conclusions.

1. From 2000 to 2019, the degree of NGP in Jiangsu Province decreased, with NG decreasing by 4.85%, the area of NGPCL decreasing from 35,991.07 km^2 to 20,270 km^2, and GPCL decreasing from 33,916.99 km^2 to 33,505.61 km^2. Meanwhile, terconversion of NGPCL and GPCL was common in Jiangsu Province. In additi the transformation of cultivated land to AS was also more common, and NGP was more likely to be transformed to AS than GPCL, which was a phenomenon th deserves the attention of land managers.
2. During the study period, Jiangsu Province exhibited a pattern of high NGPCL der ties in the central and southern regions, with lower densities in the northern areas. this paper, the NGPR of county units in Jiangsu Province was measured, and the e lution of NGPR illustrated the tendency for the level of NGP to be greater in the sou and central Jiangsu Province than in the north. The results of spatial autocorrelati illustrated the spatial correlation of NGPR in Jiangsu Province, and the tendency NGP drove and influenced the NGP in the surrounding areas. The NGPR was do nated by H-H agglomerations and L-L agglomerations, and H-H agglomerations we located in the southern part of Jiangsu Province, where the economy and urbanizati are high, while L-L agglomerations were gradually concentrating in the northeaste part of Jiangsu Province.
3. Climatic factors, GDP, and population density contributed more to NGP in Jiang Province, while distance to government departments, soil types, and distance to levels of highway also had some influence on NGP, and topographic factors dro NGP the least.

NGPCL will broadly maintain its current spatial pattern under the three scenari with denser concentrations in the south and center and sparser distribution of NGPCL the north. In the NDS, there will be a large reduction in both NGPCL and GPCL; in the C the reduction in NGPCL and GPCL will be somewhat suppressed; and in the FSS, the tre of reduction in GPCL will be reversed and the degree of NGP will be further suppresse

Author Contributions: Conceptualization, L.W. and C.J.; methodology, L.W. and C.J.; software, C A.L. and M.S.; validation, L.W. and C.J.; formal analysis, H.C., W.G. and H.O.; investigation, H.C. a W.G.; resources, C.J., L.W., H.C. and W.G.; data curation, C.J.; writing—original draft preparati L.W., C.J. and X.L.; writing—review and editing, L.W. and C.J.; visualization, L.W., C.J. and M supervision, L.W.; project administration, L.W.; funding acquisition, L.W. All authors have read a agreed to the published version of the manuscript.

Funding: This research was funded by the National Natural Science Foundation of China, gra number 42101252.

Data Availability Statement: Data available in a publicly accessible repository that does not iss DOIs: Publicly available datasets were analyzed in this study. This data can be found here: ht //www.nesdc.org.cn (accessed on 1 April 2023).

Conflicts of Interest: The authors declare no conflicts of interest.

References

1. Liu, Z.; Zhong, H.; Li, Y.; Wen, Q.; Liu, X.; Jian, Y. Change in grain production in China and its impacts on spatial supply a demand distributions in recent two decades. *J. Nat. Resour.* **2021**, *36*, 1413–1425. [CrossRef]
2. Zhang, C.; Wang, X.; Liu, Y. Changes in quantity, quality, and pattern of farmland in a rapidly developing region of China: A ca study of the Ningbo region. *Landsc. Ecol. Eng.* **2019**, *15*, 323–336. [CrossRef]
3. Wang, X.; Hao, J.; Dai, Z.; Haider, S.; Chang, S.; Zhu, Z.-Y.; Duan, J.; Ren, G.-X. Spatial-temporal characteristics of cropla distribution and its landscape fragmentation in China. *Farming Syst.* **2024**, *2*, 100078. [CrossRef]

Wu, Y.; Zhang, P.; Yu, Y.; Xie, R. Progress Review on and Prospects for Non-grain Cultivated Land in China from the Perspective of Food Security. *China Land Sci.* **2021**, *35*, 116–124.
Xie, Y.; Wang, Z.; Wang, Y.; Zheng, J.; Xiang, S.; Gao, M. Spatial-temporal variation and driving types of non-grain cultivated land in hilly and mountainous areas of Chongqing. *J. Agric. Resour. Environ.* **2024**, *41*, 15–26. (In Chinese)
Cheng, X.; Tao, Y.; Huang, C.; Yi, J.; Yi, D.; Wang, F.; Tao, Q.; Xi, H.; Ou, W. Unraveling the Causal Mechanisms for Non-Grain Production of Cultivated Land: An Analysis Framework Applied in Liyang, China. *Land* **2022**, *11*, 1888. [CrossRef]
Li, T.; Hao, D. Current situation of"non-grain production"of cultivated land in China and the research progress of retillage and fertilization technology. *Chin. J. Appl. Ecol.* **2023**, *34*, 1703–1712.
Qiu, L.; Zhu, J.; Pan, Y.; Dang, Y.; Wu, S. Distribution Characteristics, Ecological Risks, and Source Identification of Heavy Metals in Cultivated Land Under Non-grain Production. *Environ. Sci.* **2023**, *44*, 2829–2837.
Zhu, Z. The Causes and Countermeasures of Non-grain during the Process of Land Circulation. *Rural Econ.* **2009**, *25*, 13–14.
Su, Y.; Li, C.L.; Wang, K.; Deng, J.S.; Shahtahmassebi, A.R.; Zhang, L.P.; Ao, W.J.; Guan, T.; Pan, Y.; Gan, M.Y. Quantifying the spatiotemporal dynamics and multi-aspect performance of non-grain production during 2000–2015 at a fine scale. *Ecol. Indic.* **2019**, *101*, 410–419. [CrossRef]
Wang, Y.; Chen, Y.; Yi, X.; Xiao, B. The Non-Grain Problem In The Process Of Land Trabsfer And The Countermeasures. *Chin. J. Agric. Resour. Reg. Plan.* **2011**, *32*, 13–16.
Xie, P. Research on Influencing Factors of Non-agriculturalization and Non-grainization of Cultivated Land in Villages and Towns. Master's Thesis, Nanjing Agricultural University, Nanjing, China, 2021.
Zhao, X.; Liu, Z. "Non-Grain" or "Grain-Oriented": An Analysis of Trend of Farmland Management. *J. South China Agric. Univ.* **2021**, *20*, 78–87.
Chen, Y.; Li, M.; Zhang, Z. Does the Rural Land Transfer Promote the Non-Grain Production of Cultivated Land in China? *Land* **2023**, *12*, 688. [CrossRef]
Zhang, Y.; Feng, Y.; Wang, F.; Chen, Z.; LI, X. Spatiotemporal differentiation and driving mechanism of cultivated land non- grain conversion in Guangdong Province. *Resour. Sci.* **2022**, *44*, 480–493. [CrossRef]
Chen, H.; Tan, Y.; Deng, X.; Xiao, W. Progress and prospects on information acquisition methods of abandoned farmland. *Trans. Chin. Soc. Agric. Eng.* **2020**, *36*, 258–268.
Su, Y.; Qian, K.; Lin, L.; Wang, K.; Guan, T.; Gan, M. Identifying the driving forces of non-grain production expansion in rural China and its implications for policies on cultivated land protection. *Land Use Policy* **2020**, *92*, 104435. [CrossRef]
Xu, G.; Lu, L.; Yang, C.; Huang, L.; Wan, C. Identification and driving mechanisms of non-grain cultivated land in hilly and mountainous areas based on multi-temporal Sentinel-1A images. *Trans. Chin. Soc. Agric. Eng.* **2023**, *39*, 236–245.
Hao, Q.; Zhang, T.; Cheng, X.; He, P.; Zhu, X.; Chen, Y. GIS-based non-grain cultivated land susceptibility prediction using data mining methods. *Sci. Rep.* **2024**, *14*, 4433. [CrossRef] [PubMed]
Li, Q.; Chen, W.; Shi, H.; Zhang, S. Assessing the environmental impact of agricultural production structure transformation-Evidence from the non-grain production of cropland in China. *Environ. Impact Assess. Rev.* **2024**, *106*, 107489. [CrossRef]
Ran, D.; Zhang, Z.; Jing, Y. A Study on the Spatial–Temporal Evolution and Driving Factors of Non-Grain Production in China's Major Grain-Producing Provinces. *Int. J. Environ. Res. Public Health* **2022**, *19*, 16630. [CrossRef]
Tang, C.; Yi, Y.; Kuang, Y. Research on Spatio-Temporal Complexity Evolution and Influencing Factors of "Nongrain" in Guangxi. *Discrete Dyn. Nat. Soc.* **2022**, *2022*, 1181108. [CrossRef]
Chen, F.; Liu, J.; Chang, Y.; Zhang, Q.; YU, H.; Zhang, S. Spatial Pattern Differentiation of Non-grain Cultivated Land and Its Driving Factors in China. *China Land Sci.* **2021**, *35*, 33–43.
Liang, X.; Jin, X.; Liu, J.; Yin, Y.; Gu, Z.; Zhang, J.; Zhou, Y. Formation mechanism and sustainable productivity impacts of non-grain croplands: Evidence from Sichuan Province, China. *Land Degrad. Dev.* **2023**, *34*, 1120–1132. [CrossRef]
Chen, W.; Liao, Y.; Kong, X. Characteris, Drivers and Control of Non-grain Production on Permanent Basic Farmland Based on Plot Scale. *Trans. Chin. Soc. Agric. Mach.* **2023**, *54*, 114–124.
Zhang, O.; Jiang, C. Analysis on Differences of "Non-Grain" of Different Types Farmers in Transfer-In Farmland. *Finance Trade Res.* **2016**, *27*, 24–31.
Liang, X. Understanding the drivers of sustainable land expansion using a patch-generating land use simulation (PLUS) model: A case study in Wuhan, China. *Comput. Environ. Urban Syst.* **2021**, *85*, 101569. [CrossRef]
Liu, X.; Li, X.; Zhang, Y.; Wang, Y.; Chen, J.; Geng, Y. Spatiotemporal evolution and relationship between construction land expansion and territorial space conflicts at the county level in Jiangsu Province. *Ecol. Indic.* **2023**, *154*, 110662. [CrossRef]
Wang, L.; Liang, A.; Li, X.; Jiang, C.; Wu, J.; Omrani, H. Understanding Recessive Transition of Cultivated Land Use in Jilin Province, China (1990–2020): From Perspective of Productive-Living-Ecological Functions. *Land* **2023**, *12*, 1758. [CrossRef]
Li, X.; Wang, L.; Pijanowski, B.; Pan, L.; Omrani, H.; Liang, A.; Qu, Y. The Spatio-Temporal Pattern and Transition Mode of Recessive Cultivated Land Use Morphology in the Huaibei Region of the Jiangsu Province. *Land* **2022**, *11*, 1978. [CrossRef]
Luo, Y.; Zhang, Z.; Li, Z.; Chen, Y.; Zhang, L.; Cao, J.; Tao, F. Identifying the spatiotemporal changes of annual harvesting areas for three staple crops in China by integrating multi-data sources. *Environ. Res. Lett.* **2020**, *15*, 074003. [CrossRef]
Li, X.; Wu, K.; Yang, Q.; Hao, S.; Feng, Z.; Ma, J. Quantitative assessment of cultivated land use intensity in Heilongjiang Province, China, 2001–2015. *Land Use Policy* **2023**, *125*, 106505. [CrossRef]

33. Zhang, M.; Wang, J.; Feng, Y. Temporal and spatial change of land use in a large-scale opencast coal mine area: A comp network approach. *Land Use Policy* **2019**, *86*, 375–386. [CrossRef]
34. Qiao, F.; Zhang, Y.J.; Gu, Z.Y.; Wang, J.; Zhang, G.Y. Grid Task Scheduling Algorithm Based on Non-Conflict Degree. *Appl. M Mater.* **2012**, *239–240*, 1497–1500. [CrossRef]
35. Xiang, H.; Ma, Y.; Zhang, R.; Chen, H.; Yang, Q. Spatio-Temporal Evolution and Future Simulation of Agricultural Land Us Xiangxi, Central China. *Land* **2022**, *11*, 587. [CrossRef]
36. Zhu, Z.; Duan, J.; Li, R.; Feng, Y. Spatial Evolution, Driving Mechanism, and Patch Prediction of Grain-Producing Cultiva Land in China. *Agriculture* **2022**, *12*, 860. [CrossRef]
37. Jiang, Z. Spatial-temporal Change and Multi-scenario Prediction of Land Use in Urumqi. Master's Thesis, Xinjiang Norr University, Xinjiang, China, 2022.
38. Wang, Y. Study on Varation Mechanism of Land Use Pattern and its Optimization Model in Songnen High Plain. Ph.D. The Northeastern University, Shenyang, China, 2017.
39. Wu, N.; Wei, Y.; Li, L.; Yang, H. Spatial distribution of non-grain crops and formation mechanism: Empirical analysis of Ningl County, Henan Province. *Prog. Geogr.* **2023**, *42*, 1298–1310. [CrossRef]
40. Peng, H.; Song, J.; Yin, W. Multi scenario simulation of cultivated land landscape pattern in Western Hubei mountainous a based on PLUS model. *Hubei Agric. Sci.* **2023**, *62*, 51–59.
41. Pontius, R.G.; Boersma, W.; Castella, J.-C.; Clarke, K.; De Nijs, T.; Dietzel, C.; Duan, Z.; Fotsing, E.; Goldstein, N.; Kok, K.; et Comparing the input, output, and validation maps for several models of land change. *Ann. Reg. Sci.* **2008**, *42*, 11–37. [CrossR
42. Pontius, R.G.; Millones, M. Death to Kappa: Birth of quantity disagreement and allocation disagreement for accuracy assessm *Int. J. Remote Sens.* **2011**, *32*, 4407–4429. [CrossRef]
43. Yu, Z.; Zhao, M.; Gao, Y.; Wang, T.; Zhao, Z.; Wang, S. Multiscenario Simulation and Prediction of Land Use in Huaibei C Based on CLUE-S and PLUS Models. *Appl. Sci.* **2023**, *13*, 7142. [CrossRef]
44. Li, J.; Yang, D.; Wu, F.; Chen, R.; He, W. Dynamic Simulation of Land Use Chnages and Assessment of Carbon Storage in Kunmi Citu Based on Plus and InVEST Models. *Bull. Soil Water Conserv.* **2023**, *43*, 378–387.
45. Zhu, W.; Zhang, J.; Cui, Y.; Zheng, H.; Zhu, L. Assessment of territorial ecosystem carbon storage based on land use char scenario: A case study in Qihe River Basin. *Acta Geogr. Sin.* **2019**, *74*, 446–459.
46. Wang, C.; Zhi, X. "Grain-oriented" or "Non-grain-oriented": Spatial-temporal analysis of the agricultural planting structure Jiangsu Province at county level. *Hubei Agric. Sci.* **2023**, *62*, 64–70.
47. Tan, Y. Study on food demand and availability of cultivated land resource based on a balanced diet in Jiangsu Province. Maste Thesis, Nanjing University, Nanjing, China, 2020.
48. Liu, X. Research on Grain Supply-Demand Balance in Jiangsu Province. Master's Thesis, Nanjing Agricultural Univers Nanjing, China, June 2012.
49. Li, Y.; Xia, C.; Xia, Y.; Huang, Z.; Hua, C. The Evolution of the Spatio-temporal Pattern of "Non-grain" Cultivated Land in Jiang Province and Its Enlightenment to Rural Revitalization. *Sci. Technol. Ind.* **2023**, *23*, 229–238.
50. Chen, Y.; Wang, J.; Zhang, F. New patterns of globalization and food security. *J. Nat. Resour.* **2021**, *36*, 1362–1380. [CrossRef]
51. Zhou, Y.; Liu, Z. A social-ecological network approach to quantify the supply-demand-flow of grain ecosystem service. *J. Cle Prod.* **2024**, *434*, 139896. [CrossRef]
52. Li, J.; Xiao, Q.; Wu, H.; Li, J. Unpacking the Global Rice Trade Network: Centrality, Structural Holes, and the Nexus of Fo Insecurity. *Foods* **2024**, *13*, 604. [CrossRef]
53. Cohen, J. A coefficient of agreement for nominal scales. *Educ. Psychol. Meas.* **1960**, *20*, 37–46. [CrossRef]
54. Landis, J.; Koch, G. The measurement of observer agreement for categorical data. *Biometrics* **1977**, *33*, 159–174. [CrossRef]

land

ticle

Disembedding and Disentangling Grassland Valuation: Insights into Grassland Management Institutions and Ecological Research in China

ıan Zhang [1,*] and KuoRay Mao [2]

[1] Institute of Sociology, Chinese Academy of Social Sciences, No 5 Jiannei Dajie Street, Dongcheng District, Beijing 100732, China
[2] Department of Sociology, Colorado State University, Fort Collins, CO 80523, USA; kuoray.mao@colostate.edu
* Correspondence: zhangqian@cass.org.cn

Abstract: After two decades of implementing top-down grassland restoration projects focused on reducing livestock numbers and pastoralist populations, the Chinese government's well-funded efforts have not significantly reversed grassland degradation. This study reviews the institutional changes in grassland management over the past forty years, highlighting the Livestock and Grassland Double Contract Household Responsibility System of the early 1980s and the Grassland Ecological Reward and Compensation Policy introduced in 2011. It demonstrates how these institutional transformations have shaped pastoralists' evolving understanding of grassland value and reveals that commodifying grassland's economic and ecological value has led to the capitalization of nature, disembedding husbandry from grassland production, and undermining the effectiveness of conservation projects. This article also showcases the development of grassland ecology research in China, noting its increasing detachment from a holistic understanding of ecosystems and the interdisciplinary needs of management practices. The disjunction between grassland ecology research and practical management has resulted in a lack of techniques aligned with local ecological and socioeconomic contexts. This article champions active engagement with and protection of pastoralist communities to reintegrate grasslands' true economic and ecological value into management practices, thereby effectively restoring degraded grasslands and achieving sustainable management.

Keywords: grassland degradation; social–ecological system; commodification; capitalization; ecological restoration

ation: Zhang, Q.; Mao, K. embedding and Disentangling ıssland Valuation: Insights into ıssland Management Institutions l Ecological Research in China. ıd **2024**, *13*, 1218. https://doi.org/ 3390/land13081218

ademic Editor: Yongheng Gao

:eived: 5 June 2024
vised: 21 July 2024
:epted: 2 August 2024
ɔlished: 6 August 2024

1. Introduction

To address the issue of grassland degradation in China, which became evident in 2000, the central government has invested substantial funds and resources in a series of grassland protection projects [1,2]. These projects have been based on the simplified logic that overgrazing is the direct cause of grassland degradation [3]. Consequently, the restoration of degraded grasslands has been pursued through the enclosure of grassland and efforts to reduce livestock numbers [4]. The value of grassland has been monetized to facilitate this reduction, and pastoralists are supposed to receive ecological compensation to make up for their lost income [1,3]. However, despite the compensation provided to pastoralists, livestock numbers have not decreased as expected, and the degraded grasslands have not recovered as planned [1,5]. To understand why these well-intentioned projects have failed to restore degraded grasslands, it is essential to investigate pastoralists' views on the changing ecological value of grassland and, subsequently, their perceptions of grassland conservation measures. This understanding has the potential to bridge the gap between policy implementation and on-the-ground realities and holds promise for more effective and sustainable grassland management strategies.

1.1. Simplified Logic of the Recovery Method: Grazing Ban and the Invalid Results

Since 2000, the issue of grassland degradation has garnered significant public attenti due to several widespread sandstorms originating from degraded grasslands and extendi southward and eastward, even reaching Korea and Japan [6]. These events highlight the severity of the problem at that time. According to the Third National Grassla Resource Survey, approximately 38.67 million hectares of grassland in the Inner Mongo Autonomous Region, accounting for 56.9% of the available grassland area, are moderat or severely degraded. The productivity of natural grasslands has generally decreased 30% to 70% [7]. Overgrazing, attributed to pastoralist activities, has been identified as t primary cause of grassland degradation [8].

In response to this pressing issue, the Chinese central government has spearheade series of projects aimed at restoring degraded grasslands. The flagship initiative, the Retu ing Grazing Land to Grassland (RGLG) project, was launched in 2003. This comprehensi project is designed to curb overgrazing through various measures, including year-rou and spring grazing bans. The government has committed significant resources, includi manpower, fencing materials, and financial investments, to this project. By 2018, the cent government had poured a total of USD 3.84 billion into the RGLG project [9]. In addition the RGLG project, the Beijing and Tianjin Sandstorm Source Control Project and the Pub Welfare Forest Conservation Project were rolled out in 2002 and 2004, respectively. In 20 the Grassland Ecological Reward and Compensation Policy (GERCP) was introduced, w the total investment exceeding USD 19.5 billion to compensate pastoralists for reduci their livestock through grazing bans, making it the largest Payment for Ecosystem Servi (PESs) project in China [10].

However, despite the substantial efforts and investments, the situation on the grou does not seem to be improving as expected. The number of livestock has not decreas even with the compensation allocated to alleviate grazing pressure on grasslands. In fa in Inner Mongolia, the livestock numbers have continued to rise, from 99.3 million she units in 2015 to 102.3 million in 2017 and a staggering 109.3 million in 2022. This trend 'partial improvement but overall deterioration' in terms of China's grassland degradati has not been fundamentally reversed [11]. This begs the question: despite the significa investments in terms of fiscal transfer funds and human power, why has the situation terms of grassland degradation not been fundamentally addressed?

1.2. Grassland Ecology in China: Payment for Ecosystems as a Solution

Amplified concerns regarding escalating environmental crises have led to evolvi strategies to meticulously quantify the cost of environmental degradation (value lost) a monetize the benefits (value gained) derived from nature, which are increasingly bei integrated into decision-making frameworks. These approaches serve as potential to to enhance both social and ecological sustainability through a proper valuation of the sources [12–14]. The ecosystem services model supposedly signifies a shift from traditio hierarchical, directive management to more decentralized, ostensibly voluntary decisi frameworks that enable policymakers and resource managers to integrate environmental ternalities into market transactions [13,15,16]. In recent decades, an intriguing phenomen has emerged among Chinese grassland scholars, who have become increasingly captivat by the concept of payment for ecosystem services as an innovative policy remedy for t rampant degradation of grasslands. By financially incentivizing local communities a stakeholders to engage in sustainable practices, PESs schemes present a novel paradig that supposedly harmonizes ecological stewardship with economic benefits, mitiga environmental degradation, and fosters a synergistic relationship between human activ and ecological preservation.

Conservation experts from various fields support assigning a monetary value to t environmental benefits that contribute to human well-being. This approach can be inc porated into policy decisions alongside traditional market values and thereby, lead to t creation of market-based tools, such as PES schemes. These tools provide compensation natural resource managers for implementing environmentally friendly management practices [15,17]. Furthermore, these schemes may include conditional agreements, payments only if a service is successfully delivered, or penalties for failure to adhere to agreed-upon service provisions [16,18–20]. Although these schemes are theoretically appealing, they often encounter practical difficulties, including high transaction costs and power imbalances between the negotiating parties, and often involve the state mediating to ensure service delivery through a mix of voluntary and regulatory mechanisms [15,16,21].

The practical implementation of PESs often significantly diverges from idealized models. A substantial body of critical literature contends that PESs schemes may perpetuate social and environmental injustices by altering traditional livelihoods and excluding Indigenous and long-tenured groups from their lands [16,22]. Moreover, these schemes have been criticized for commodifying nature, which oversimplifies the intricate ecological and social dynamics involved. This could result in policy decisions that prioritize economic efficiency at the expense of ecological health and human well-being [16]. As PESs has become a cornerstone of global environmental governance, the social and ecological implications of these models warrant careful consideration and adjustment to ensure their positive contribution to a just transition and environmental sustainability.

In China, researchers are increasingly attracted to PESs as a policy solution to overgrazing due to its potential to foster long-term ecological resilience and socioeconomic prosperity, while aligning with the country's highly interventionist environmental regulatory framework [23]. This model regards environmental governance and ecology as a state-led development project and is a pivotal element of the broader ecological modernization initiative [23,24]. This regulatory approach positions the state as a mediator between environmental and economic concerns, incorporating critical imperatives, such as revenue generation, poverty alleviation, environmental protection, and legitimation, into comprehensive national ecological reforms [25,26].

The state's top-down implementation model internalizes environmental issues by emphasizing the technocrats' quantification of natural resources and environmental externalities, which, although effective, may inadvertently perpetuate social inequities [27]. The advent of remote sensing technology has further reduced the costs associated with environmental monitoring and has improved the accountability of policy outcomes by providing large-scale, continuous, and repeatable observations. Remote sensing offers significant advantages, including monitoring extensive and inaccessible areas, delivering real-time data, and minimizing the necessity for extensive fieldwork [28]. The synergy between earth system sciences and grassland ecologists has bolstered the implementation of this top-down, interventionist conservation strategy [29].

However, the top-down environmental regulatory regime is not merely an arbitrary administratively managed system, but a comprehensive system that merges legal, economic, and political instruments to enhance regulatory enforcement [25,30]. This interventionist state approach, driven by the development of ecological sciences, provides an epistemological foundation for legitimizing state-led technocratic practices involving socioenvironmental engineering and naturalizing social inequalities between urban and rural populations [31]. Consequently, the perpetuation of PESs has become central to environmental governance, state formation, and the continued uneven valuation and stratification in the use of land resources in the grassland regions of China [23]. One example of how this technocratic approach may result in the uneven utilization of land resources in China is the limitation of relying solely on remote sensing technology to assess the progress of grassland restoration projects. A primary critique of remote sensing is the absence of ground truth data, which is crucial for validating and calibrating remote sensing measurements [32]. Without ground-based verification, the measurement accuracy and reliability of satellite-derived data remain questionable. Moreover, relying exclusively on remote sensing data neglects the disconnect between the stated goal of grassland protection against overgrazing and the on-the-ground techniques local officials use to assess and monitor grassland degradation [33]. While remote sensing and test plot measurements are explicitly geared towards monitoring grass production, the carrying capacity of lo grasslands cannot be accurately assessed without systematically coupling data on fora growth with data on livestock numbers.

Additionally, information gathered locally is often compiled solely to fulfill sh term policy goals, which frequently change according to the formation of different gr development agendas [34,35]. This information blockage results in the viewpoints rational planners and technicians employed by the central state taking precedence o those of local pastoralists. Consequently, the goals of conservation programs shift towa fulfilling program quotas set by the bureaucracy, rather than achieving long-term ecologi sustainability and economic viability for local communities [30,35]. As such, it is import to hear the perspectives of pastoralists on the causes of grassland degradation and value of grassland ecology, which will be explored in the following sections.

1.3. Pastoralists' Perceptions on Grassland Value: A Missing Perspective

Undoubtedly, the perceptions of pastoralists regarding grassland conservation m sures are crucial to the success of these policy programs. Grassland serves as a ho production resource, or base for compensation requests for pastoralists. The differing vie on the functions of grassland and the value they generate are decisive in determini whether grasslands can be truly and sustainably protected. It should be noted that change in valuation by pastoralists for grassland conservation is not solely triggered PESs-based policy projects. In fact, pastoralists' perceptions of the value of grassland ha been gradually transforming since the implementation of the Livestock and Grassla Double Contract Household Responsibility System (LGDCHRS) in the early 1980s, wh brought fundamental changes to the grassland property rights regime.

Various scholars have provided explanations from different perspectives for the f ure of grassland restoration projects, which can be categorized into the following th aspects. The first and mainstream explanation, as articulated by the Deputy Director the Grassland Supervision Center of the State Forestry and Grassland Administration 2018, is overgrazing. Consequently, the foremost conservation measure is to elimin overgrazing [8]. Overgrazing is seen as a characteristic of traditional nomadic anin husbandry. More than 80% of China's natural grassland has degraded due to this pract leading to low productivity and production methods that, at the expense of ecology, ha rendered the grassland an unsustainable, severely degraded, and economically deple ecosystem [36].

The second explanation posits that these conservation projects oversimplify the co plexity of the social–ecological system of grasslands. For example, the appraisal mec nisms at the national level fall short of capturing most of the value of pastoral syste because they operate with simplifications that focus on market exchanges and the form economy [37]. These projects do not consider the variability and uncertainty of the dryla environment but tend to use impractical and generalizable solutions by simplifying analysis of complex problems, such as imposing a unified carrying capacity standard each pastoral household within a township [23,38].

The third explanation is the "tragedy of commodity", whereby the implementation PESs has altered local internal conservation mechanisms [1,39]. These mechanisms inclu disrupting the mutual dependence between livestock and vegetation, excluding pastorali from grassland conservation and supervision, and encouraging both pastoralists a outsiders to engage in activities harmful to grasslands, such as digging for desert stor medicinal herbs, and catching *Mesobuthus eupeus* to sell [40]. One criticism of PESs that it may shift people's decision-making logic from considering actions that benefit social–ecological system to those that are most advantageous for individuals [41].

Over the past four decades, a series of institutional and policy changes have sign cantly altered pastoralists' perceptions of the value and significance of grasslands. Howev few scholars have explored the reasons for grassland degradation and slow recovery fr this perspective. This article aims to elucidate the shifts in the perceived value of gra lands among pastoralists and examine how these shifts influence the implementation of grassland conservation policies in China. By analyzing these changes, this study seeks to provide a deeper understanding of the complex interplay between local knowledge systems and policy frameworks. The insights gained from this analysis will offer more effective strategies for sustainable grassland management by integrating local knowledge and practices within ecological studies. The objective of this study is to emphasize the necessity of holistic and adaptive management strategies that account for the variability and complexity of grassland ecosystems. Engaging pastoralists in the decision-making process and ensuring their active participation in conservation efforts are critical factors for achieving effective outcomes.

2. Methods and Data

This study utilizes data from government documents, policy papers, and official statistics to examine the implementation of LGDCHRS since the 1980s and PESs in the Inner Mongolia Autonomous Region of China, from 2011 to 2022. The policy on grassland management has transitioned from a decentralized governance model to a more top-down administrative approach imposed by the Chinese central government. We selected and analyzed the planning and implementation guidelines of grassland management policies on the Xilingol grassland of Inner Mongolia. This analysis aids in understanding the formal approaches to ecological zoning and grassland preservation, as well as the local practices used to deliver compensation for ecosystem services and enforcement.

We conducted over 200 in-depth interviews with pastoralists, village cadres, and banner husbandry officials in the Xilingol grassland (Figure 1), during multiple field trips between 2001 and 2022. These interviews, conducted with consent, were recorded for subsequent analysis. The survey questionnaire addresses several key issues: the implementation process of grassland contracting, changes in water resource utilization, impacts of climate change, costs and benefits of pastoral livestock management, strategies and costs of pastoralists in coping with natural disasters, changes in cooperative and reciprocal practices among pastoralists in terms of livestock management and disaster resilience, the impact of fencing on livestock management and grassland protection, the effects of payments for ecosystem services (PESs) on pastoral livelihoods, and the perceptions of pastoralists towards PESs. The qualitative data from these interviews reflect pastoralists' evolving perspectives on grassland conservation programs and the varying informal practices that grassroots cadres employ to enforce compliance. The qualitative longitudinal interviews comprehensively portray how grassland management policies have affected pastoralists' perceptions of grassland value and modified their herding practices. This, in turn, has shaped the evolving socioeconomic and human–nature relations in the grassland regions of Inner Mongolia.

Figure 1. Research Area in Inner Mongolia, China.

To extend the robustness of our analysis, we employed a mixed methods approa Quantitative data from official statistics published in the Animal Husbandry Yearbo (1985–2022) were triangulated with qualitative insights from policy documents (See Tabl and interviews to ensure a comprehensive understanding of the policy impacts. categorized different codes to identify recurring themes and patterns within the qualitat data, providing a nuanced view of the local implementation practices and the resulti socio-ecological changes. This methodology allows for a detailed examination of both formal policy frameworks and the grassroots-level responses, offering insights into effectiveness and challenges of grassland conservation efforts in Inner Mongolia.

Table 1. Grassland Management Policies and Guidelines in China.

Year	Policy and Guidance Documents	Issuing Agencies
1982	Summary of the National Rural Work Conference	Central Committee Document No. 1
1983	Several Issues on Current Rural Economic Policies	Central Committee Document No. 1
1996	Regulations on Further Implementing and Improving the "Double Rights and One System" of Grasslands in Inner Mongolia Autonomous Region	Inner Mongolia Autonomous Region, No. 168
2011	Guidance on the Implementation of the Grassland Ecological Protection Subsidy and Reward Mechanism Policy for 2011	Ministry of Agriculture, Ministry of Finance
2016	Notice on Issuing Opinions on Improving the Separation of Ownership Rights, Contracting Rights, and Management Rights of Rural Land	General Office of the Central Committee of the Communist Party of China, General Office of the State Council, No. 67
2016	Guidance on the Implementation of the New Round of Grassland Ecological Protection Subsidy and Reward Policy (2016–2020)	Ministry of Agriculture, Ministry of Finance
2021	Guidance on the Implementation of the Third Round of Grassland Ecological Protection Subsidy and Reward Policy	Ministry of Finance, Ministry of Agriculture and Rural Affairs, National Forestry and Grassland Administration
2021	Opinions on Deepening the Reform of the Ecological Protection Compensation System	General Office of the Central Committee of the Communist Party of China, General Office of the State Council

3. Results

3.1. An Overview of Grassland Management Institutions in Inner Mongolia

Since the establishment of the People's Republic of China in 1949, there have been two significant changes in grassland management institutions in Inner Mongolia. The first major change occurred with the implementation of LGDCHRS in 1984, which transformed the collective property rights of grassland into private use rights [38] (pp. 96–97) [42] p. 12. This shift led to the transition from traditional nomadic livestock husbandry to a sedentary system, effectively halting seasonal movement [38] (pp. 96–97). The second major change was the introduction of PESs in 2011. This program provides pastoralists with relatively high compensation, encouraging them to reduce their livestock numbers or even cease animal husbandry and relocate to urban areas [43,44]. The specific grassland management institutions during these three stages are detailed below.

3.1.1. Before the LGDCHRS

For a long history, the ownership of grassland in Inner Mongolia was held by feudal nobles, temples, or the state. Generally, pastoralists did not possess individual land ownership; however, they had certain recognized rights to use grassland for grazing their own livestock and the herds of local nobles [38] p. 93. During the collective economy period from the 1950s to the 1970s, when grassland was collectively owned, pastoralists established mutual assistance groups. Households voluntarily moved together to jointly use grassland and manage livestock based on labor division and cooperation [38] (pp. 95–96). Subsequently, some groups were combined into larger communes, each with its own allocated grassland. Livestock grazing was restricted by the boundaries of the commune's grassland, and the transhumance of animals throughout the four seasons was arranged by the commune [38] p. 96. Due to the common use of grassland, the daily grazing radius of livestock was relatively extensive, with horses, cattle, and sheep having a grazing radius of 30 km, 20 km, and 10 km, respectively [42] p. 37. In 1980, Xilinhot, located in the center of the Xilingol grassland, had established five communes, one of which comprised 503 pastoralists and 87,999 livestock. To establish this commune, each household had to sell their livestock at a discount of 20–30% of the total price to the commune, with the remaining 70–80% of the livestock value used as shares to join the commune. At the end of each year, pastoralists received 3% of the value of their livestock as a dividend [38] p. 96.

3.1.2. The Implementation of LGDCHRS

Following the implementation of the household contract responsibility system in agricultural areas, LGDCHRS was introduced in the 1980s in pastoral areas of Inner Mongolia. This system distributed livestock and grassland to individual pastoral households. LGDCHRS aimed to address the "*Daguofan*", a term describing the tragedy of the commons, where every livestock grazes on communal grassland. Grassland management was reorganized into smaller, contracted areas for each pastoral household. The most direct result was the cessation of seasonal movement, leading to a significant reduction in the daily grazing radius of livestock, for example, 3–3.5 km for horses, 2 km for cattle, and 1.5–2 km for sheep, in eastern Inner Mongolia [42] p. 37.

These changes have been recognized as progress in the discourse, which claims that backward nomadism, characterized by "following water and grass to breed livestock", should be replaced by intensified livestock breeding dependent on infrastructure, including sheds, wells, artificial grassland, high-yield feed fields, and storage facilities. Consequently, the grassland landscape was gradually fragmented by fences. The risk of grass availability, previously mitigated by mobility, has increased sharply due to the cessation of seasonal movement. Grazing impacts caused by the same number of livestock on grassland during droughts exemplify and exacerbate overgrazing and grassland degradation. Moreover, the labor division and cooperation among pastoralists diminished, requiring pastoralists to become "versatile" in managing their own livestock and building infrastructure [45]. The roles of village leaders and experienced pastoralists in arranging grassland utilization and livestock breeding at the village scale have been abandoned. This includes managi seasonal movements, improving livestock breeds, and responding to natural disast Consequently, one of the key principles of adaptive governance of grasslands, action the ecological scale, has been lost [46].

3.1.3. The Implementation of PESs

In 2011, the Grassland Ecological Reward and Compensation Policy (GERCP) w initiated across the entire pastoral area of China, encompassing 13 provinces. This pol operates on a five-year cycle and aims to encourage pastoralists to reduce their livestc numbers and protect grasslands by providing compensation for grazing bans and rewa for maintaining livestock numbers within the grassland carrying capacity [3]. The GER project recognizes pastoralists' efforts to provide grassland ecosystem services and off compensation as payment for these services. The policy measures include: (1) for gra lands with very harsh living environments, severe degradation, and unsuitable conditic for grazing, a grazing ban and enclosure were implemented, with the central governm subsidizing pastoralists at the rate of USD 11.7/ha/year; (2) for usable grasslands outsi the grazing ban area, based on the approved reasonable livestock carrying capacity, the c tral government provided rewards for maintaining a balance between grass and livestc to pastoralists who did not exceed the carrying capacity, at a rate of USD 2.9/ha/year [This mechanism attempts to compensate pastoralists who contribute to grassland conser tion through government subsidies. During the first round of policy implementation fr 2011 to 2015, satisfactory results were achieved, such as poverty elimination and a red tion in grassland exploitation [10]. However, the livestock numbers did not decrease expected. For instance, in Inner Mongolia, during the first three years of PESs, the numl of livestock decreased from 96.94 million sheep units in 2010 to 93.4 million sheep unit decrease of 4%. However, by 2014, the number of livestock had increased to 97.7 milli sheep units, surpassing the 2010 levels before the implementation of PESs.

3.2. Changes in the Pastoralists' Perceptions of Grassland Value

Based on the above analysis, the implementation of LGDCHRS and PESs has sign cantly changed grassland management in Inner Mongolia [1]. It is crucial to recognize t these policies and projects have also profoundly influenced pastoralists' attitudes towa grassland, affecting the effectiveness of these initiatives [23,39]. This section will explc the changes in pastoralists' perceptions of grassland value from three perspectives: nature of grassland, the ownership of grassland, and the purpose of grassland conservati Although there may be an element of romanticism in describing the collective period, statements in regard to these three aspects are fundamentally accurate.

3.2.1. What Is Grassland for Pastoralists?

Before the implementation of LGDCHRS, the grassland was both the homeland a the most important asset for pastoralists. Mongolian singer Tengger encapsulates t sentiment in his song "Heaven," where he describes the grassland as "home" with its b sky, clear lake, and green expanses. From an external perspective, pastoralists rely grassland because they are dispersed in remote areas, isolated from political and econor centers, and have no other industries to depend on. However, from the pastorali perspective, the grassland resembles a mother, a metaphor frequently found in Mongoli songs and proverbs [48]. A Mongolian proverb states, "The first-class rich have frien the second-class rich have knowledge and wisdom, and the third-class rich have cattle a sheep, but the greatest wealth is still the grassland" [45]. This comparison implies t wealth is not merely economic but is foundational to the pastoralists' way of life.

After grassland was contracted to individual households, it transformed into a mea of production. Once pastoralists left livestock husbandry, the grassland became an as that could be rented out for income. This transformation did not occur overnight. Initia during the implementation of LGDCHRS, pastoralists found it difficult to accept construction of fences on grassland. As an elderly Mongolian woman remarked, "If the grassland was divided like a spider web, then there must be a problem with the grassland". However, driven by the market economy, such as the sharp rise in cashmere prices in the 1980s, pastoralists were motivated to increase their goat herds, believing "it meant losing money if we did not increase goats" [49] (pp. 153–157). Consequently, grassland became a means of production that facilitated income generation. In the collective economy, pastoralists had to consider grassland and livestock use value. Post-LGDCHRS, the value of grassland and livestock became intertwined with market transactions. In essence, exchange value became the core concern for pastoralists after the implementation of LGDCHRS.

Following the implementation of PESs in 2011, the grassland became a voucher for some pastoralists to receive compensation [1,23]. On the one hand, the high compensation prompted some pastoralists to completely abandon the grassland and move to towns; on the other hand, it attracted a few pastoralists who had left grassland and livestock breeding for years, to return and reclaim their abandoned grassland. These individuals primarily sought a paper certificate, which they could use to apply for compensation, regardless of the location or health of the grassland [39]. The previously mentioned exchange value required pastoralists to actively work on their grassland. Under the concept of exchange value, pastoralists had to manage their livestock well, while protecting the use value of the grassland. Even when renting the grassland to others, they needed to consider its health. However, under PESs, pastoralists can receive compensation merely for having a nominal contracted area of grassland and a contract certificate [1,35]. Thus, aside from exchanging compensation for PESs, the exchange value of grassland has diminished further in the eyes of some pastoralists.

3.2.2. Who Does the Grassland Belong to?

As stated above, the grassland is often compared to "Mother Earth". Before the implementation of LGDCHRS, pastoralists believed that the grassland was jointly owned by all pastoralists. This common ownership was guaranteed by a series of institutional arrangements. There were many beliefs and rules emphasizing conservation in the use of grassland [45]. For example, Prof. Liu Shurun's research in the Hulunbuir grassland in Inner Mongolia revealed the role of a "*nutugchin*", a coordinator of grassland management and livestock breeding, who played a crucial role in conserving grassland and local knowledge. "Nutug" means "hometown" in Mongolian, and "chin" means "the person of...". Thus, it can be literally translated as "the person who guards the grassland hometown". Before the implementation of LGDCHRS, "nutugchins" were nominated at all administrative levels of leagues, banners, and *sumus* (counties). In some areas, they were referred to as *galin ah*, especially where camel breeding was prevalent. "*Nutugchins*" were primarily responsible for arranging grassland use and livestock movement [45]. Importantly, during this period, leaders at all administrative levels and pastoralists respected and obeyed the commands of "*nutugchins*". As the Mongolian proverb says: "The horse who eats *artemisiae herba* will brim with energy and vitality, and the person who listens to the old man will have wisdom growth". With these experienced individuals guarding the grassland, the "grassland mother" could be protected, while nourishing the material and spiritual lives of pastoralists [45] (pp. 126–131).

At the beginning of LGDCHRS implementation in the 1980s, even though the use rights of grassland were distributed to individual households, most pastoralists did not immediately change their mindset. This is also the reason why Inner Mongolia implemented the second round of grassland contracting in 1995. For herders, using grasslands on a household basis presents many difficulties. It was difficult for them to imagine how to graze livestock within a fixed, fenced, and much smaller grassland area [38] p. 101. However, under various influences, they gradually adopted the "my grassland" concept and put up fences around their contracted areas to protect their use rights. A primary factor was the increase in livestock, which led pastoralists to realize the necessity of using fences to "protect their own grassland from others' livestock" [45] (pp. 126–131). From 1984 to

1999, favorable weather conditions and market prices, particularly the rapid increase cashmere prices, led to a significant increase in livestock numbers in Inner Mongolia, w sheep numbers increasing by 56% from 23.77 million in 1984 to 37.03 million in 1999 [5 The widespread growth in livestock numbers made pastoralists understand the need build fences to keep neighbors' cattle and sheep out, or else they would suffer ecologi and economic losses. Different regions in Inner Mongolia experienced varying modificati processes. For instance, Zhang [49] (pp. 127–132) noted in her research on the Ordos gra land in western Inner Mongolia that local pastoralists began building fences around 19 gradually shifting the meaning of grassland from "collective" and "public" to "privat However, Li and Zhang [38] p. 101 found that most fences in central Inner Mongoli Xilingol grassland were constructed after 2000. With the entry of state-owned or priv capital for activities such as mining, road construction, agricultural planting, and later wi and solar power development, which occupied grassland, pastoralists needed fences assert their use rights and receive compensation. Consequently, under internal and exter pressures, pastoralists established and exercised their grassland use rights by constructi fences. However, in response to external pressures, pastoralists often used fences merely obtain compensation rather than to genuinely protect their use rights.

Although the use rights of grassland still belong to pastoralists, the implementati of PESs has generally increased the perceived instability of these rights. The top-dov policy enables every pastoralist to participate in the project, with the price of receivi compensation being the imposition of certain restrictions, such as a seasonal or year-rou grazing ban. PESs further promotes the diversification of pastoralists, who can be rougl divided into four categories. The first category comprises households with no livestock a few livestock, who leave the grassland and move to towns, relying on the compensati to make a living [1]. Under relatively loose supervision, some of them rent out th grassland to earn additional income. The second category includes outsiders who take t opportunity to use the grassland to raise livestock [1,39]. The third category consists moderately wealthy and wealthy households, who continue to engage in animal husbanc with increasing costs, loans, and risks [51]. The fourth category includes a small numb of pastoralists, who have left the grassland for years but return to claim their share of t grassland due to the compensation benefits.

The question of who owns the grassland has become more ambiguous. Pastorali who leave the grassland with compensation or reclaim their share of grassland sol to obtain compensation do not care about the grassland beyond their contract certifica which allows them to earn compensation. They do not consider the health status or even t specific location of their grassland. Outsiders who rent grassland are primarily interest in the immediate benefits of animal husbandry, often leading to overuse of the contract grassland. If this grassland becomes unusable, they simply rent another piece. Only t local pastoralists who continue to engage in animal husbandry care about the health their grassland. However, under the pressures of climate change, grassland degradati and loan burdens, they also tend to raise more livestock than their grassland's sustainal carrying capacity [1,23].

3.2.3. For Whom Is Grassland Conservation?

Before the implementation of LGDCHRS, pastoralists primarily relied on grassla animal husbandry for their livelihood and believed that sustainable grassland use w essential for their own well-being. The pastoral areas were remote from agricultural zor and even more distant from industrial centers. At that time, life was simple; there was lit need for cash, and local income came almost entirely from animal husbandry, with t grassland serving as its foundation [52] p. 7. The transhumance system, which involv the seasonal migration of livestock, operated like a complex mechanism, with numero interlocking components. Each pastoralist household or group functioned as a small ge contributing to the overall operation of the system. This complete dependence on grassla resources compelled every pastoralist to engage in grassland conservation.

After the grassland was contracted to individual households, pastoralists initially had a strong desire to protect it. However, various restrictions rendered them increasingly powerless [52] p. 23. Firstly, since different households were allocated specific grassland parcels, moving livestock to other pastures became difficult unless arrangements and payments for rent were made. Secondly, prior to settlements, when pastoralists lived in yurts, there were few paths, and both people and livestock could move freely without the constraints of fences. After settlements, multiple roads emerged in front of each house, and numerous routes between fences became severely degraded due to over-trampling. Thirdly, compared to the cooperative and straightforward tasks arranged during the collective era, pastoralists had to become versatile after LGDCHRS [52] p. 23. Their adaptive governance, which respected natural grassland laws, was replaced by unreasonable practices that ignored the grassland ecosystem's characteristics. Traditional grazing knowledge became obsolete, necessitating the acquisition of new skills, such as driving, feeding livestock with forage, and using the Internet [52] p. 33. Additionally, the cost of purchasing fodder and forage has increased, forcing pastoralists to increase their livestock numbers to repay bank loans. Some even resort to usurious loans, creating a vicious cycle. Despite knowing that the degraded grassland cannot support so many livestock, pastoralists must maintain high livestock numbers to meet their financial obligations. Without sufficient funds to buy fodder and forage, both livestock and livelihoods are jeopardized [45] (pp. 126–131).

The implementation of PESs further transformed pastoralists' moral sentiment towards grassland conservation. For some, the grassland's meaning shifted from being a homeland to a voucher for obtaining compensation [1]. The ecological service function of "grassland as an ecological barrier in northern China" gained prominence, and various conservation projects were implemented, primarily through grazing bans [2]. However, the role of "grasslands as a homeland for pastoralists" was neglected. Firstly, PESs directly compensated pastoralists, fostering dependency and increasing the number of individuals who relied on compensation and grassland rents rather than engaging in livestock rearing [52]. This attracted outsiders seeking short-term benefits, exacerbating overgrazing issues. Secondly, PESs schemes distributed compensation to individuals rather than collectives, undermining the collective action needed for effective grassland ecological function maintenance. This led to the fragmentation of local social systems and the abandonment of community cooperative rules for rational grassland use [39]. Lastly, compensation encouraged irrational consumption and disrupted family unity. For instance, a village leader in eastern Inner Mongolia noted that over twenty SUVs appeared within a month of compensation distribution. Many pastoralists lacked plans for reasonable use of the compensation. The implementation of PESs also led to the division of large families into smaller units, further fragmenting the ecosystem and creating more social problems [23,39]. Overall, while LGDCHRS and PESs schemes aimed to improve grassland management and pastoralist livelihoods, they introduced new challenges and complexities that require careful consideration and adaptation to ensure long-term sustainability and social cohesion in pastoral communities.

3.3. Characteristics of Grassland Ecological Research in China

As mentioned above, the commodification of both ecosystem services and economic value has profoundly altered the social–ecological system in pastoral areas, including relationships among pastoralists and between humans and nature. Despite continued state investment, the resulting social disorganization and grassland degradation persist [35,44]. However, reflection on this issue should not end here; it must extend into the field of grassland ecology as an academic discipline. Grassland ecology in China, developed since the 1950s, has played a crucial role in grassland management. Reflecting on this field can provide a foundation for the future management of pastoral areas. As Beck emphasized in his "Risk Society" [53] p. 79, while the misapplication of scientific rationality has caused numerous problems, it should still be used to correct errors. Therefore, this section further analyzes the dilemma faced by grassland management and pastoral areas, by exploring the development of grassland ecology in China over the past four decades, focusing potential solutions.

Research on grassland ecology in China began in the 1950s and has gained increa attention since the late 1970s. According to interviews with pioneers in livestock a grassland science, including Zhang Zutong, Fu Xiangqian, Chen Shan, and Liu Shu (July 2008), early grassland ecological research in China was closely linked to livest and animal husbandry studies. The first grassland major in China, established in 1 was established by the Department of Animal Husbandry and Veterinary Medicine Inner Mongolia University. Later, influenced by Prof. Wang Dong and Prof. Ren Jizho grassland major was established at Gansu Agricultural University, maintaining a str connection with animal husbandry [38] p. 40. The Department of Ecology at Inner Mong University was established in 1957, led by Prof. Li Jidong, one of the founders of mod Chinese ecology, who relocated the ecology and geobotany team from Peking Univer to Inner Mongolia University. In 1977, Inner Mongolia University established Chin first undergraduate major in ecology. Field research has been a significant focus in development of grassland ecology. An ecological research station was established Xilingol grassland in 1979, concentrating on monitoring, research, and demonstration had natural advantages in terms of conducting grassland research, because it was closest research station to grassland and pastoralists at that time. Based on an analysis the annual meeting records from 1981 to 1998 at the research station, as well as 11 collec paper publications during this period [54], early grassland ecological research exhibi two main characteristics. These early efforts laid the groundwork for understanding a addressing the complex challenges in grassland management, providing essential insig that remain relevant in developing effective strategies for sustainable pastoralism a ecological conservation in China's grassland regions.

3.3.1. Losing the Understanding of the Complexity and Integrity of Grassland Ecosyste

Early research from the ecological research station reveals a shift in research directi with a decreased focus on the overall grassland ecosystem and an increased focus detailed studies, such as plant succession, plant physiology, and photosynthesis. This tre aligns with Lin and Fyles' findings, which highlight the limitations of an overempha on technical knowledge [55]. While this approach has facilitated a deep understandi of the technical aspects of plants and land, it falls short of comprehending the holis complexity of the ecosystem. When research began at the station in 1980, 28 topics w identified and categorized into seven areas: plants, animals, soil, water, material cyc grassland utilization, and database construction. For instance, the fifth edition of "Resea on Grassland Ecosystem," published in 1997, included detailed studies on the effects grazing on soil physical properties, soil nutrients, soil microorganisms, and the loc community response. However, notably, it lacked articles addressing the overall impac grazing on vegetation [56].

3.3.2. Taking an Engineering Approach to Grassland

These studies also reflect an increasingly distant approach from nature and a focus artificial interventions. Since 1981, the annual meetings have proposed establishing a studying artificial grassland. Based on the author's analysis of the summaries from working meetings of the positioning station from 1981 to 1998, by 1984, various exp ments on artificial grassland began to appear in the collection of papers. These experime included studies on the biological characteristics of *Leymus chinensis* under cultivation, establishment of *Leymus chinensis* artificial grassland, winter-snow-layer fertilization *Artemisia frigida* communities, background investigations on cutting grass succession t areas in *Leymus chinensis* grassland, and cutting grass succession tests on artificial *Leym chinensis* grassland. Since then, studies on natural grassland have significantly diminish The few studies that focus on natural grassland often emphasized its backwardness a lack of artificial modification, critiquing the unreasonable utilization of natural grassla

(1985) [57] and its management practices (1981) [58]. These studies continued to advocate for an eco-developmental logic underpinning systems science approaches to environmental management as not just techniques of managing natural resources, but as the very foundation of the progress of China's grassland region, a notion of utmost importance to development and legitimacy [23]. Emphasizing human control of nature through technical management, these studies advocate for the technoscientific control of socio-natural systems and portray ecology as both malleable and that it can be altered through human intervention and the correct calculation of its ecological functionality [23,30,59]. This logic posits that ecological engineering through environmental modeling and functional land zoning will transform a herding society into an ecologically modern society and pave the way for a "healthier" grassland region and community [26,59].

4. Discussion

4.1. The Capitalization of Nature and the Two Disembedding Processes in Regards to Grassland Value

As we examine the evolving significance of grassland for pastoralists, it becomes apparent that nature is increasingly "cut" and "extracted" from the social life of local people within the developed market mechanism. Implementing the LGDCHRS and PESs has precipitated two disembedding processes within the grassland social–ecological system. The first process involves the commodification of the grassland's economic value, a direct consequence of the LGDCHRS. This commodification leads to the economic value being disembedded from the grassland's ecological, social, and cultural dimensions. The second process, stemming from the implementation of PESs, concerns the commodification of the grasslands' ecological value, where only a portion of this value is harnessed to provide ecological barrier services in central China. This selective commodification alters the moral inclinations of pastoralists towards grassland protection.

4.1.1. Disembedding Process of Grassland Economic Value

In her research conducted in western Inner Mongolia, Zhang ([49] p. 191) introduces the concept of "Disembedding 'Nature'" to elucidate the transformation in the relationship between nature and humans following the implementation of the LGDCHRS. This paradigm shift is marked by the transition from a relationship characterized by integration and symbiosis to one of division and opposition, with nature increasingly objectified by human actions. This transformation is primarily evident in the subdivision of grasslands into smaller, tradable plots, a change that many pastoralists find difficult to accept. This resistance stems from alterations to their longstanding grazing practices and, more significantly, from a profound conflict of beliefs. As pastoralists articulate, "The grassland is our mother earth and cannot be dismembered and divided by her children" [45] (pp. 126–131).

Despite the challenges, grassland contracting has been implemented across Inner Mongolia, facilitated by the region's relatively flat terrain and the absence of insurmountable natural barriers. Typically, grasslands are divided based on family size and livestock numbers according to specific standards, whether applied to pastoralist groups or individual households. In 1996, Inner Mongolia launched a second round of LGDCHRS implementation, during which grassland previously contracted to pastoralist groups was further distributed among individual households [38] p. 71. While some areas retain common grassland, the more prevalent arrangement now involves fragmented grassland utilized by individual households. Consequently, the relationship between pastoralists and the grassland has evolved from adaptation to active construction and transformation [38] p. 71. The increasing prevalence of fences, motor-pumped wells, forage planting, silage pits, warm sheds, and various mechanical facilities, underscores this shift. Pastoralists are compelled to become multifaceted workers to cope with these changes in livestock management post-LGDCHRS [45] (pp. 126–131).

Meanwhile, the relationships among pastoralists have also changed accordingly. First, contradictions have emerged within families and between neighbors. These include disputes over the fairness of grassland division, livestock encroaching on neighbors' grassla and fences blocking access routes. Secondly, income from renting grassland, especia to outsiders, weakens the moral incentive for pastoralists to abide by social norms a promote common interests [60]. Thirdly, the original reciprocal support among pastoral in the face of disasters has been replaced by economic transactions. Renting other peopl grassland at high prices and even paying for livestock's drinking water from wells d ing droughts has made pastoralists more calculating. Finally, in the era of the collect economy, pastoralists could rely on the leadership of *nutugchins* and village leaders winter and disaster relief. Now, pastoralists must rely on themselves. Facing increas uncertainty, they feel confused and uncertain about their future, making them more focu on short-term benefits rather than long-term sustainability. All these changes have to the gradual physical and emotional division of pastoralists by fences, which crea physical distance, contradictions, and mutual distrust. This has given rise to a "fenci society" that physically and emotionally divides pastoralists [61].

The emergence of a "fencing society" has significantly altered the relationship betw people and grassland. As previously mentioned, nature is increasingly objectified human actions. As a result of the increasingly developed market mechanism, natur "cut" and "extracted" from the social life of local people, endowed with a standardiz market value, becoming a profitable economic production factor, while losing its vibr vitality and original rich connotation. This process is the capitalization of nature. Mongolian society, this represents a profound shift in the cultural significance of natu accompanied by environmental consequences [49] p. 153. For pastoralists, the grassla is now seen primarily as a productive resource. Elvin noted that failing to exploit t resource economically results in lost potential wealth. The term "the cash-in imperati encapsulates this notion [62] p. 2. This perspective also sheds light on why pastorali despite knowing that outsiders might overexploit the grassland, still choose to rent their contracted land. Furthermore, the cashmere price surge in the 1980s led pastorali to increase their goat herds, operating under the belief that reduced grazing equates financial loss.

4.1.2. Disembedding Process of Grassland Ecosystem Service Value

Based on our research in a township in the south-central Xilingol grassland in Inn Mongolia, it is demonstrated how the implementation of PESs in the 2010s led to t collapse of the local co-management mechanism that had protected grazing lands 30 years [1]. This collapse precipitated a tragedy of the commons, characterized by outsid bringing in their livestock and local pastoralists indefinitely increasing their herds. Pr to the implementation of PESs, even households with no or a few livestock depended livestock breeding and pastoral communities for their livelihood. The future of pastorali hinged on the sustainable use of grassland, reinforced by local regulations that prohibit renting grassland to outsiders. During this time, community support mechanisms w also in place, such as earning income through shepherding or protecting grassland mowing. However, post-PES implementation, households with minimal or no livesto received compensation based on their grassland contracting certificates, despite often bei unaware of the specific locations of their grazing lands, since these were commonly us After completing market transactions for ecosystem services, these households relocated towns, severing ties with the community's mutual support systems. Under lax supervisi they began renting their grazing lands to outsiders, without regard for the consequenc exacerbating the livestock influx. This situation led other households to abandon the nor on controlling livestock numbers and protecting grassland, instead using compensati funds to secure loans for increasing their livestock numbers.

Therefore, the commodification of ecosystem services fails to reflect the broader val of the ecosystem as it strips away the social and ecological dimensions embedded in th services, which are crucial at varying scales [63] (pp. 172–174). In this case, the origir grazing land co-management and livestock quantity control mechanisms were the m directly and significantly affected elements. PESs may shift people's decision-making rationale from considering what practices fit the community social-ecological system to considering what is most beneficial for individuals [1]. The process of commodification has led to the disorder of social metabolism, resulting in unsustainable social and ecological outcomes, thus leading to the tragedy of commodification [63] (pp. 172–174).

4.2. *Research on Grassland Ecology and Its Detachment from Grassland Management*

The studies on grassland ecology often focus on technological and technocratic solutions, while overlooking traditional grassland management practices. The research can be divided in regard to the following two aspects.

4.2.1. Disentangling Research on the Natural Grassland Ecosystem

This research typically assumes that traditional animal husbandry, which relies entirely on natural grassland, is backward and lacks stable, high-quality, and high-yielding grazing land. For example, a 1988 study argued that traditional livestock herds grazed year-round lack a comprehensive scientific breeding and management system, impeding increased production in animal husbandry. Ironically, the "movement" of herds allows pastoralists to maintain stable, high-quality, and high-yielding grazing land, requiring "no supplemental feeding" [64] p. 47. While "no supplemental feeding" was assessed as a lack of scientific management systems, it indicates the full and rational use of natural grassland, reducing costs. Therefore, supplemental feeding does not represent advanced methods, but rather signifies grassland degradation and increased costs. The belief that supplemental feeding creates "improved livestock" overlooks the fact that livestock raised on supplemental feeding may be unfit for natural conditions, weak, and sickly, thus requiring supplemental feeding. Pursuing more meat, milk, and cashmere production based on "improved" livestock often ignores other significant drawbacks. Based on these misconceptions, it was proposed that "if China's animal husbandry wants to significantly improve production efficiency, it must combine artificial grassland with natural grassland and transition to semi-domesticated grazing on the basis of planting artificial pasture and concentrated feed" [65]. However, decades of practice have shown that abandoning natural advantages and transforming landscapes into artificial spaces for animal production creates numerous problems [38] p. 110; [66] (pp. 47–54).

4.2.2. Disentangling Research on the Pastoralists' Knowledge and Practice

Grassland contracting, in essence, is a "mandated marketization" reform based on two erroneous assumptions: the tragedy of the commons and the backwardness of nomadism [38] (pp. 62–69). From the perspective of the tragedy of the commons, traditional grassland management and animal husbandry were misunderstood as a situation of "no owner of the grassland, no boundary for grazing, no rules for management, no responsibility for construction, and no guilt for destruction" [38] p. 68. This view contributed to the belief that it caused grassland degradation and desertification. Those who assert the backwardness of nomadism claim that traditional animal husbandry relied mainly on "raising livestock by relying on nature" and "extensive management methods", which depend solely on natural conditions and focus only on livestock quantity, rather than quality [49] (pp. 214–216). Researchers from outside often fail to understand why herds are constantly relocated and why pastoralists do not plant forage grass. If there had been better communication with local pastoralists, these misunderstandings could have been resolved, revealing that pastoralists manage their animals within their ecosystems more holistically.

Both scholars and policymakers have underestimated the knowledge and experience of traditional pastoralists, valuing technical interventions instead. While data are valuable, we should also seek insights from modern Leopold foresters, farmers, and indigenous elders, who have spent their lives observing the land, and ask them what they see [55]. As previously discussed, pastoralists have always been practitioners of land ethics, respecting, relying on, and protecting the grassland. They learn from livestock, understanding whether forage is toxic, or delicious, and nutritious, and mastering livestock management skills grassland [45] (pp. 126–131). Ecology provides a model for a more vibrant and harmoni human community, and ecological theory must be holistic rather than simplified, neglect the system's complexity [67] (pp. 493–499). This article emphasizes that local people, w have long been involved in the operation of this system, cannot be ignored.

This article illustrates the changing perceptions of pastoralists regarding the soc economic, and ecological value of grassland, by presenting the implementation of grassla contracting, which has led to the dominance of anthropocentrism in grassland managem and replaced the original ecological ethics of pastoralists. This view is entirely based on c siderations of technical production improvements, ignoring the diversity and complexity ecosystems and the traditional knowledge on those systems. Faced with widespread gr land degradation, scientific rationality has once again prevailed, asserting that degrad grassland can be restored if livestock grazing is reduced. To this end, various conser tion projects and the largest PESs project have been implemented. The overall trend ecological development has been to abandon a holistic viewpoint and an interdisciplin approach. The narrative in this article is just one example of humanity's advocacy scientific rationality and its immersion in conquering and transforming nature. This arti emphasizes the importance of the pastoralists' perspective by highlighting how their vie on grassland's social, economic, and ecological value have evolved. Only through activ understanding, learning from, and protecting the grassland can we re-embed and real the true economic and ecological value of the grassland.

5. Conclusions

Our findings underscore the critical importance of integrating local knowledge i policy frameworks for grassland conservation. Institutional transformations in Chin grassland management policies have significantly shaped pastoralists' evolving und standing of grassland value. The commodification of grassland's economic and ecologi value has led to the capitalization of nature, disembedding husbandry from grassla production and undermining conservation efforts. This article chronicles the traject of grassland ecology research over the past four decades, highlighting its increasing tachment from a holistic understanding of ecosystems and the interdisciplinary needs management practices. This study reveals that the disjunction between grassland ec ogy research and practical management has resulted in a lack of techniques aligned w local ecological and socioeconomic contexts. Consequently, this research advocates a c laborative approach involving all stakeholders to develop and implement effective a sustainable solutions to grassland degradation in China.

This research is unique in two significant ways. First, it offers a continuous histori perspective, comprehensively reviewing policy changes over the years, rather than foc ing on a single period. Second, it highlights pastoralists' perspectives, particularly th attitudes and actions regarding recent ecological compensation policies, which are oft overlooked. This study examines the transformation of pastoralists' perceptions on e logical compensation within the context of successive conservation programs. Due to d constraints, this study focuses on one region in Inner Mongolia as a case study. Howev the diverse pastoral regions of China may experience varying effects, and generalizing insights and suggestions requires caution. This research emphasizes an evidence-bas approach rather than starting from theoretical assumptions, contributing significantly understanding pastoral areas, pastoralists, and pastoralism.

Our findings demonstrate that the concept of "my grassland" among pastoralists c not form immediately after grassland contracting but developed gradually due to the ne to prevent others' livestock from encroaching on their grassland. Analyzing the changi value of grasslands from the pastoralist's perspective is crucial, especially given their la of voice in policy formulation. This article suggests that policies should be formed th create space for local pastoralists to make informed decisions about the rational use grasslands, emphasizing confidence in their cultural knowledge, rather than relying sol on externally imposed technology. Policymakers should understand how pastoralists form ecological value instead of relying on technocrats' abstract reification. This study suggests that policymakers consider how subsidies are distributed and utilized, giving local stakeholders a voice to address regional needs and improve fund utilization efficiency. PESs schemes require complementary measures, such as restricting the entry of outsiders and providing compensation for ecological services generated by community-level collective actions. It is crucial to align ecological protection with pastoralists' livelihoods rather than view them as opposing forces.

This article is an exploratory study to provide a localized understanding of how pastoralists in Inner Mongolia perceive ecological value. However, there is much to consider regarding how capital and social networks shape specific implementation methods. Addressing these issues is fundamental, especially under conditions of climate change and socioeconomic development, which increase the complexity and uncertainty. Further research is needed to explore situations in other regions of China. In conclusion, actively engaging with and learning from pastoralists is vital to realizing the true economic and ecological value of grasslands. This article suggests more nuanced and effective strategies for sustainable grassland management, through comprehensive and collaborative efforts.

Author Contributions: Conceptualization, Q.Z. and K.M.; methodology, Q.Z.; validation Q.Z. and K.M.; formal analysis, Q.Z. and K.M.; investigation, Q.Z.; resources, Q.Z.; data curation, Q.Z.; writing—original draft Q.Z. and K.M. writing—review and edition K.M. and Q.Z.; visualization, Q.Z.; supervision, Q.Z.; project administration, Q.Z. All authors have read and agreed to the published version of the manuscript.

Funding: This research received no external funding.

Data Availability Statement: No new data were created and interview data is unavailable due to privacy and ethical restrictions.

Conflicts of Interest: The authors declare no conflict of interest.

ferences

Zhang, Q. Tragedy of Commodity: The reason for grassland overuse under payment of ecosystem services. *J. Soc. Dev.* **2017**, *4*, 181–195. (In Chinese)

Li, W.; Li, Y.; Gongbuzeren. Rangeland Degradation Control in China: A Policy Review. In *The End of Desertification?* Behnke, R.H., Mortimore, M., Eds.; Springer Earth System Science: Berlin, Germany, 2016. [CrossRef]

Li, Z.; Su, B.; Liu, M. Research Progress on the Theory and Practice of Grassland Eco-Compensation in China. *Agriculture* **2022**, *12*, 721. [CrossRef]

Hu, B.; Wen, Q.; Xi, F.; Li, M.; Wang, L.; Ren, Y. Effects of enclosure on grassland ecological and economic benefits in Northern China. *Chin. Geogr. Sci.* **2024**, *34*, 67–78. [CrossRef]

Gao, L.; Kinnucan, H.W.; Zhang, Y.; Qiao, G. The effects of a subsidy for grassland protection on livestock numbers, grazing intensity, and herders' income in inner Mongolia. *Land Use Policy* **2016**, *54*, 302–312. [CrossRef]

Wang, S.; Tong, C.; Zuo, X. International Influence of Dust Storms and Global Cooperative Treatment. *J. Nantong Teach. Coll. (Soc. Sci. Ed.)* **2008**, *4*, 111–115. (In Chinese)

Joint Research Group on Grassland Ecology. Control desertification, prevent desertification, and building green barriers: Research on Grassland Ecology and Governance in Inner Mongolia Autonomous Region. *Res. World* **2003**, *3*, 14–18. (In Chinese)

Pan, Q.; Sun, J.; Yang, Y.; Liu, W.; Li, A.; Peng, Y.; Xue, J.; Xia, H.; Huang, J. Issues and Solutions on Grassland Restoration and Conservation in China. *Bull. Chin. Acad. Sci.* **2021**, *36*, 666–674. (In Chinese)

China Net (Grassland Channel). Available online: http://grassland.china.com.cn/2018-08/22/content_40472470.html (accessed on 20 March 2024). (In Chinese)

Hou, L.; Xia, F.; Chen, Q.; Huang, J.; He, Y.; Rose, N.; Rozelle, S. Grassland ecological compensation policy in China improves grassland quality and increases herders' income. *Nat. Commun.* **2021**, *12*, 4683. [CrossRef]

Website of National Forestry and Grassland Administration & National Park Administration. Available online: https://www.forestry.gov.cn/main/586/20210820/165512066802631.html (accessed on 17 March 2024). (In Chinese)

Costanza, R.; Groot, R.; Braat, L.; Kubiszewski, I.; Fioramonti, L.; Sutton, P.; Farber, S.; Grasso, M. Twenty years of ecosystem services: How far have we come, and how far do we still need to go? *Ecosyst. Serv.* **2017**, *28*, 1–16. [CrossRef]

Gómez-Baggethun, E.; Muradian, R. In markets we trust? Setting the boundaries of market-based instruments in ecosystem services governance. *Ecol. Econ.* **2015**, *117*, 217–224. [CrossRef]

14. Schröter, M.; Zanden, E.H.; Oudenhoven, A.; Remme, R.P.; Serna-Chavez, H.M.; Groot, R.S.; Opdam, P. Ecosystem services a contested concept: A synthesis of critique and counter-arguments. *Conserv. Lett.* **2014**, *7*, 514–523. [CrossRef]
15. Gómez-Baggethun, E.; Groot, R.; Lomas, P.L.; Montes, C. The history of ecosystem services in economic theory and practice: Fr early notions to markets and payment schemes. *Ecol. Econ.* **2010**, *69*, 1209–1218. [CrossRef]
16. Kaiser, J.; Haase, D.; Krueger, T. Payments for ecosystem services: A review of definitions, the role of spatial scales, and critiq *Ecol. Soc.* **2021**, *26*, 12. [CrossRef]
17. Costanza, R.; d'Arge, R.; De Groot, R.; Farber, S.; Grasso, M.; Hannon, B.; Limburg, K.; Naeem, S.; O'neill, R.V.; Paruelo, J.; e The value of the world's ecosystem services and natural capital. *Nature* **1997**, *387*, 253–260. [CrossRef]
18. Swallow, B.M.; Kallesoe, M.F.; Iftikhar, U.A.; Noordwijk, M.; Bracer, C.; Scherr, S.J.; Raju, K.V.; Poats, S.V.; Duraiappah, A Ochieng, B.O.; et al. Compensation and Rewards for Environmental Services in the Developing World: Framing Pantropi Analysis and Comparison. *Ecol. Econ.* **2009**, *14*, 26. [CrossRef]
19. Wunder, S. Revisiting the concept of payments for environmental services. *Ecol. Econ.* **2015**, *117*, 234–243. [CrossRef]
20. Wunder, S. The efficiency of payments for environmental services in tropical conservation. *Conserv. Biol.* **2007**, *21*, 48– [CrossRef]
21. Coase, R.H. The Problem of Social Cost. *J. Law Econ.* **2013**, *56*, 837–877. [CrossRef]
22. Apostolopoulou, E.; Chatzimentor, A.; Maestre-Andrés, S.; Requena-i-Mora, M.; Pizarro, A.; Bormpoudakis, D. Reviewi 15 years of research on neoliberal conservation: Towards a decolonial, interdisciplinary, intersectional and community-engag research agenda. *Geoforum* **2021**, *124*, 236–256. [CrossRef]
23. Mao, K.; Zhang, Q.; Truslove, M. Grassland conservation and environmental inequality in Inner Mongolia, China. In *Handbook Inequality and the Environment*; Edward Elgar Publishing: Cheltenham, UK, 2023; pp. 429–448.
24. Muldavin, J. The politics of transition: Critical political ecology, classical economics, and ecological modernization theory China. In *The Political Geography Handbook*; Cox, K., Lowe, M., Robinson, J., Eds.; Sage: London, UK, 2007; pp. 247–262.
25. Li, Y.; Shapiro, J. China goes green: Coercive environmentalism for a troubled planet. *Environ. Politics* **2020**, *30*, 678–680.
26. Zinda, J.A. Unpacking Authoritarian Environmental Governance: Gauging Authoritarian Elaboration in China and Beyo *Sociol. Dev.* **2023**, *9*, 195–216. [CrossRef]
27. Geall, S.; Ely, A. Narratives and pathways towards an ecological civilization in contemporary China. *China Q.* **2018**, *236*, 1175–1 [CrossRef]
28. Iftikhar, A.; Cawkwell, F.; Dwyer, E.; Barrett, B.; Green, S. Satellite remote sensing of grasslands: From observation to managem *J. Plant Ecol.* **2016**, *9*, 649–671.
29. Masenyama, A.; Mutanga, O.; Dube, T.; Bangira, T.; Sibanda, M.; Mabhaudhi, T. A systematic review on the use of remote sensi technologies in quantifying grasslands ecosystem services. *GIScience Remote Sens.* **2022**, *59*, 1000–1025. [CrossRef]
30. Zinda, J.A.; He, J. Ecological civilization in the mountains: How walnuts boomed and busted in southwest China. *J. Peasant. St* **2020**, *47*, 1052–1076. [CrossRef]
31. Rodenbiker, J. Making ecology developmental: China's environmental sciences and green modernization in global context. *A Am. Assoc. Geogr.* **2021**, *111*, 1931–1948. [CrossRef]
32. Marsett, R.C.; Qi, J.; Heilman, P.; Biedenbender, S.H.; Watson, M.C.; Amer, S.; Weltz, M.; Goodrich, D.; Marsett, R. Remote sensi for grassland management in the arid southwest. *Rangel. Ecol. Manag.* **2006**, *59*, 530–540. [CrossRef]
33. Kolås, Å. Degradation discourse and green governmentality in the Xilinguole grasslands of Inner Mongolia. *Dev. Chang.* **20** *45*, 308–328. [CrossRef]
34. Harlan, T. Conservation or decarbonization? Small hydropower and state logics of green development in China. *Ann. Am. As Geogr.* **2020**, *110*, 1464–1482. [CrossRef]
35. Mao, K.; Zhang, Q.; Weeks, N. Authoritarian environmentalism as just transition? A critical environmental justice examination state environmental intervention in northwestern China. In *Environmental Justice in the Anthropocene*; Routledge: London, U 2021; pp. 175–187.
36. Zhang, X.S.; Tang, H.P.; Dong, X.B.; Li, B.; Huang, Y.; Gong, J. The dilemma of steppe and it's transformation in China. *Bull. Ch Acad. Sci.* **2016**, *61*, 165–177. (In Chinese)
37. Krätli, S. If Not Counted Does Not Count? A Programmatic Reflection on Methodology Options and Gaps in Total Econon Valuation Studies of Pastoral Systems. IIED Issue Paper London. 2014. Available online: https://www.iied.org/sites/defau files/pdfs/migrate/10082IIED.pdf (accessed on 16 June 2024).
38. Li, W.J.; Zhang, Q. *Understanding Rangeland Challenges: Discovery of Rangeland Use and Management Issues at Arid and Semi-a Regions*; Economic Science Press: Beijing, China, 2009; p. 210. (In Chinese)
39. Fan, M.; Zhang, Q. Who is providing ecological compensation?—Reflection on grassland ecological protection subsidy a reward policies based on scale issues. *Acad. Bimest.* **2018**, *4*, 46–52. (In Chinese)
40. Zhang, Q.; Fan, M. Whether payment for ecosystem services can protect grassland ecology? —Case study of rangeland ecologi policy in Alxa Left Banner. *J. China Agric. Univ. (Soc. Sci.)* **2020**, *3*, 36–46. (In Chinese)
41. Vatn, A. An institutional Analysis of Payments for Environmental Services. *Ecol. Econ.* **2010**, *69*, 1245–1252. [CrossRef]
42. Wang, J. *Agricultural-Pastoral Ecology and Traditional Mongolian Society*; Shandong People's Publishing House: Jinan, China, 20
43. Li, W.; Gongbuzeren. *Pastoralism: The Custodian of China's Grasslands*; IIED: London, UK, 2013. Available online: https://ww iied.org/10042iied (accessed on 15 July 2024).

Yang, L.; Qiao, G. Grassland Ecological Compensation, Income Level and Policy Satisfaction: An Empirical Analysis Based on a Survey of Herders in Ecological Protection Redline Areas. *Sustainability* **2023**, *15*, 1664. [CrossRef]
Liu, S.R.; Liu, S.T. *The Reflection of Grassland Degradation*; Inner Mongolia Culture Publishing House: Hulunbier, China, 2020; pp. 126–131. (In Chinese)
Zhang, Q.; Ai, L. Adaptive governance and climate change: Case analysis of Inner Mongolia grassland and counterplan study. *Clim. Chang. Res.* **2018**, *4*, 411–422. (In Chinese)
Ministry of Agriculture and Rural Development and Ministry of Finance, Notification on Issuing the "Guiding Opinions on the Implementation of the 2011 Grassland Ecological Protection Subsidy and Reward Mechanism Policy". 2011. Available online: http://www.moa.gov.cn/nybgb/2011/dqq/201805/t20180522_6142764.htm (accessed on 15 July 2024).
Ulijibat, B. "Grassland Mother"—An Exploration of the Theme of Maternal Love in Mongolian Songs. *Trends in Future.* **2021**, *13*, 1–3. (In Chinese)
Zhang, W. *Disembedding "Nature": Environmental and Social Change of a Pastoral Area in Inner Mongolia since 1949*; Intellectual Property Publishing House: Beijing, China, 2016; pp. 191, 214–216. (In Chinese)
Gotohui. Available online: https://www.gotohui.com/ndata/show-92662 (accessed on 15 March 2024). (In Chinese).
Dalintai; Chimeddorji, J. Analysis on the value chain of grassland animal husbandry: Findings based on a survey of twelve typical, pure animal husbandry banners in Inner Mongolia. *J. Minzu Univ. China (Philos. Soc. Sci. Ed.)* **2020**, *1*, 70–78. (In Chinese)
Zheng, H. *The Logic of Grassland: Field Notes in Pastoral Areas*; The Ethnic Publishing House: Beijing, China, 2018; pp. 23, 33. (In Chinese)
Beck, U. *Risk Society: Towards a New Modernity*; Yilin Press: Beijing, China, 2022; p. 79. (In Chinese)
Duzhanchi' Blog. Available online: https://blog.sciencenet.cn/blog-39664-1213097.html (accessed on 5 March 2024). (In Chinese).
Lin, Q.; Fyles, J.W. Following in Aldo Leopold's footsteps: Humans-in-ecosystem and implications for ecosystem health. In *Ecological Economics for the Anthropocene*; Brown, P.G., Timmerman, P., Eds.; Columbia University Press: New York, NY, USA, 2015; pp. 208–232.
Li, Y.; Zhong, W.; Kang, L.; Liu, Y.; Liao, Y.; Guan, S.; Jia, S. Research on the Impact of Grazing on Biodiversity and Environment of Grassland. In *Research on Grassland Ecosystem*; Science Press: Beijing, China, 1997; pp. 1–11. (In Chinese)
Zhong, Y.K. The Current Situation and Problems of Natural Mowing Field Utilization in Baiyin Xile Pasture. In Research on Grassland Ecosystem: The Fifth Collection. 1986; *Unpublished material.* (In Chinese)
Zhao, X.Y. The Management and Utilization of Natural Grassland Resources in the Xilin River Basin: The Perspective of Ecosystem. In *Research on Grassland Ecosystem: The First collection.* 1981; *Unpublished material.* (In Chinese)
Pan, J. *China's Environmental Governing and Ecological Civilization*; Springer: Berlin/Heidelberg, Germany, 2013.
Lam, W.F. Coping with change: A study of local irrigation institutions in Taiwan. *World Dev.* **2001**, *29*, 1569–1592. [CrossRef]
Menggengdalai. The Rise of fence society: Institutional change, fence construction and social transformation in pastoral areas. *J. China Agric. Univ. (Soc. Sci.)* **2023**, *40*, 39–55. (In Chinese)
Elvin, M. *The Retreat of the Elephants: An Environmental History of China*; Jiangsu People's Publishing House: Nanjing, China, 2021; p. 2. (In Chinese)
Longo, S.B.; Clausen, R.; Clark, B. *The Tragedy of the Commodity: Oceans. Fisheries, and Aquaculture*; Rutgers University Press: New Brunswick, USA, 2015; pp. 172–174.
IIED. *Valuing Variability: New Perspectives on Climate Resilient Drylands Development*; Environmental Science Press: Beijing, China, 2017; p. 47. (In Chinese)
Jiang, S. Discussion on Reasonable Utilization Strategies of Grasslands—Taking the Baiyin Xile Area in Central Xilingol, Inner Mongolia as an Example. In *Research on Grassland Ecosystem: The Second Collection*; Science Press: Beijing, China, 1988. (In Chinese)
Han, N.Y. *The Logic of Grassland: Report of Grassland Ecology and Pastoralists' Livelihood*; The Ethnic Publishing House: Beijing, China, 2018; pp. 47–54. (In Chinese)
Worst, D. *Nature's Economy: A History of Ecological Ideas*; The Commercial Press: Beijing, China, 1999; pp. 493–499. (In Chinese)

land

MDI

Article

Contribution of Rare Earth Elements Is Key to the Economy of the Future

Juan-Ramón Cuadros-Muñoz [1], Juan-Antonio Jimber-del-Río [1,*], Rafael Sorhegui-Ortega [2], Michelle Zea-De la Torre [3] and Arnaldo Vergara-Romero [1]

[1] Department of Agricultural Economics, Finance and Accounting, Faculty of Law and Financial Economy Universidad de Córdoba, 14071 Córdoba, Spain; et2cumuj@uco.es (J.-R.C.-M.); z82veroa@uco.es (A.V.-R.
[2] Research Departement, Universidad Bolivariana del Ecuador, Durán 092405, Ecuador; rasorheguio@ube.edu.ec
[3] School of Economics, Universidad Ecotec, Samborondón 092302, Ecuador; mzea@mgs.ecotec.edu.ec
* Correspondence: jjimber@uco.es

Abstract: An econometric model was developed to analyze the contribution of various factor the gross value added to the agricultural, manufacturing, and service sectors in the United Sta The study found that variables such as rare earth element mining, the employment it generates, domestic consumption, imports, and prices of certain elements significantly impact economic sect The models showed a good fit and met the necessary statistical assumptions. Rare earth eleme are essential for a wide range of technological products, with China being the leading produ and consumer. This has raised concerns about the dependence on other countries. These eleme significantly impact the economy's primary, secondary, and tertiary sectors used in agricult manufacturing, and services. Rare earth elements' mining and processing are complex and expens processes, and demand is expected to continue to increase in the future.

Keywords: rare earth elements; critical elements; next-generation markets; economy

Citation: Cuadros-Muñoz, J.-R.; Jimber-del-Río, J.-A.; Sorhegui-Ortega, R.; Zea-De la Torre, M.; Vergara-Romero, A. Contribution of Rare Earth Elements Is Key to the Economy of the Future. *Land* **2024**, *13*, 1220. https://doi.org/10.3390/land13081220

Academic Editor: Nick B. Comerford

Received: 24 May 2024
Revised: 19 July 2024
Accepted: 22 July 2024
Published: 7 August 2024

1. Introduction

Rare earth elements are 17 chemical elements found in the Earth's crust in minin quantities, between 10 and 100 parts per million. Rare earth elements are found in a wi range of rocks but are concentrated in a few minerals, such as monazite, bastnäsite, a xenotime [1]. Recent years have seen a surge in research on how geopolitical events in ence energy and resource prices [1,2]. Nassar et al. [2] were among the first to demonstr that world governance, political stability, the absence of violence, and global supply conc trations have a more significant impact on rare earth elements' prices when global supply limited. Building on this, Ref. [3] found that periods of economic turmoil and geopoliti unrest lead to increased "spillover" effects, meaning that price fluctuations in oil mark can have a greater impact on financial markets, and vice versa.

The principal rare earth elements' deposits are in China, Australia, the United Sta Brazil, and Russia [4]. China is the leading producer and consumer of rare earth e ments. These are essential elements for manufacturing a wide range of products, includi electronic devices, automobiles, defense equipment, and clean energy [4].

The importance of rare earth elements has increased in recent years due to the grow of the global economy and the increasing demand for products that depend on them Rare earth elements are essential for developing critical energy, transition technologies, a digitalization technologies, such as electric motors, wind turbines, and batteries [5].

High-tech devices, medical equipment, and military systems all rely heavily on ra earth elements (REEs) [6]. These elements are becoming even more critical as clean ener technologies develop [6]. As the demand for clean energy surges in the coming decades, will the demand for REEs, placing a strain on the current supply chain [6].

Rare earth elements (REEs) are like wonder materials for modern technology. Their unique electronic, optical, catalytic, and magnetic properties allow them to solve a range of technological challenges [7]. While there is no single way to categorize their uses, REEs are generally employed in nine main sectors as catalysts, polishing compounds, glass, phosphors and pigments, metallurgy, batteries, magnets, ceramics, and others [7]. In 2015, the global consumption of REEs reached an estimated 119,650 metric tons of REO, with their biggest use as catalysts, followed by magnets, polishing, and others (Figure 1). Experts predict a 5% annual growth rate in global REE demand by 2020, with clean energy's booming market expected to further propel this growth for years to come [8]. This surge in demand puts a strain on the global REE supply chain, creating a significant challenge [8].

The push for cleaner energy through decarbonization and electrification is skyrocketing demand for neodymium (Nd). The International Energy Agency (IEA) predicts demand could double or even triple in the next few decades, with clean energy becoming a major consumer of rare earth elements (REEs), potentially exceeding 40% of total demand [9]. This economic importance, coupled with potential supply disruptions, has countries scrambling for sustainable REE sources. As a result, we have seen a recent surge in REE exploration projects and processing plants [9].

Nations around the world have come together to address climate change by setting a goal of limiting global warming to be below 2 °C compared to pre-industrial levels. This effort is called the Paris Agreement. To achieve this goal, countries are implementing strategies to reduce emissions' "mitigation" and prepare for the effects of climate change "adaptation" [10].

Mitigation efforts include transitioning to cleaner energy sources like solar and wind power, which are becoming increasingly cost-effective. Rare earth elements, which have unique properties that make them essential for many renewable energy technologies, are also in high demand [10].

The transition to renewable energy is happening rapidly, with the share of renewables in electricity generation expected to reach over 60% by 2030 and 88% by 2050. Solar and wind power are expected to become the leading sources of electricity [11].

Transportation is another major source of emissions, and electric vehicles are seen as a key part of the solution. In a scenario where emissions are reduced to net zero, electricity is projected to be the primary fuel for transportation by 2040 [12].

Given this evolving landscape, a timely analysis of global REE resources and potential supply chains is critical for stakeholders. This study provides a comprehensive overview, including a compilation of ongoing rare earth element projects, innovative supply chain designs, and highlights regions with the most promising REE potential [12].

China currently dominates the global REE market, producing roughly 85% in 2016, with Australia following at 10% [6]. However, there is good news: despite limited production, the world has a wealth of REE resources. As of 2017, there were 178 identified REE deposits scattered around the globe, containing an estimated total of 478 million tonnes (Mt) of rare earth oxides (REOs). Over half (58%) of these deposits hold more than 0.1 Mt of REOs, and nearly 60 have undergone technical assessments [6].

Rare earth elements, a critical strategic resource, have become a subject of intense research due to concerns about China's influence. Scholars like [13,14] worry that China might leverage its dominant position to gain geopolitical advantages.

Beyond physical disruptions, research by [15] highlights how societal and geopolitical factors like speculative markets, export bans [16], and environmental regulations can impact supply [16]. Since downstream applications hold more power in the global market [17], China's policies significantly influence prices. The authors of [18] even found that increased trade policy uncertainty in the US benefits the rare earth element market by potentially boosting demand, while such uncertainty in China might restrict supply [12].

Further research by [19] suggests a positive correlation between geopolitics and import prices, but a negative correlation with overall import value. This complexity highlights the interplay of factors like rising anti-globalization sentiment and the COVID-19 pandem influencing the rare earth element market in unpredictable ways [20].

At the current production rate, these resources could supply the world for ove century. However, there is a catch. Clean technologies are expected to require a signific amount of specific REEs, particularly neodymium (Nd) and dysprosium (Dy). By 20 these two elements are projected to make up 75% and 9% of the clean tech REE dema respectively [19,21]. This is concerning because they only represent 15% and 0.52% global REE resources. This imbalance means that neodymium and dysprosium will lik be major factors driving the development of new REE exploration projects and clean ene technologies in the coming years.

The extraction and processing of rare earth elements is a complex and expensi process. The extraction and processing processes of rare earth elements can be divided i the following stages [22]:

- Prospecting: This stage consists of searching for rare earth element deposits.
- Exploitation: This stage consists of extracting rare earth elements from deposits.
- Concentration: This stage consists of the concentration of rare earth elements.
- Purification: This stage consists of the purification of rare earth elements.

The scarcity of deposits around the world in which it is profitable to extract th materials is a source of tensions between the central technology-producing countries.

This article aims to evaluate the implications of rare earth elements on the gr value added and break it down for primary, secondary and tertiary sectors. Also, i expected that this analysis can evaluate the global potential of REE resources, evalu the economic viability of current advanced REE projects, and analyze the medium- a long-term demand (2016–2030) for REEs from clean energy technologies. The relevar of this research is to establish a foundation to quantitatively analyze future opportunit challenges, and constraints within the global REE supply chain and demand landscape

This article follows the following structure. The introduction is followed by a theor cal framework that puts us in the situation of the current discussion in the scientific fie specifying its contribution to the three main sectors of the economy, the primary, second and tertiary sectors.

The materials and methods are explained below, including data collection and t econometric model used, followed by the results obtained, the discussion, and conclusio This study ends with the bibliographic references cited in the article.

1.1. Theoretical Framework

The 17 chemical elements that make up rare earth elements are Lanthanum (L Cerium (Ce), Praseodymium (Pr), Neodymium (Nd), Promecium (Pm), Samarium (S Eudimium (Eu), Gadolinium (Gd), Terbium (Tb), Dysprosium (Dy), Holmium (Ho), Erbiu (Er), Thulium (Tm), Ytterbium (Yb), Lutecide (Lu), Scandium (Sc) and Yttrium (Y). The elements are found in the periodic table in group 3 and are characterized by their magne optical, and catalytic properties [23].

Rare earth elements are crucial materials in electronics, optics, or magnetism and a currently irreplaceable. Small amounts of up to 16 or 17 components are usually present our mobile phones, except for the average. In short, they are used in circuits, speakers, a screens in all device parts [24]. They are vital for electric vehicles and electronic devic but we also find their use is transcendental in the weapons industry, such as in miss guidance systems, radar-invisible aircraft, or nuclear submarines [24]. This industry currently on the rise due to the Russia–Ukraine conflict, Chinese rearmament, and the fe that the United States will lose its role in world hegemony [25].

The use of rare earth elements is not only used to take lives but also to save them. T role of these chemical elements is fundamental in medical applications. For example, th are found in contrast injections to see the functioning of an organ or an X-ray [26].

Scandium is a non-lanthanide rare earth element used to strengthen metal bon We can find it in low-consumption lights and modern televisions. Scandium, a valua element used in applications, is typically difficult to extract [27]. This research explores a promising method for obtaining scandium from bauxite residue, a waste product from aluminum production. The technique utilizes high-pressure sulfuric acid leaching to achieve high yields of scandium while minimizing the extraction of iron and aluminum, which are also present in the residue [27,28].

Yttrium is a non-lanthanide rare earth element used in superconductors, pulsed light lasers, drugs in chemotherapy and rheumatoid arthritis treatment, and surgical supplies. It is also used in low-consumption light bulbs or camera lenses. Beyond traditional mining, lanthanides and yttrium can be extracted as valuable byproducts from processing other minerals [29]. These include apatite mining and uranium tailings. Notably, red mud, a waste product from the Bayer process for aluminum production, is enriched with lanthanides and yttrium. In Greece alone, 5 million tons of red mud are produced annually, and this material contains a significant concentration of these elements—exceeding 0.1% in total [29].

Lanthanum is used in the manufacturing of lenses for cameras and telescopes. It can also be used in wastewater treatment and oil refining. Lanthanum has a strong attraction to phosphate, binding it to form insoluble $LaPO_4$. This compound sinks to the bottom and cannot be absorbed by algae or aquatic plants. One major advantage of lanthanum is its effectiveness across various environmental conditions. Traditionally, aluminum (Al) or iron (Fe) were used for phosphorus control, but their success heavily depended on factors like water pH, oxygen levels, alkalinity, and the presence of other minerals and organic matter. Lanthanum's broader applicability makes it a more reliable solution [27,30].

Cerium is the most abundant rare earth element. It has many uses, including being used as a catalyst in catalytic converters in vehicle exhaust systems to reduce emissions. It is widely used to manufacture magnets and create alloys with iron, magnesium, and aluminum. Cerium (Ce), a member of the lanthanide group on the periodic table, made its debut in 1803. Discovered in its oxide form by researchers in both Sweden and Germany, the oxide was later named "ceria" by Swedish scientist Jons Jacob Berzelius [29,31].

Praseodymium is used in pigmentation for glass and gems and in the manufacturing of magnets. We can find it when creating high-strength metals, such as those used in aircraft engines [27]. Praseodymium stands out among the rare earth elements due to its unique ability to exist in multiple chemical states. This versatility translates to praseodymium-based catalysts excelling at both capturing and activating CO_2. Additionally, these catalysts boast superior stability compared to their copper and iron counterparts, making them a promising option for CO_2 conversion technologies [27]. The decrease in the size of electronic devices has been possible thanks to the magnetic properties of ytterbium and terbium.

Another element that is part of the rare earth elements group is neodymium. It is present on mobile phones or headphones, not only on those devices. It is a colorant in ceramic glazes and manufactures' welding glasses and lasers. Neodymium allows us, through alloys with boron and iron, to create a neodymium-iron-boron magnet, which is currently the most powerful magnet. The magnetic properties of rare earth elements make a high demand for them [32]. This magnet type represents one of the most powerful permanent magnets used in electric motors, vital for the transition toward electric vehicles (see Tables 1 and 2). In a couple of decades, more than a million electric vehicles are expected to be on the roads of the United States, increasing the demand for neodymium exponentially. Demand exceeds production by around 2 or 3 thousand tons annually. Neodymium is a strategically important resource and an essential element in modern societies. It is a key enabler of the energy transition due to its application in electric motors and wind turbines [33].

Table 1. Uses of rare earth elements (La, Ce, Pr, Nd, Pm, Sm, Eu, Gd, and Tb).

	La	Ce	Pr	Nd	Pm	Sm	Eu	Gd	Tb
Permanent Magnets	✗	✗	✗	✗		✗		✗	
Ceramics	✗	✗	✗	✗	✗	✗	✗	✗	✗
Batteries	✗	✗	✗	✗		✗		✗	
Construction Materials	✗	✗	✗	✗	✗		✗		✗
Lasers					✗		✗		✗
Aircraft Alloys									
Optical Glasses									
Catalysis									
Magnetism									

Note: "✗" are the different use of rare earth elements, according to the "World Production of Rare Earths" rep by the United States Geological Survey [27].

Table 2 shows us the following rare earth elements.

Table 2. Uses of rare earth elements (Dy, Ho, Er, Tm, Yb, Lu, Sc, and Y).

	Dy	Ho	Er	Tm	Yb	Lu	Sc	Y
Permanent Magnets	✗							
Ceramics	✗	✗	✗	✗	✗	✗		
Batteries	✗							
Construction Materials		✗	✗	✗	✗	✗		
Lasers		✗	✗	✗	✗	✗		
Aircraft Alloys							✗	
Optical Glasses							✗	✗
Catalysis							✗	✗
Magnetism								✗

Note: "✗" are the different use of rare earth elements, according to the "World Production of Rare Earths" rep by the United States Geological Survey [27].

According to the "World Production of Rare Earths" report by the United Sta Geological Survey [27], the central producing countries of rare earth elements in 2023 w the following (see Table 3).

Table 3. Countries producing or exploiting rare earth elements.

Country	Production (Metric Tons)
China	210,000
United States	43,000
Australia	18,000
Myanmar	12,000
Thailand	7100
Vietnam	4300
India	2900
Russia	2600
Madagascar	960
Brazil	80
Malaysia	80

China has been the leading producer of rare earth elements for decades, accounti for more than 69% of global production. The authors of [23] report that the United Sta is the second largest producer in the world (14%), followed by Australia (6%), Myanm (4%), and Thailand (2%).

The production of rare earth elements has increased in recent years due to the increas demand for these elements to manufacture technological products. Demand for rare ea elements is expected to continue to rise in the coming years as technology develops a the global economy expands [34].

China's dependence on rare earth element production is a matter of growing concern. China has used its control of the rare earth element market as an instrument of political pressure. It has led to efforts to diversify the supply of rare earth elements and reduce dependence on China [25,26].

In recent years, significant investments have been made in rare earth elements exploration and development projects in countries outside of China. These projects aim to increase rare earth element production in these countries and reduce dependence on China [32].

However, diversifying the supply of rare earth elements is a complex and challenging process. Rare earth element deposits are relatively scarce and difficult to extract. Additionally, the mining and processing of rare earth elements can have a negative impact on the environment [22].

1.2. Contribution of Rare Earth Elements to the Sectors of the Economy

Rare earth elements significantly impact the primary sector, as they are used in various products and applications essential for agriculture, forestry, and fisheries. Among the main uses that are addressed in previous research are the following:

Neodymium is used to manufacture phosphate fertilizers, an important source of phosphorus for crops. Phosphorus is an essential nutrient for plant growth. Praseodymium is used to manufacture pesticides such as insecticides and fungicides. These pesticides control pests that attack crops [27,35].

Terbium is used to manufacture electric motors for agricultural machinery, such as tractors and combines. Electric motors are more efficient than internal combustion engines and produce fewer polluting emissions [29,36].

Gadolinium manufactures sensors for forestry equipment, such as distance and motion sensors. These sensors are used to monitor the operation of forestry equipment. Dysprosium is used to manufacture some pesticides, such as insecticides and fungicides [27,37].

These pesticides are used to control pests that attack forests. Erbium is used to manufacture wood products such as furniture and construction boards. These wood products are used in construction and furniture. Thulium is used to manufacture electric motors for fishing equipment such as boats and nets. Electric motors are more efficient than internal combustion engines and produce fewer polluting emissions [29].

Yttrium is used to manufacture navigation instruments, such as Global Positioning Systems (GPSs). These instruments are used to navigate the sea. Scandium is used in the manufacture of fileting machines for fish processing. These machines cut fish into filets [27,38].

Rare earth elements are the invisible raw materials that power many of the technologies that shape our world. Their unique properties drive innovation and enable the development of a more sustainable future and a modern manufacturing industry [39]. Their main uses include the following:

Neodymium is the element most used in the manufacturing of permanent magnets. These magnets are used in various applications, such as electric motors for automobiles, hard drives for computers, and speakers for audio equipment [40].

Cerium is used as a catalyst in the production of sulfuric acid, an important chemical used in a wide range of applications, such as fertilizers, detergents, and batteries [41].

Yttrium is used in the manufacturing of optical glasses for lenses and filters. These lenses are used in cameras, telescopes, glasses, and other optical devices. Scandium is used in alloys with aluminum to improve its strength and ductility. These alloys are used to manufacture airplanes, automobiles, ships, and other structures [39].

The main applications to the third sector of the economy include the following:

Neodymium is used to manufacture permanent magnets for electric motors in mobile phones, computers, and televisions. Gadolinium is used to manufacture MRI contrast agents, which are used to diagnose diseases [42].

Dysprosium is used to manufacture permanent magnets for wind turbines, essen for renewable energy generation. Terbium is used to manufacture electric motors for elec vehicles, which are more efficient and produce fewer polluting emissions than inter combustion vehicles. Yttrium is used in the manufacturing of missiles, which are used armies to attack targets [43].

2. Materials and Methods

The research objectives are met by applying the following methodology and verifyi the empirical assumptions analyzed for the object of study.

2.1. Data Source and Variables

The data were obtained from the Rare Earth Statistics and Information issued by National Minerals Information Center. This information is regulated by the U.S. Geologi Survey acts as the scientific arm of the Department of the Interior, providing up-to-da scientific information relevant to an ever-changing world. It responds to society's evolvi needs by offering a wide range of data and expertise in geology, water, biology a cartography. This information supports decision-making on environmental, resource, a public safety issues in the United States [44].

The reports were reviewed and the information was extracted from the Mine Commodity Summaries and Minerals Yearbook from 1990 to 2022, thus providing observations. The prices of the leading rare earth elements and the gross value added the primary (agricultural), secondary (manufacturing), and tertiary (services) sectors of United States of America were extracted (see Table 4).

The trade restriction variable was a dummy variable where 0 was used for the yea in which there were no restrictions and 1 was used for the year's corresponding to t restrictions initially imposed by the administration of President Donald Trump.

Table 4. Model variables.

Symbology	Variable	Measurement	Variable Type
Y_1	Primary Gross Added Value	Log (% Total GDP)	Explained-Endogenous
Y_2	Secondary Gross Added Value	Log (% Total GDP)	Explained-Endogenous
Y_3	Tertiary Gross Added Value	Log (% Total GDP)	Explained-Endogenous
X_1	Rare earth elements' exploitation	Log (Metric Tons)	Explanatory-Exogenous
X_2	Employment generated by rare earth elements	Log (Annual average)	Explanatory-Exogenous
X_3	Domestic consumption of rare earth elements	Log (Metric Tons)	Explanatory-Exogenous
X_4	Import of rare earth elements	Log (Metric Tons)	Explanatory-Exogenous
X_5	Dysprosium	Log (Average Price)	Explanatory-Exogenous
X_6	Europium	Log (Average Price)	Explanatory-Exogenous
X_7	Neodymium	Log (Average Price)	Explanatory-Exogenous
X_8	Terbium	Log (Average Price)	Explanatory-Exogenous
X_9	Trade restriction	Dummy	Explanatory-Exogenous

The data for this study were processed using JASP v.0.17.2.1 [45], a comprehensive a user-friendly statistical software package developed by the JASP Team at the University Amsterdam. JASP offers a wide range of statistical analyses suitable for various resear designs, from basic descriptive statistics to complex modeling techniques. In this study, utilized JASP version 0.17.2.1, which incorporates the latest advancements and bug fixes ensure the accuracy and reliability of our statistical analyses.

JASP is used because it uniquely offers both Bayesian and Frequentist statical a proaches, allowing users to compare and contrast results from different perspectives a gain a more comprehensive understanding of their data. This flexibility is particula valuable in fields where both approaches are prevalent.

Figure 1 shows the behavior over time of the gross value added to each sector the economy. It can be seen that the primary GVA is maintained over time, while t manufacturing sector has a slight decrease and in the case of the tertiary sector, an increa is seen.

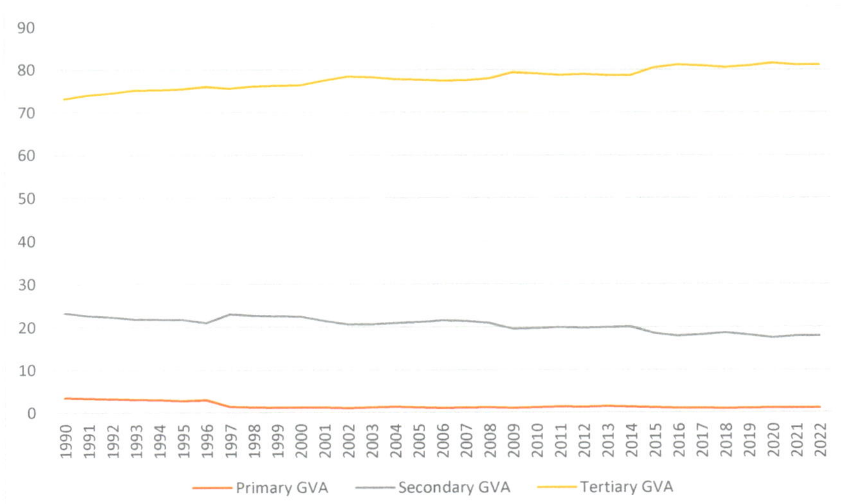

Figure 1. GVA sector.

Figure 2 shows the behavior of the rare earth element industry from exploitation to consumption or export, depending on the case. It can be seen that both the exploitation of rare earth elements and their exportation have a structural upward trend in the last seven years.

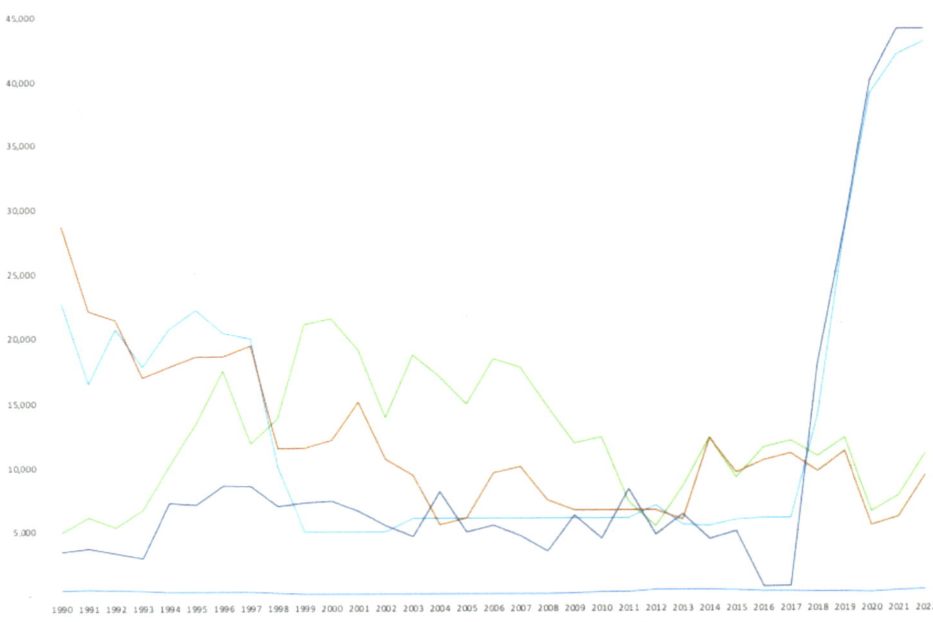

Figure 2. Rare earth elements' market behavior.

In Figure 3, you can see the prices of rare earth elements over time. It can be seen that the prices decreased in the nineties, but in recent years they have been on the rise.

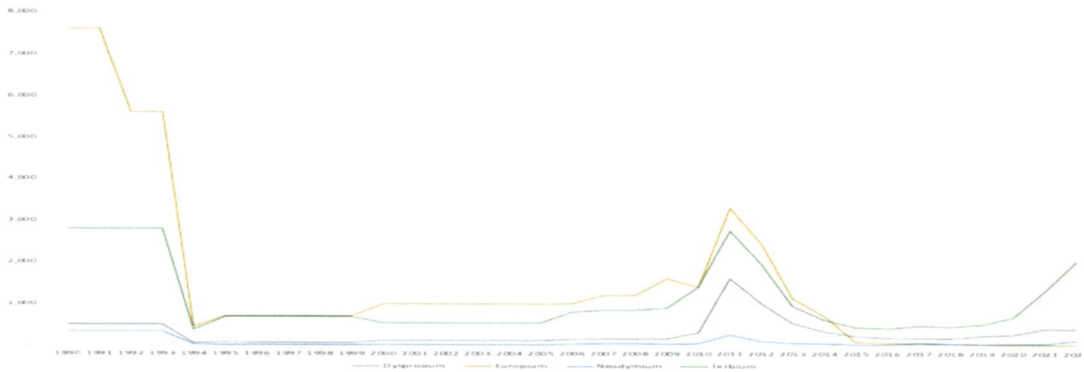

Figure 3. Dysprosium, Europium, Neodymium, and Terbium prices.

2.2. Econometric Model

The econometric model meets the article's objective to analyze the contribution of v ious factors in the gross value added to the economic sectors in agriculture, manufacturi and services. The mathematical expression of the model is shown in Equation (1).

$$Y_t = f(X_1, X_2, \ldots X_n)$$

where Y_t represents an endogenous or explained variable, while the X_n are the variab that indicate the explanatory or exogenous factors.

Consequently, to correctly explain the empirical content, three models are propos that are expressed in Equations (2)–(4).

$$Y_1 = b_0 + b_1X_1 + b_2X_2 + b_3X_3 + b_4X_4 + b_5X_5 + b_6X_6 + b_7X_7 + b_8X_8 + b_9X_9 + \mu_i$$

$$Y_2 = b_0 + b_1X_1 + b_2X_2 + b_3X_3 + b_4X_4 + b_5X_5 + b_6X_6 + b_7X_7 + b_8X_8 + b_9X_9 + \mu_i$$

$$Y_3 = b_0 + b_1X_1 + b_2X_2 + b_3X_3 + b_4X_4 + b_5X_5 + b_6X_6 + b_7X_7 + b_8X_8 + b_9X_9 + \mu_i$$

where each model represents each economic sector of the economy of the United Sta of America. Now, the term disturbance or random error is μ_i. This term groups all variables not included in the model but can explain the object of study [46]. Each represents the explanatory variables described in Table 4.

3. Results

Table 5 shows the description of the variables and their main statistics. It allows to analyze the absence of outliers and evaluate the normality of the data and the numb of observations.

Table 5. Descriptive statistics.

Variable	N	Mean	SE
Primary GVA (Y_1)	33	0.304	0.078
Secondary GVA (Y_2)	33	3.020	0.015
Tertiary GVA (Y_3)	33	4.355	0.005
Exploitation (X_1)	33	9.232	0.127
Employment (X_2)	33	5.270	0.101

Table 5. *Cont.*

Variable	N	Mean	SE
Consumption (X_3)	33	9.272	0.079
Import (X_4)	33	9.335	0.073
Dysprosium (X_5)	33	5.313	0.144
Europium (X_6)	33	6.410	0.285
Neodymium (X_7)	33	4.031	0.152
Terbium (X_8)	33	6.755	0.112
Trade restriction (X_9)	33	0.152	0.364

Table 6 shows the coefficients of the models explained in the Materials and Methods, with the standard deviation and the plausible significance of the variable in the models in parentheses. The latter is represented as a measure of probability that the coefficient of the variables is different from zero. For this, the significance is represented when it is less than 0.00 with (***) and 0.05 with (**).

Table 6. Model coefficients.

Variable	Model 1 (Y_1)	Model 2 (Y_2)	Model 3 (Y_3)
Intercept	−3.896 (0.823) ***	1.854 (0.357) ***	4.678 (0.039) **
Exploitation (X_1)	0.346 (0.077) ***	0.021 (0.014) **	
Employment (X_2)	0.302 (0.075) ***		
Consumption (X_3)		0.061 (0.017) ***	−0.028 (0.005) ***
Import (X_4)		0.032 (0.027) **	
Dysprosium (X_5)	−0.211 (0.092) ***		0.013 (0.004) ***
Europium (X_6)	0.281 (0.086) ***	0.034 (0.019) ***	−0.007 (0.004) ***
Neodymium (X_7)	0.196 (0.115) ***	−0.022 (0.015) ***	
Terbium (X_8)	−0.307 (0.129) ***		−0.013 (0.005) ***
Trade restriction (X_9)	0.088 (0.085) *	0.045 (0.058) **	0.012 (0.013) ***

Note: *, **, *** are the significance levels of 10%, 5% and 1% or less respectively.

Table 7 shows the additional adjustments to validate the model, showing acceptable collinearity indices. The last two adjustments were made to verify that the models do not present problems in which two or more variables are highly correlated.

Table 7. Models' goodness of fit.

Fit	Model 1 (Y_1)	Model 2 (Y_2)	Model 3 (Y_3)
R^2	0.896	0.884	0.919
RMSE	0.161	0.032	0.009
Tolerance (Colinearity)	***	***	***
VIF (Colinearity)	***	***	***

Note: *** is the significance level of 1% or less.

The models meet the statistical parameters and assumptions of the variables' linearity, exogeneity, homoscedasticity, normality, and independence.

This plot compares the quantiles (percentiles) of two datasets. In this case, one dataset represents the quantiles of your standardized residuals, and the other represents the quantiles of a theoretical normal distribution.

For the specified model 1, the Q–Q plot of standardized residuals presents strong evidence supporting the normality of the distribution of the residuals. The points on the graph are aligned approximately along a straight diagonal line, indicating a notable similarity between the observed distribution of standardized residuals and a theoretical normal distribution (see Figure 4).

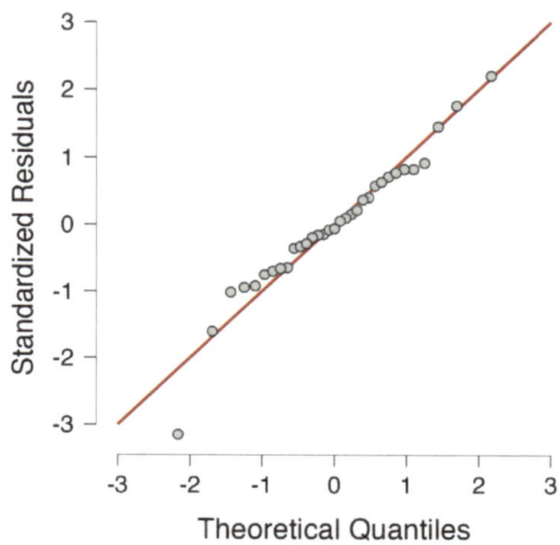

Figure 4. Q–Q plot of standardized residuals for Model 1.

Importantly, no curved patterns or significant deviations from the diagonal line a observed in the Q–Q chart. This reinforces the idea that there is no evidence to sugges non-normal distribution of the residuals (see Figure 5).

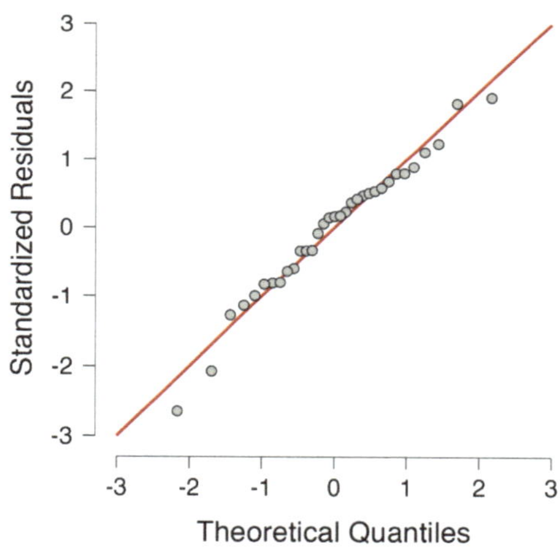

Figure 5. Q–Q plot of standardized residuals for Model 2.

Based on the evidence provided by the Q–Q plot, we can analyze that the residu of the specified model follow a normal distribution (see Figure 6). This conclusion h important implications for the reliability of the statistical analysis performed.

The plot of residuals versus predicted values for the specified model provides stro evidence supporting the adequacy of model fit. In this graph, the points are randon distributed around a horizontal line in the center, exhibiting no discernible patterns. Tl random distribution suggests that the model effectively captures the relationship betwe the dependent and independent variables and that there are no significant violations of t assumptions of the linear regression model (see Figure 7).

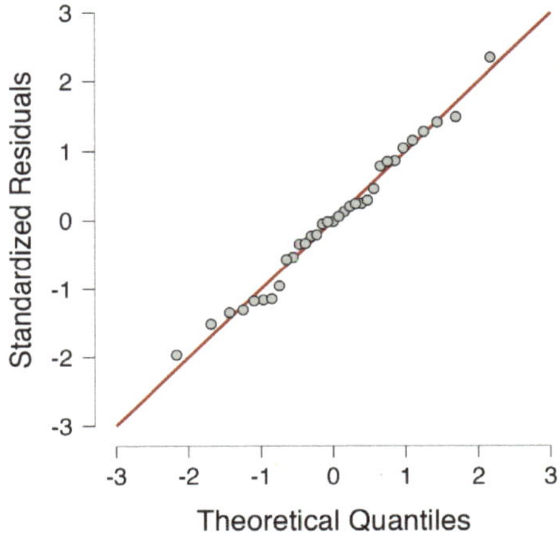

Figure 6. Q–Q Plot Standardized Residuals Model 3.

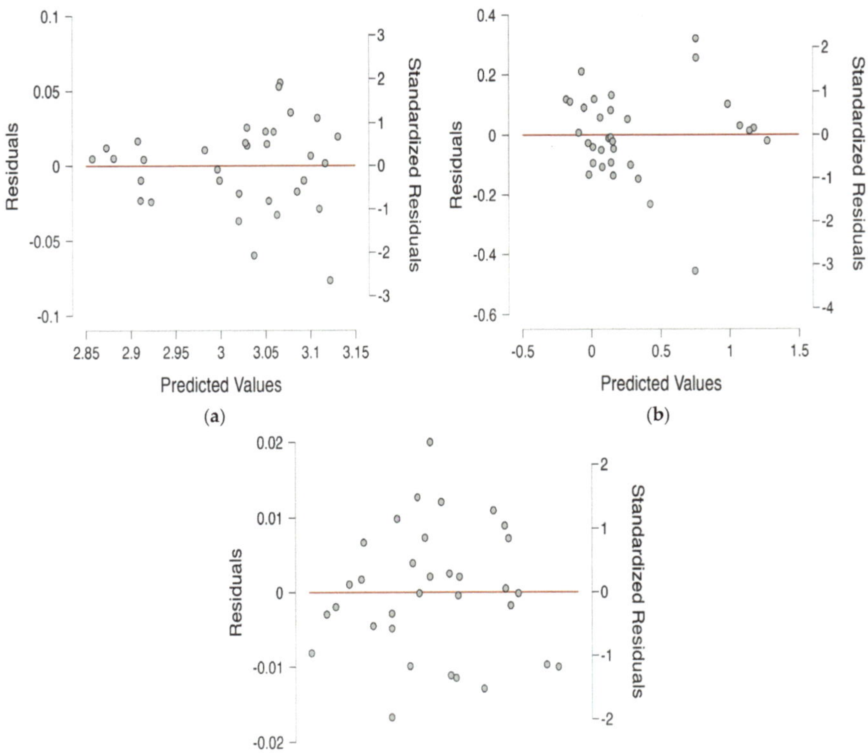

Figure 7. Plot of residuals vs. predicted values. (**a**) Model 1, (**b**) Model 2, (**c**) Model 3.

The plot of residuals vs. predicted values provides compelling evidence that the specified model adequately fits the data. The random distribution of residuals supports the validity of the model assumptions, strengthening confidence in the predictions, inferen and interpretations obtained from the analysis. This graphical analysis complements evaluation of the model and contributes to the robustness of the conclusions derived fr the study.

4. Discussion

This study employed an econometric model to systematically investigate the influe of various factors on the gross value added (GVA) to the agricultural, manufacturing, a service sectors in the United States. The model's findings unveil statistically significant pacts of rare earth element (REE) exploitation, utilization, domestic consumption, impo and the prices of specific REEs (e.g., Dysprosium, Europium, Neodymium) on secto GVA. These results corroborate prior research highlighting the pivotal role of REEs modern economies, stemming from their extensive application in technological advan ments [26,46]. Notably, the control variable representing restrictions on REE trade sugge a stronger influence of REEs on the primary (agricultural) sector compared to the ma facturing sector. This aligns with prevailing concerns regarding overreliance on a sin source, particularly China's dominant position in REE production and consumption [47, While potential alternatives to mitigate this concentration are acknowledged, long-te monitoring and evaluation are imperative to assess their effectiveness and implication

The econometric analysis extends beyond the manufacturing sector, revealing t REEs exert a pervasive influence across the primary and service sectors as well. Th diverse applications in agricultural practices, manufacturing processes, and various serv industries underscore the multifaceted economic importance of REEs [49,50]. However, model also sheds light on the challenges associated with REE extraction and processi including their complex and costly nature [51,52]. These factors, coupled with the projec surge in REE demand [53], necessitate the development of robust strategies to address sustainability and security of supply concerns. Ensuring a stable and sustainable supply these critical elements is paramount for fostering long-term economy growth, resilien and environmental stewardship.

The findings of this study have significant implications for policymakers, indus leaders, and researchers engaged in the sustainable management of REE resources. P cymakers can leverage the study's insights to formulate informed policies that prom the efficient and environmentally responsible utilization of REEs while mitigating risks associated with overreliance on a single supplier. Industry leaders can utilize the search findings to guide their strategic decision-making processes, ensuring the sustaina sourcing, processing, and application of REEs throughout their value chains. Research can build upon the study's methodology and findings to further investigate the intric dynamics between REEs, economic development, and environmental sustainability.

The study's contributions extend beyond the realm of resource economics, offeri valuable insights for broader economic discourse. The identification of REEs as criti factors influencing sectoral GVA highlights the interconnectedness of modern econom and the need for holistic approaches to economic development. Moreover, the study und scores the importance of considering environmental and social sustainability dimensic alongside economic growth objectives. By recognizing the multifaceted impacts of REE lization, policymakers and industry leaders can strive for a more sustainable and equita distribution of the benefits associated with these critical resources.

Future research directions can further enhance our understanding of the comp relationship between REEs, economic development, and sustainability. One avenue exploration lies in conducting comparative studies across different countries and regio to identify factors that influence the effectiveness of REE-related policies and practic Additionally, in-depth analyses of specific REE-intensive industries can provide granu insights into the economic and environmental implications of REE utilization. Furtherm research examining the potential for substitution of REEs with alternative materials c inform strategies for reducing reliance on these critical resources.

This study unveils the profound influence of rare earth elements on the gross value added of various economic sectors, highlighting their critical role in modern economies. The findings underscore the need for comprehensive strategies that promote the sustainable and responsible management of REE resources, ensuring a balance between economic growth, environmental stewardship, and social equity. By fostering international collaboration and research, we can collectively navigate the challenges and opportunities presented by these critical elements, shaping a more sustainable and resilient future for all.

5. Conclusions

The econometric analysis conducted in this study has demonstrated the importance of rare earth elements in the agricultural, manufacturing, and service economic sectors in the United States. It has been identified that factors such as the exploitation of rare earth elements, the employment they generate, domestic consumption, imports, and the prices of these elements significantly impact the economy. The concentration of rare earth element production and consumption in China poses challenges regarding dependence on other countries and the security of supply.

Furthermore, it has been highlighted that rare earth elements are essential for manufacturing technological products and have a transversal impact on the economy's primary, secondary, and tertiary sectors. Mining and processing rare earth elements are complex and expensive, underscoring the importance of addressing the sustainability and security of supply issues.

In this sense, it is essential to properly understand and manage the impact of rare earth elements on the economy to promote sustainable and resilient economic development. Special attention is required to diversify supply sources, innovate extraction and processing technologies, and promote sustainable practices using these essential elements. Additionally, it is recommended to analyze the behavior of the price of Terbium since this harms the primary and tertiary sectors.

Although the findings are important for future research, it should be mentioned that they also have limitations. Among the limitations of this study are the data and the potential for biases or missing information related to the temporal dimension.

For future research, the long-term analysis and evaluation of public policies aimed at the exploitation of rare earth elements, and the research and development of products made using these elements should be considered. Likewise, it is recommended to analyze panel data with information from other producing countries and thus take advantage of the temporal nature of the variables.

Author Contributions: Conceptualization, J.-R.C.-M. and R.S.-O.; methodology, J.-R.C.-M. and J.-A.J.-d.-R.; software, A.V.-R.; validation, J.-R.C.-M., J.-A.J.-d.-R., R.S.-O. and M.Z.-D.l.T.; formal analysis, R.S.-O.; investigation, J.-R.C.-M.; resources, J.-A.J.-d.-R.; data curation, R.S.-O.; writing—original draft preparation, J.-R.C.-M.; writing—review and editing, J.-A.J.-d.-R. and A.V.-R.; visualization, A.V.-R.; supervision, J.-A.J.-d.-R.; project administration, A.V.-R. and M.Z.-D.l.T.; funding acquisition, A.V.-R. All authors have read and agreed to the published version of the manuscript.

Funding: This research was funded by the Universidad Bolivariana del Ecuador, grant number PROY-UBE-2023-017, and the Universidad Bolivariana del Ecuador funded the APC under the project "Community Economic Resilience through Citizen Satisfaction in the Province of Guayas".

Data Availability Statement: The database is available in the Helvia repository of the Universidad de Córdoba at http://hdl.handle.net/10396/27658, accessed on 1 January 2024.

Conflicts of Interest: The authors declare no conflicts of interest.

eferences

Li, Z.Z.; Meng, Q.; Zhang, L.; Lobont, O.R.; Shen, Y. How do rare earth prices respond to economic and geopolitical factors? *Resour. Policy* **2023**, *85*, 103853. [CrossRef]

Nassar, N.T.; Du, X.; Graedel, T.E. Criticality of the rare earth elements. *J. Ind. Ecol.* **2015**, *19*, 1044–1054. [CrossRef]

3. Antonakakis, N.; Chatziantoniou, I.; Filis, G. Oil shocks and stock markets: Dynamic connectedness under the prism of rec geopolitical and economic unrest. *Int. Rev. Financ. Anal.* **2017**, *50*, 1–26. [CrossRef]
4. Filho, W.L.; Kotter, R.; Özuyar, P.G.; Abubakar, I.R.; Eustachio, J.H.P.P.; Matandirotya, N.R. Understanding Rare Earth Eleme as Critical Raw Materials. *Sustainability* **2023**, *15*, 1919. [CrossRef]
5. Ilankoon, I.M.S.K.; Dushyantha, N.P.; Mancheri, N.; Edirisinghe, P.M.; Neethling, S.J.; Ratnayake, N.P.; Rohitha, L.P.S.; I sanayake, D.M.D.O.K.; Premasiri, H.M.R.; Premasiri, H.M.R.; et al. Constraints to rare earth elements supply diversificati Evidence from an industry survey. *J. Clean. Prod.* **2022**, *331*, 129932. [CrossRef]
6. Zhou, B.; Li, Z.; Chen, C. Global potential of rare earth resources and rare earth demand from clean technologies. *Minerals* **2017** 203. [CrossRef]
7. De Boer, M.A.; Lammertsma, K. Scarcity of rare earth elements. *ChemSusChem* **2013**, *6*, 2045–2055. [CrossRef]
8. Trench, A.; Sykes, J.P. Rare earth permanent magnets and their place in the future economy. *Engineering* **2020**, *6*, 115–1 [CrossRef]
9. Kim, T.Y. *The Role of Critical Minerals in Clean Energy Transitions*; International Energy Agency: Washington, DC, USA, 20 pp. 70–71. Available online: https://static.sif.it/SIF/resources/public/files/energy2023/Kim.pdf (accessed on 10 October 20
10. United Nations. Climate Action. The Paris Agreement. 2024. Available online: https://www.un.org/en/climatechange/pa agreement (accessed on 11 November 2023).
11. World Economic Forum. Energy Transition. The World Added 50% More Renewable Capacity Last Year Than in 2022. 20 Available online: https://www.weforum.org/agenda/2024/02/renewables-energy-capacity-demand-growth/ (accessed on December 2023).
12. Almassalkhi, M.R.; Kundu, S. Intelligent Electrification as an Enabler of Clean Energy and Decarbonization. *Curr. Sustain./Ren Energy Rep.* **2023**, *10*, 183–196. [CrossRef]
13. Pawar, G.; Ewing, R.C. Recent advances in the global rare-earth supply chain. *MRS Bull.* **2022**, *47*, 244–249. [CrossRef]
14. Kalantzakos, S. The race for critical minerals in an era of geopolitical realignments. *Int. Spect.* **2020**, *55*, 1–16. [CrossRef]
15. Dushyantha, N.; Batapola, N.; Ilankoon, I.M.S.K.; Rohitha, S.; Premasiri, R.; Abeysinghe, B.; Ratnayake, N.; Dissanayake, K. T story of rare earth elements (REEs): Occurrences, global distribution, genesis, geology, mineralogy and global production. *Geol. Rev.* **2020**, *122*, 103521. [CrossRef]
16. Mancheri, N.A.; Sprecher, B.; Bailey, G.; Ge, J.; Tukker, A. Effect of Chinese policies on rare earth supply chain resilience. *Reso Conserv. Recycl.* **2019**, *142*, 101–112. [CrossRef]
17. Hu, X.; Sun, B.; Wang, C.; Lim, M.K.; Wang, P.; Geng, X.; Yao, C.; Chen, W.Q. Impacts of China's exports decline in rare ea primary materials from a trade network-based perspective. *Resour. Policy* **2023**, *81*, 103321. [CrossRef]
18. Aziman, E.S.; Ismail, A.F.; Rahmat, M.A. Balancing economic growth and environmental protection: A sustainable approach Malaysia's rare-earth industry. *Resour. Policy* **2023**, *83*, 103753. [CrossRef]
19. Xia, Q.; Du, D.; Cao, W.; Li, X. Who is the core? Reveal the heterogeneity of global rare earth trade structure from the perspecti of industrial chain. *Resour. Policy* **2023**, *82*, 103532. [CrossRef]
20. Hau, L.; Zhu, H.; Yu, Y.; Yu, D. Time-frequency coherence and quantile causality between trade policy uncertainty and rare ea prices: Evidence from China and the US. *Resour. Policy* **2022**, *75*, 102529. [CrossRef]
21. Liu, S.L.; Fan, H.R.; Liu, X.; Meng, J.; Butcher, A.R.; Yann, L.; Yang, K.-F.; Li, X.C. Global rare earth elements projects: N developments and supply chains. *Ore Geol. Rev.* **2023**, *157*, 105428. [CrossRef]
22. Shuai, J.; Zhao, Y.; Shuai, C.; Wang, J.; Yi, T.; Cheng, J. Assessing the international co-opetition dynamics of rare earth resour between China, USA, Japan and the EU: An ecological niche approach. *Resour. Policy* **2023**, *82*, 103446. [CrossRef]
23. Hoatson, D.M.; Jaireth, S.; Miezitis, Y. The major rare-earth-element deposits of Australia: Geological setting, exploration, a resources. *Geosci. Aust.* **2011**, *204*, 72–128. [CrossRef]
24. Uysal, T. Economic analysis of rare earth element processing methods for Mountain Pass ore. *J. S. Afr. Inst. Min. Metall.* **2022**, 1 407–412. [CrossRef]
25. Fishman, T.; Myers, R.J.; Rios, O.; Graedel, T.E. Implications of emerging vehicle technologies on rare earth supply and dema in the United States. *Resources* **2018**, *7*, 9. [CrossRef]
26. Lee, Y.; Dacass, T. Reducing the United States' risks of dependency on China in the rare earth market. *Resour. Policy* **2022**, 102702. [CrossRef]
27. Holthaus, D. The Rare Earths Economy: Can Supply Keep up with Growing Demand? *Am. Ceram. Soc. Bull.* **2022**, *3*, 38–41. Availa online: https://ceramics.org/wp-content/uploads/2022/03/CGM-April2022_Feature.pdf (accessed on 13 December 2023).
28. USGS. United States Geological Survey 2022. Rare Earths Statistics and Information. Available online: https://www.usgs.go centers/national-minerals-information-center/rare-earths-statistics-and-information (accessed on 1 March 2024).
29. Chen, Y.; Zheng, B. What happens after the rare earth crisis: A systematic literature review. *Sustainability* **2019**, *11*, 1288. [CrossR
30. USGS. United States Geological Survey 2024. Historical Statistics for Mineral and Material Commodities in the United States. Av able online: https://www.usgs.gov/centers/national-minerals-information-center/historical-statistics-mineral-and-materi commodities (accessed on 1 March 2024).
31. Jowitt, S.M. Mineral economics of the rare-earth elements. *MRS Bull.* **2022**, *47*, 276–282. [CrossRef]
32. Kim, Y.; Kim, S.; Lee, J. Economic evaluation of rare earth elements contained in coal ash. *Resour. Recycl.* **2019**, *28*, 26– [CrossRef]

Ma, J.; Li, M. The feedback effect of rare earth development policy on economy and society. *E3S Web Conf.* **2021**, *228*, 01014. [CrossRef]
Financial Times. Rare Earths. Available online: https://www.ft.com/rare-earths (accessed on 1 March 2024).
Ortiz, C.E.A.; Viana Júnior, E.M. Rare earth elements in the international economic scenario. *Rem Rev. Esc. Minas* **2014**, *67*, 361–366. [CrossRef]
Bisaka, K.; Thobadi, I.C.; Pawlik, C. Extraction of rare earths from iron-rich rare earth deposits. *J. S. Afr. Inst. Min. Metall.* **2017**, *117*, 731–739. [CrossRef]
Zhang, S. Study on Economic Significance of Rare Earth Mineral Resources Development Based on Goal Programming and Few-Shot Learning. *Comput. Intell. Neurosci.* **2022**, *2022*, 7002249. [CrossRef] [PubMed]
Yu, G.; Xiong, C.; Xiao, J.; He, D.; Peng, G. Evolutionary analysis of the global rare earth trade networks. *Appl. Math. Comput.* **2022**, *430*, 127249. [CrossRef]
Jaroni, M.S.; Friedrich, B.; Letmathe, P. Economical feasibility of rare earth mining outside China. *Minerals* **2019**, *9*, 576. [CrossRef]
Wang, J.; Guo, M.; Liu, M.; Wei, X. Long-term outlook for global rare earth production. *Resour. Policy* **2020**, *65*, 101569. [CrossRef]
Yinbu, S. Social Elements: A Summary of Social Science Research on Rare Earth Industry in Baotou. *Int. Sci. Res.* **2022**, *2*, 7–20. [CrossRef]
Kryukov, V.A.; Yatsenko, V.A.; Kryukov, Y.V. Rare Earth Industry—How to take advantage of opportunities. *Gornaya Promyshlennost Russ. Min. Ind.* **2020**, *5*, 68–84. (In Russian) [CrossRef]
Vekasi, K. Politics, markets, and rare commodities: Responses to Chinese rare earth policy. *Jpn. J. Political Sci.* **2019**, *20*, 2–20. [CrossRef]
Yatsenko, V. Specific Aspects in Assessing Economic Feasibility of Rare Earth Mining. *World Econ. Manag.* **2020**, *20*, 138–152. (In Russian) [CrossRef]
JASP Team. *JASP*; Version 0.17.2.1; JASP Team: Amsterdam, The Netherlands, 2023.
Apergis, E.; Apergis, N. What is extracted from earth is gold: Are rare earths telling a new tale to economic growth? *J. Econ. Stud.* **2018**, *45*, 177–192. [CrossRef]
Kryukov, I.V.; Samsonov, N.I.; Iatsenko, V.A. The Russian Rare-Earth Industry: Should China's Experience Be Adopted? *Probl. Econ. Transit.* **2020**, *62*, 95–107. [CrossRef]
Zou, B.; Poncin, S.; Bertinelli, L. The US-China Supply Competition for Rare Earth Elements: A Dynamic Game View. *Environ. Model. Assess.* **2022**, *27*, 883–900. [CrossRef]
Gholz, E.; Hughes, L. Market structure and economic sanctions: The 2010 rare earth elements episode as a pathway case of market adjustment. *Rev. Int. Political Econ.* **2021**, *28*, 611–634. [CrossRef]
Vekasi, K. The geoeconomics of critical rare earth minerals. *Georget. J. Int. Aff.* **2021**, *22*, 271–279. [CrossRef]
He, Y. Reregulation of China's rare earth production and export. *Int. J. Emerg. Mark.* **2014**, *9*, 236–256. [CrossRef]
Proelss, J.; Schweizer, D.; Seiler, V. The economic importance of rare earth elements volatility forecasts. *Int. Rev. Financ. Anal.* **2020**, *71*, 101316. [CrossRef]
Guo, Q.; You, W. A comprehensive evaluation of the international competitiveness of strategic minerals in China, Australia, Russia and India: The case of rare earths. *Resour. Policy* **2023**, *85*, 103821. [CrossRef]

MDPI AG
Grosspeteranlage 5
4052 Basel
Switzerland
Tel.: +41 61 683 77 34

Land Editorial Office
E-mail: land@mdpi.com
www.mdpi.com/journal/land

www.ingramcontent.com/pod-product-compliance
Lightning Source LLC
Chambersburg PA
CBHW041401210926
51850CB00013B/3144
* 9 7 8 3 7 2 5 8 2 5 5 1 6 *